"十三五"江苏省高等学校重点教材(编号:2020-1-069)

高等学校土木工程专业系列规划教材

U0731873

弹塑性力学基础（双语版）
（第2版）

丁建国　编著

范　进　主审

WUHAN UNIVERSITY PRESS
武汉大学出版社

图书在版编目(CIP)数据

弹塑性力学基础:双语版/丁建国编著. —2版.—武汉:武汉大学出版社,2022.9
"十三五"江苏省高等学校重点教材　高等学校土木工程专业系列规划教材
ISBN 978-7-307-23233-4

Ⅰ.弹…　Ⅱ.丁…　Ⅲ.①弹性力学—高等学校—教材　②塑性力学—高等学校—教材　Ⅳ.O34

中国版本图书馆 CIP 数据核字(2022)第 136726 号

责任编辑:刘小娟　　责任校对:李嘉琪　　装帧设计:吴　极

出版发行:**武汉大学出版社**　(430072　武昌　珞珈山)
　　　　　(电子邮箱:whu_publish@163.com　网址:www.stmpress.cn)
印刷:广东虎彩云印刷有限公司
开本:880×1230　1/16　印张:19　字数:563 千字
版次:2014 年 4 月第 1 版　　2022 年 9 月第 2 版
　　　2022 年 9 月第 2 版第 1 次印刷
ISBN 978-7-307-23233-4　　　定价:65.00 元

丛 书 序

土木工程涉及国家的基础设施建设,投入大,带动的行业多。改革开放后,我国国民经济持续稳定增长,其中土建行业的贡献率达到1/3。随着城市化的发展,这一趋势还将继续呈现增长势头。土木工程行业的发展,极大地推动了土木工程专业教育的发展。目前,我国有500余所大学开设土木工程专业,在校生达40余万人。

2010年6月,中国工程院和教育部牵头,联合有关部门和行业协(学)会,启动实施"卓越工程师教育培养计划",以促进我国高等工程教育的改革。其中,"高等学校土木工程专业卓越工程师教育培养计划"由住房和城乡建设部与教育部组织实施。

2011年9月,住房和城乡建设部人事司和高等学校土建学科教学指导委员会颁布《高等学校土木工程本科指导性专业规范》,对土木工程专业的学科基础、培养目标、培养规格、教学内容、课程体系及教学基本条件等提出了指导性要求。

在上述背景下,为满足国家建设对土木工程卓越人才的迫切需求,有效推动各高校土木工程专业卓越工程师教育培养计划的实施,促进高等学校土木工程专业教育改革,2013年住房和城乡建设部高等学校土木工程学科专业指导委员会启动了"高等教育教学改革土木工程专业卓越计划专项",支持并资助有关高校结合当前土木工程专业高等教育的实际,围绕卓越人才培养目标及模式、实践教学环节、校企合作、课程建设、教学资源建设、师资培养等专业建设中的重点、亟待解决的问题开展研究,以对土木工程专业教育起到引导和示范作用。

为配合土木工程专业实施卓越工程师教育培养计划的教学改革及教学资源建设,由武汉大学发起,联合国内部分土木工程教育专家和企业工程专家,启动了"高等学校土木工程专业系列规划教材"建设项目。该系列教材贯彻落实《高等学校土木工程本科指导性专业规范》《卓越工程师教育培养计划通用标准》和《土木工程卓越工程师教育培养计划专业标准》,力图以工程实际为背景,以工程技术为主线,着力提升学生的工程素养,培养学生的工程实践能力和工程创新能力。该系列教材的编写人员,大多主持或参加了住房和城乡建设部高等学校土木工程学科专业指导委员会的"土木工程专业卓越计划专项"教改项目,因此该系列教材也是"土木工程专业卓越计划专项"的教改成果。

土木工程专业卓越工程师教育培养计划的实施,需要校企合作,期望土木工程专业教育专家与工程专家一道,共同为土木工程专业卓越工程师的培养作出贡献!

是以为序。

2014年3月于同济大学四平路校区

特别提示

　　教学实践表明，有效地利用数字化教学资源，对于学生学习能力以及问题意识的培养乃至怀疑精神的塑造具有重要意义。

　　通过对数字化教学资源的选取与利用，学生的学习从以教师主讲的单向指导模式转变为建设性、发现性的学习，从被动学习转变为主动学习，由教师传播知识到学生自己重新创造知识。这无疑是锻炼和提高学生的信息素养的大好机会，也是检验其学习能力、学习收获的最佳方式和途径之一。

　　本系列教材在相关编写人员的配合下，逐步配备基本数字教学资源，主要内容包括：

　　文本：课程重难点、思考题与习题参考答案、知识拓展等。

　　图片：课程教学外观图、原理图、设计图等。

　　视频：课程讲述对象展示视频、模拟动画，课程实验视频，工程实例视频等。

　　音频：课程讲述对象解说音频、录音材料等。

数字资源获取方法：

① 打开微信，点击"扫一扫"。

② 将扫描框对准书中所附的二维码。

③ 扫描完毕，即可查看文件。

更多数字教学资源共享、图书购买及读者互动敬请关注"开动传媒"微信公众号！

Foreword

Elastic-plastic mechanics is a discipline branch that studies the elastic and plastic deformation characteristics and pursues rigorous reasoning and accurate solution. Elastic mechanics started from the equilibrium differential equation and geometric deformation equation of deformable body proposed by Navier and Cauchy in 1821-1822, and plastic mechanics started from the maximum shear stress yield criterion proposed by Tresca in 1864. So far, elastic-plastic mechanics has been researched for 200 years and widely applied in engineering. It is an essential foundation and basis for studying and solving the structural strength issues of aeronautics, astronautics, weapon, civil engineering, machinery, and hydraulic engineering. Professor Jianguo Ding has been engaged in the teaching and research of elastic-plastic mechanics for nearly 40 years with considerable experience and achievements. In 2014, he wrote the English book *Fundamentals of Elastic and Plastic Mechanics* (*Edition for Bilingual Teaching*) (Wuhan University Press), which is widely welcomed by readers for its concise content and pure English. This revised bilingual book adds the related theories and engineering application examples based on the first edition. *Fundamentals of Elastic and Plastic Mechanics* (2nd edition) not only progressively explains the profound in simple terms but also forms its own system in a bilingual way. This book is scientific, readable, rigorous, and normative, keeping the essence of classic English books on elastic-plastic mechanics. Furthermore, the chapter arrangement, knowledge system, case analysis, and Chinese-English comparison have distinctive features.

Bilingual teaching is urgently needed to cultivate talents with an international academic vision, while bilingual books are the critical basis of bilingual teaching. This book fills in the gaps of Chinese-English bilingual books on elastic-plastic mechanics at home and abroad, which is of great significance for promoting bilingual teaching of elastic-plastic mechanics in China and motivating the construction and development of basic disciplines.

Professor Xiaoting Rui

Presiding Professor of Mechanics in Nanjing University of Science and Technology

The member of the Chinese Academic of Science

February 2022

序

弹塑性力学是研究弹性和塑性物体变形规律的学科分支，追求推理严谨及求解精确。弹性力学始于1821—1822 年 Navier 和 Cauchy 提出的变形体平衡微分方程及几何变形方程，塑性力学始于 1864 年Tresca 提出的最大剪应力屈服准则，至今弹塑性力学已有 200 年的研究历史，在工程中得到了广泛的应用，是研究解决现代航空、航天、兵器、土木、机械、水利等工程结构强度问题的重要基础和依据。丁建国教授从事弹塑性力学教学和科研工作近 40 年，经验与成果丰富。2014 年，丁建国教授编著的《弹塑性力学基础(双语教学版)》(武汉大学出版社)，因其内容精练、英文纯正而广受读者欢迎。本书是在第一版基础上，增加了有关理论、工程应用实例。本书内容深入浅出、循序渐进，双语自成体系，科学性强，可读性好，保留了弹塑性力学经典英文教材的精髓，内容严谨、规范。章节编排、知识体系、实例分析、中英文对照别具一格。

双语教学是培养具有国际学术视野人才的迫切需要，双语教材则是双语教学的重要基础。本书填补了国内外弹塑性力学中英文双语教材空白，对推动我国弹塑性力学双语教学工作，促进基础学科建设和发展具有重要意义。

南京理工大学力学学科首席学科带头人
中 国 科 学 院 院 士

2022 年 2 月

Preface

Engineers often encounter elastic-plastic mechanics problems in engineering practice. The seismic performance of structures under earthquakes must be analyzed and estimated when some civil engineering structures are designed. Engineers have to study the working state of mechanical components under various external loads to make safe and economical designs. Therefore, the basic theories and analytic methods for elastic-plastic mechanics are essential. Because the plastic state does not necessarily refer to a loss of load-bearing capacity, engineers allow plastic deformation in some components to dissipate earthquake energy to increase the overall structural seismic capability. It is of great theoretical and practical significance to study and master the "Fundamentals of Elastic and Plastic Mechanics".

To implement the Ministry of Education of PRC's promotion of Bilingual Teaching in colleges and universities, the author had the opportunity to bilingually teach a "Fundamentals of Elastic and Plastic Mechanics" course. After years of teaching practice and experience, following the principle of concise content, pure English, and system integrity, we compiled *Fundamentals of Elastic and Plastic Mechanics* (*Edition for Bilingual Teaching*) in 2014. This book is suitable for students majoring in civil engineering, hydraulic engineering, mechanical engineering, and ordnance engineering in China. There are many differences between the contents of a foundational course in elastic and plastic mechanics in our country and other developed Western countries, such as the United States. The original foreign books are expensive and are not suitable for Chinese students. After referring to some classic English books, we chose the suitable parts and rearranged them according to the content system used in China. Therefore, this book maintains the quintessence of the original foreign books while making the content suitable for Chinese students.

This book supplements the following contents based on the *Fundamentals of Elastic and Plastic Mechanics* (*Edition for Bilingual Teaching*) (2014): strain terms in two-dimensional polar coordinates, equilibrium differential equation in two-dimensional polar coordinates, total strain theory, simultaneous solution method with both equilibrium differential equation and yield criterion, and principal stress method of calculating the deformation force for cylinder upsetting. Meanwhile, the corresponding Chinese content is added. The newly revised bilingual book *Fundamentals of Elastic and Plastic Mechanics* (2nd edition) is more conducive to students' learning and improves students' ability to solve engineering problems. This book contains eight chapters: Force Systems, Stress, Strain, Stress-strain Relations, Basic Equations of Elasticity, Problems in Plane Strain and Plane Stress from the Theory of Elasticity, The Theory of Plasticity, and The Examples of Elastoplastic Analysis. A large number of corresponding exercises in both Chinese and English are provided.

Throughout writing this book, I have obtained the support and help of Professor Fan Jin, Associate Professor Xiao Feng, and some graduate students from the School of Science in Nanjing University of Science and Technology. I extend my heartfelt thanks to them.

Deficiencies are inevitable because of the limited level of the author, and I appreciate your comments.

Jianguo Ding
February 2022

前　言

工程技术人员在工程实践中经常遇到弹塑性力学问题,当设计土木工程结构时,需要对结构进行地震作用下的抗震性能分析和评估,当机械构件受各种外载作用时,要判断其工作状态,并做出既安全又经济的设计,这些都需要弹塑性力学的基本理论和分析方法。由于塑性状态下的结构并没有完全丧失承载能力,当进行工程结构设计时,有时还需让一部分结构先进入塑性状态消耗地震能量,以确保整体结构有足够的抗震能力。因此,学习并掌握"弹塑性力学基础"具有重要的理论意义和现实意义。

为了落实教育部关于在高等学校中推广双语教学课程的要求,编著者有幸承担了"弹塑性力学基础"的双语教学工作。美国等西方发达国家在弹塑性力学基础课程的内容安排方面与我国存在较大的差异,并且国外原版教材比较昂贵,要求我国学生直接选用国外原版教材并不合适,因此,编著者在经过了多年对该课程的双语教学实践并在经验总结的基础上,按照内容精练、英文纯正及体系完整的原则,通过参考多本英文版经典教材,从中选用合适的部分并按我国的内容体系重新编排,在2014年出版了可供我国土木工程、水利工程、机械工程、兵器工程等本科专业学生使用的《弹塑性力学基础(双语教学版)》。这样既保留了国外原版教材的精髓,又使教材内容体系符合我国国情。

本书在2014年版《弹塑性力学基础(双语教学版)》的基础上补充了以下内容:二维极坐标下的应变项、二维极坐标下的平衡微分方程、塑性力学全量理论、平衡微分方程和屈服准则联合应用及圆柱体镦粗变形力计算的主应力法实例等内容,同时增加了与英文内容相对应的中文内容。总之,《弹塑性力学基础(双语版)(第2版)》更加有利于学生的学习,更有利于培养学生解决工程问题的能力。本书共分8章,其内容分别为力系(Force Systems)、应力(Stress)、应变(Strain)、应力-应变关系(Stress-strain Relations)、弹性力学基本方程(Basic Equations of Elasticity)、弹性理论中平面应变和平面应力问题(Problems in Plane Strain and Plane Stress from the Theory of Elasticity)、塑性理论(The Theory of Plasticity)、弹塑性分析实例(The Examples of Elastoplastic Analysis)。此外,本书还配有大量相互对应的中英文习题。

在本书编写过程中得到了南京理工大学理学院范进教授、肖枫副教授及部分研究生的支持和帮助,特向他们致以衷心的感谢。

由于编著者水平有限,书中难免有不足之处,欢迎读者批评指正。

丁建国

2022年2月

目 录
Contents

英 文 篇

中　文　篇

英　文　篇

1　Force Systems

1.1　Introduction

In our studies of rigid-body mechanics, the deformation of bodies was of no significance in the problems we were able to solve. You will recall that, in such problems, Newton's law was all we needed in order to compute certain unknown forces acting on bodies in equilibrium. The study of statics did lead, however, to problems where the use of Newton's law alone was insufficient for the handling of the problem even though the bodies involved seemed quite "rigid" from a physical point of view. For those problems, called statically indeterminate problems, the deformation, however small, was significant for the determination of the desired forces. As an illustration, consider the simply supported beam shown in Figure 1-1(a) with the free body diagram of the beam shown in Figure 1-1(b). You may readily solve for the supporting forces A, B_x, and B_y by the method of rigid-body mechanics, provided that there is little change of position of the external loads as they are applied to the beam. We can accomplish this because we know that the resultant force system on the free body is of zero value. By setting this resultant equal to zero while using the undeformed geometry of the problem we can easily solve for the three unknowns. Suppose next that there are three supports for the beam instead of two as shown in Figure 1-2. Clearly, the deformation of the beam can be expected to be even smaller in extent than for the two-support system, so there should be no difficulty in this problem arising from a changing geometry. The total supporting force system from the two supports is in accordance with the dictates of rigid-body mechanics. However, we cannot, in this latter problem, determine the value of supporting forces A, C, B_x and B_y uniquely since there are an infinite number of combinations of values that will give the precise total resultant force required by rigid-body mechanics. To choose the proper combination of values requires the consideration of the deformation of the beam, small as this deformation might be. We see that the rigid-body considerations afford us a necessary requirement for the resultant of the supporting force system whereas the deformation analysis supplies the additional information sufficient to determine the values of each supporting force. Statically indeterminate problems akin to the one discussed will be one of the classes of problems that we shall undertake to solve.

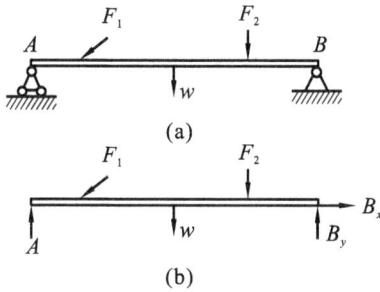

Figure 1-1   The simply supported beam and its force view

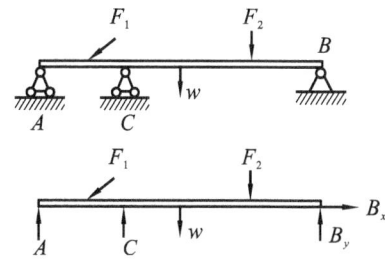

Figure 1-2   The beam of three supports and its force view

Next, consider some arbitrary solid in equilibrium as shown in Figure 1-3. Suppose we pass a hypothetical plane M through the body as shown in the diagram. We wish to determine the force distribution that is transmitted from one portion of the body A to the other portion B through this interface. Considering part B as a free body (Figure 1-4), we can find by the methods of rigid-body mechanics a force and couple at some position in the section that is the correct resultant force system for the desired distribution, provided, of course, that the applied forces have not appreciably changed their initial known orientation as a result of deformation. But just like the supports of the indeterminate beam problem, there are an infinite number of distributions that can yield this resultant force system. We must investigate the deformation of the entire body in order to obtain sufficient additional information for establishing a unique force distribution.

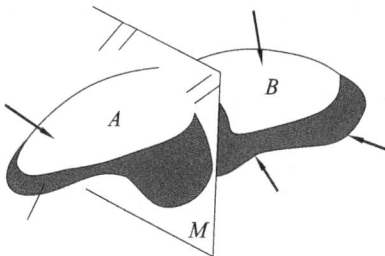

Figure 1-3   Some arbitrary solid in equilibrium

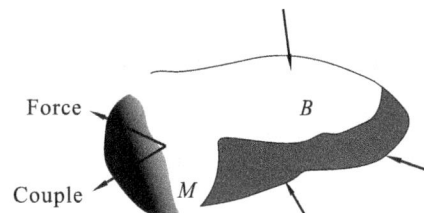

Figure 1-4   The free body diagram of some arbitrary solid

The knowledge of force distribution in solids is of vital importance to the design of most systems. It might be pointed out that the spectacular increase in the range of single-stage chemical rockets in recent years from the 200-mile range of the German V-2 to the 1000-mile-plus range of the IRBM has been the result largely of structural improvements rather than of propulsion gains.

We shall therefore be interested in internal distribution of forces in solids as well as the computation of certain discrete forces in statically indeterminate structural problems like the beam problem.

Until now,we have been centering our attention on the computation of certain discrete forces and force distributions for problems requiring the consideration of the deformation of the body. It should be clear that sometimes the information of the deflection itself may be of paramount interest and not that of the force distributions. Because we shall limit ourselves to problems involving small deformation,we shall be able to determine the deflection of the statically determinate two-support beam by first computing the supporting forces using rigid-body mechanics,and then proceeding with considerations of deformation. And in the statically indeterminate three-support beam,we shall compute the supporting forces as well as the deflection at the same time since the deformation is intrinsically connected with both computations.

Unlike your studies of rigid-body continua in which you consider statics and dynamics of rigid bodies,we shall,in our studies of deformable solids,restrict ourselves essentially to statical problems. However,there is an important class of non-statical problems for which the formulations of this book can be applied. Suppose that,in response to a given force system,a body moves while undergoing little change in shape so that by rigid-body mechanics we can compute the motion of the body at any instant using the undeformed shape of the body. We then may employ the methods of this book to compute the deformation of the body resulting from the combined action of the given forces and computed inertial forces. However,there is an important proviso that the aforementioned forces vary slowly with time. This requirement is imposed because we shall compute the deformation of the body at time t as if the applied force system and the inertial force distribution computed from the rigid-body motion of the body at time t were static loads on the body. Clearly such an approach for a rapid time-varying force distribution would have little meaning. The case of the rotating disk is an important example of the kind of problem that may be suitable for the approach described here.

The dynamics of deformable solids is reserved for more advanced course,where you may consider the vibrations of beams,plates,and shells; or wave propagation in solids; or possibly,shock loading of structures. Others may study the stresses in machines such as jet engines,or the vibrations induced in rockets by the propulsion system. These are fascinating problems beyond the scope of this book. They require the understanding of the fundamentals that we shall stress in our studies here.

Since we shall be dealing with forces to a great extent,it is now useful to make certain classification.

1.2 Types of Force Distributions

We shall at this time set forth classifications of force distributions. Force fields which act throughout a body, i. e. , force fields which exert influence on the mass distributed throughout the body, are called body-force distributions. The force of gravity and inertial forces are the chief examples of such distributions. Body-force distributions are expressed per unit mass or per unit volume of the body elements they directly influence. Thus if $\boldsymbol{B}(x,y,z,t)$ is a body-force distribution per unit mass, the force on element of mass dm would be

$$\mathrm{d}\boldsymbol{F} = \boldsymbol{B}(x,y,z,t)\,\mathrm{d}m \tag{1-1}$$

In addition to body-force distributions, we also have surface-force distributions. These force distributions act on the boundary of a body that we may want to consider. The surface-force distribution is given per unit of area of the boundary acted on. A surface-force distribution might consist of a force distribution acting on the outside surface of the body shown in Figure 1-3 or might equally well include also the distribution on a surface exposed by a hypothetical plane, such as M in the diagram. The force and couple shown in Figure 1-4 are the resultant force system of such a distribution. We sometimes wish to distinguish between surface-force distribution at actual boundaries and surface-force distribution at hypothetically exposed boundaries, such as plane M. When this is the case, we call the surface-force distribution at the actual boundary the surface traction. We shall have occasion later in the book to wish to use this distinction.

1.3 Closure

In this chapter, we have attempted to show some connection between earlier rigid-body mechanics and studies to be undertaken in this book. Also we presented certain definitions of force distribution that will help us in our study of deformable solids. We shall next consider carefully force distributions on internal surfaces which are hypothetically exposed by the use of free-body diagrams.

2 Stress

2.1 Introduction

In Chapter 1, we discussed two types of force distributions, namely body-force distributions and surface-force distributions. We shall now consider the latter in greater detail. You will recall that surface-force distributions may be found on an actual boundary, in which they are called surface tractions, or they may be considered at internal sections of a body exposed by the technique of free bodies. Clearly any internal interface of a body may in this way be exposed so as to have a surface-force distribution associated with it. By this reasoning we can think of surface-force distributions pervading the entire body. Indeed, it is by such a viewpoint that we are able to describe quantitatively how external loads are transmitted through a body.

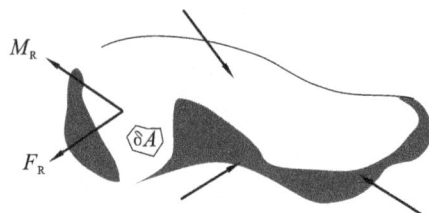

Figure 2-1   A body in equilibrium

Consider a small area δA of a hypothetical interface of a body in equilibrium, as shown in Figure 2-1. Notice that the rigid-body resultant force F_R and couple M_R have been shown for the whole interface. For the area element there will also be a resultant force δF and a resultant couple δM as indicated in Figure 2-2.

If this area element is decreased to infinitesimal size, the couple can be neglected because the force distribution across the area then approaches that of a uniform and parallel distribution which can be replaced by a single force, as we learned in rigid-body mechanics. We will then delete the couple δM from the ensuing remarks since we shall soon go to the infinitesimal limit. It is convenient to decompose δF into a set of orthogonal components, as shown in Figure 2-3 in which one of the components δF_n is normal to the area element whereas the other components δF_{s_1} and δF_{s_2} are tangent to the area element.

We may now define the normal stress σ and the shear stress τ_s by the following limiting processes

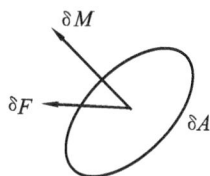

Figure 2-2   The area element of a body

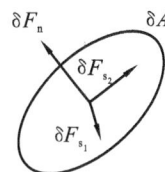

Figure 2-3   A set of orthogonal components
in the area element of a body

$$\begin{cases} \sigma = \lim\limits_{\delta A \to 0} \dfrac{\delta F_{\mathrm{n}}}{\delta A} = \dfrac{\mathrm{d}F_{\mathrm{n}}}{\mathrm{d}A} \\[2mm] \tau_{\mathrm{s}_1} = \lim\limits_{\delta A \to 0} \dfrac{\delta F_{\mathrm{s}_1}}{\delta A} = \dfrac{\mathrm{d}F_{\mathrm{s}_1}}{\mathrm{d}A} \\[2mm] \tau_{\mathrm{s}_2} = \lim\limits_{\delta A \to 0} \dfrac{\delta F_{\mathrm{s}_2}}{\delta A} = \dfrac{\mathrm{d}F_{\mathrm{s}_2}}{\mathrm{d}A} \end{cases} \qquad (2\text{-}1)$$

We see that normal and shear stresses are intensities of force components given per unit area. Note that they are scalar quantities. Giving normal and shear stress distributions is our way of describing a force distribution over a plane interface and, as we shall soon see the distribution of force through a body.

2.2   Stress Notation

In Section 2.1, we showed how we could describe a force distribution over a plane interface inside a body. We can in this way set forth normal and shear stresses for any interface at a point, hence, we can describe the manner in which external forces are transmitted throughout a body. With three stress components set forth for each interface at a point, it becomes imperative to formulate an effective and meaningful notation to identify the stresses. For this purpose, consider in a deformed body an infinitesimal element with faces parallel to Cartesian reference as shown in Figure 2-4 where each of these faces forms infinitesimal rectangular parallelepiped. Stresses on three faces have been shown. Notice that a double index scheme has been utilized. The first subscript indicates the direction of the normal to the plane on which the stress is considered; the second subscript denotes the direction of the stress itself. The normal stress must then have one indice, since the stress direction and the normal to the plane on which the stress acts are collinear. The shear stresses, on the other hand, have mixed indices. For example, τ_{yx} is the shear stress acting on the plane parallel to the xz plane whose normal proceeds in the y direction whereas the stress itself is oriented in the x direction.

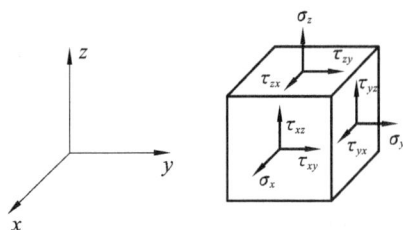

Figure 2-4   An infinitesimal element with faces parallel to Cartesian reference

As for the sign convention, we shall follow these rules: a stress acting on an area element whose outward normal points in the positive direction of any coordinate axis will be taken as positive if the stress itself also points in the positive direction of any coordinate axis(The axes for the area vector and stress need not be the same axes). A stress is positive also if both the area vector and the stress point in the negative direction of the same or different coordinate axes. Thus you will note on inspection that the stresses shown in Figure 2-4 are all positive stresses. If now the area vector and stress are not directed simultaneously in either positive or negative coordinate directions, the stress is negative.

We shall see, in Section 2.3, that knowing the stresses on three orthogonal interfaces at a point, we can determine, by employing transformation formulations for stress, stresses on any interface at that point. Specifying distributions corresponding to interfaces parallel to Cartesian set of axes is tantamount to specifying the stress distribution throughout the entire body. The notation that we have presented will then prove to be extremely helpful.

It should be clearly understood that stress is not restricted to solids. Our conclusions here and indeed throughout the entire chapter apply to any continuous medium exhibiting viscosity or rigidity.

We now proceed to develop transformation formulations for stress.

2.3   Transformation Formulations for Stress

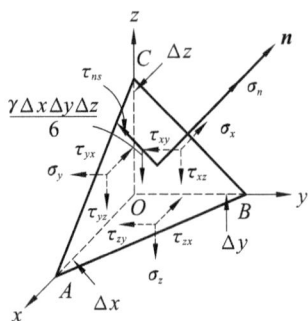

Figure 2-5   A small tetrahedron of

a continuous medium

Let us consider a small tetrahedron of a continuous medium as shown in Figure 2-5. The orthogonal edges of the tetrahedron are of length Δx, Δy and Δz respectively. Positive shear and normal stresses have been shown on the faces parallel to the reference planes. On the inclined surface whose outward normal has been indicated as $\boldsymbol{n}$, we have shown the normal stress σ_n and the total shear stress τ_{ns}. It is convenient to denote the direction cosines of $\boldsymbol{n}$ with respect to the x, y and z axes as $l, m,$ and n respectively. With this notation, we now express the areas of the coordinate faces of the tetrahedron in terms of the area of the inclined face ABC in a manner given by Equation (2-2).

$$\begin{cases} AOC = ABCm \\ BOA = ABCn \\ COB = ABCl \end{cases} \qquad (2\text{-}2)$$

Next we write Newton's law in the direction of $\boldsymbol{n}$. Thus

$$\sigma_n ABC - \sigma_x COBl - \tau_{xx}COBn - \tau_{xy}COBm - \sigma_y AOCm - \tau_{yz}AOCn - \tau_{yx}AOCl -$$

$$\sigma_z BOAn - \tau_{zx}BOAl - \tau_{zy}BOAm - \gamma \frac{\Delta x \Delta y \Delta z}{6}n = \rho \frac{\Delta x \Delta y \Delta z}{6}a_n$$

$$(2\text{-}3)$$

where a_n is the acceleration in the direction $\boldsymbol{n}$, γ is the specific weight, and ρ is the density.

We now replace areas AOC, BOA and COB in Equation (2-3), using Equation(2-2). Next we divide through by ABC. Finally, we take the limit of each term as the quantities $\Delta x, \Delta y$ and Δz approach zero. Clearly, the last two terms vanish, since ABC is the order of magnitude of a product of two of the terms $\Delta x, \Delta y$ and Δz, leaving in both cases a $\dfrac{\Delta x \Delta y \Delta z}{ABC}$ term which vanishes in the limit. We thus have for σ_n at a point

$$\sigma_n = \sigma_x l^2 + \sigma_y m^2 + \sigma_z n^2 + \tau_{xy}lm + \tau_{xz}ln + \tau_{yz}mn + \tau_{yx}ml + \tau_{zx}nl + \tau_{zy}nm \qquad (2\text{-}4)$$

We see that the normal stress on any plane at a point depends only on the stresses on an orthogonal set of planes at the point and the direction cosines associated with the desired normal stress, where these direction cosines are measured relative to coordinate axes parallel to the aforementioned set of orthogonal planes.

We now proceed to compute shear stress on the inclined surface ABC of the tetrahedron by a similar computation as was performed for the normal stress. Accordingly, in Figure 2-6 we have shown the tetrahedron with a shear stress τ_{ns} on the inclined face. The direction cosines of this stress are given as l', m' and n'. Since the normal-stress direction and the shear-stress direction are at right angles to each other, the following equation must be satisfied by the two sets of direction cosines

$$ll' + mm' + nn' = 0 \qquad (2\text{-}5)$$

If we now write Newton's law in the direction of the shear stress, denoted as $\boldsymbol{s}$ in Figure 2-6, we may proceed in a manner paralleling the development of Equation (2-4) to form the following equation

$$\tau_{ns}=ll'\sigma_x+mm'\sigma_y+nn'\sigma_z+lm'\tau_{xy}+ml'\tau_{yx}+ln'\tau_{xz}+nl'\tau_{zx}+mn'\tau_{yz}+nm'\tau_{zy} \quad (2\text{-}6)$$

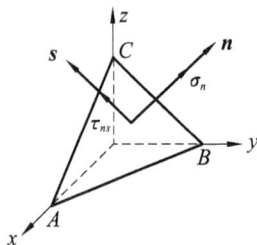

Figure 2-6   The tetrahedron with a shear stress on the inclined face

Transformation Equations (2-4) and (2-6) thus permit us to compute all stresses at a point provided that we know the nine stresses on a set of orthogonal faces at the point. You will recall that a vector quantity requires the specification of only three components at a point. The concept of stress, obviously, is more complicated. Quantities which transform at a point in a manner given by Equations(2-4) and (2-6) are called second-order tensors. In addition to the stress tensor, we shall also find in the following chapter that strain is a tensor. And in the study of rigid-body dynamics one deals with the inertia tensor. Because these quantities transform in a certain way when we change coordinates at a point, they have certain distinct characteristics which set them apart from other quantities. We shall shortly investigate some of these properties and we shall devise a more efficient notation for handling these quantities. The stress tensor is usually represented as follows

$$\begin{bmatrix} \sigma_x & \tau_{xy} & \tau_{xz} \\ \tau_{yx} & \sigma_y & \tau_{yz} \\ \tau_{zx} & \tau_{zy} & \sigma_z \end{bmatrix} \quad (2\text{-}7)$$

Notice that the first subscript can be considered to identify a row of the array of terms whereas the second subscript can be considered to identify a column of the array. Also note that the downward left-to-right diagonal, called the principal or main diagonal, is composed only of normal stresses. The sum of the terms in the principal diagonal is termed the trace of the tensor.

As a first step in shortening the notation, we now introduce the indicial or Cartesian tensor notation where we express a vector V having components V_x, V_y and V_z as V_i, where it is assumed that i takes on all values x, y, z. Thus using the subscript i is tantamount to denoting three scalar components and hence sufficient to specify a vector. For the stress tensor, we may merely give τ_{ij} and σ_i in place of the array in Equation (2-7) with the tacit understanding that i and j can take on all values x, y, z in all possible permutations. We shall later devise very useful operations between vector and tensor quantities by setting forth certain procedures using the indices.

2.4 The Stress Tensor Is Symmetric

We shall now show that shear stresses at a point with reversed indices must be equal to each other, i. e. , $\tau_{ij} = \tau_{ji}$. This means that the corresponding shear stresses on each side of the principal diagonal of the array of stresses given by Equation (2-7) are equal. To do this most simply consider an infinitesimal element of a body in equilibrium shown in Figure 2-7(a). If we take moments about the edge O-x only those stresses shown in Figure 2-7(b) are involved and, accordingly, dropping higher order terms we may arrive readily at the conclusion that

$$\tau_{yx} = \tau_{xy} \tag{2-8}$$

We can also show that this relation holds even when the element has motion.

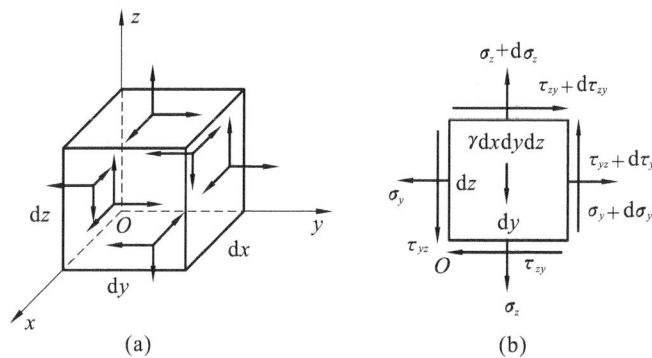

Figure 2-7   An infinitesimal element of a body in equilibrium

Not only are the shear stresses τ_{xy} and τ_{yx} equal at a point, but also they are limited to certain senses relative to each other. The two possibilities are shown in Figure 2-8. In short, you may notice that the shear stresses must always point toward, or away from a corner. By considering moments about the various corners of the element while observing Figure 2-8, you can see that only the orientations shown are permitted by Newton's law.

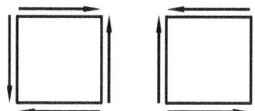

Figure 2-8   The orientations of shear stresses

By similar considerations for the other coordinate edges of the element, we may generalize our conclusions to the three-dimensional case of stress. Thus using Cartesian tensor notation we have

$$\tau_{ij} = \tau_{ji} \tag{2-9}$$

Notice furthermore that we have properly oriented the shear stresses relative to each other in Figure 2-4 and in Figure 2-5.

Many of the tensors that the engineer deals with are symmetric tensors. These include the strain tensor (to be discussed in the next chapter), the inertia tensor of rigid-body mechanics, and the quadrupole tensor of electromagnetic theory.

Using the symmetry property of the stress tensor, we can rewrite the transformation Equation (2-4) in the following form

$$\sigma_n = \sigma_x l^2 + \sigma_y m^2 + \sigma_z n^2 + 2(\tau_{xy}lm + \tau_{xz}ln + \tau_{yz}mn) \tag{2-10}$$

Also the transformation Equation (2-6) may similarly be rewritten as follows

$$\tau_{ns} = ll'\sigma_x + mm'\sigma_y + nn'\sigma_z + (lm' + ml')\tau_{xy} + (ln' + nl')\tau_{xz} + (mn' + nm')\tau_{yz} \tag{2-11}$$

2.5　Evaluation of Principal Stress; Tensor Invariants

We have shown an infinitesimal tetrahedron in Figure 2-9 having three faces with known stresses on the reference planes. We assume that the inclined face ABC is a principal plane and we shall employ the letter σ without subscripts to be the principal stress on this plane. The direction of σ, which is of course the direction of the normal to the inclined surface ABC, has been given by the unit vector $\boldsymbol{n}$ whose direction cosines are denoted as l, m and n.

To ascertain the principal stress σ, we employ Newton's law in the coordinate directions x, y and z. Considering the z component first, we have

$$\sigma ABCn - \sigma_z AOB - \tau_{yz}AOC - \tau_{xz}BOC - \gamma \frac{\Delta x \Delta y \Delta z}{6} = \rho \frac{\Delta x \Delta y \Delta z}{6}a_z \tag{2-12}$$

Dropping the gravity and inertia terms as higher-order terms and employing Equation (2-2), we arrive, after we cancel ABC and rearrange terms, at the following equation

$$\tau_{xz}l + \tau_{zy}m + (\sigma_z - \sigma)n = 0$$

where we have made use of the complementary property of shear stress. By considering Newton's law in other coordinate directions in a similar manner, we arrive at two other equations like the foregoing. We give the three resulting equations in the following form

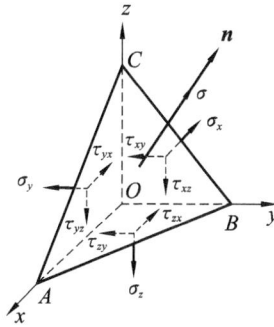

Figure 2-9   The stresses on all faces of an infinitesimal tetrahedron

$$(\sigma_x - \sigma)l + \tau_{xy}m + \tau_{xz}n = 0 \tag{2-13a}$$

$$\tau_{yx}l + (\sigma_y - \sigma)m + \tau_{yz}n = 0 \tag{2-13b}$$

$$\tau_{zx}l + \tau_{zy}m + (\sigma_z - \sigma)n = 0 \tag{2-13c}$$

We now wish to ascertain σ and the corresponding direction cosines l, m and n. At a first step, we can adopt the viewpoint that the direction cosines are the unknowns in the preceding set of equations and are to be solved in terms of the stresses. Thus, using Cramer's rule, we have for l

$$l = \frac{\begin{vmatrix} 0 & \tau_{xy} & \tau_{xz} \\ 0 & \sigma_y - \sigma & \tau_{yz} \\ 0 & \tau_{zy} & \sigma_z - \sigma \end{vmatrix}}{\begin{vmatrix} \sigma_x - \sigma & \tau_{xy} & \tau_{xz} \\ \tau_{yx} & \sigma_y - \sigma & \tau_{yz} \\ \tau_{zx} & \tau_{zy} & \sigma_z - \sigma \end{vmatrix}} \tag{2-14}$$

Clearly l will be zero, as will the other direction cosines, unless the denominator in Eequation (2-14) is zero so as to permit an indeterminate result for l. But the direction cosines cannot all be zero since we know that

$$l^2 + m^2 + n^2 = 1 \tag{2-15}$$

Thus a necessary condition required for the solution of our problem is

$$\begin{vmatrix} \sigma_x - \sigma & \tau_{xy} & \tau_{xz} \\ \tau_{yx} & \sigma_y - \sigma & \tau_{yz} \\ \tau_{zx} & \tau_{zy} & \sigma_z - \sigma \end{vmatrix} = 0 \tag{2-16}$$

By expanding out the determinant, we get a cubic equation in the unknown σ, i. e.

$$\sigma^3 - (\sigma_x + \sigma_y + \sigma_z)\sigma^2 + (\sigma_x\sigma_y + \sigma_y\sigma_z + \sigma_z\sigma_x - \tau_{xy}^2 - \tau_{yz}^2 - \tau_{zx}^2)\sigma - (\sigma_x\sigma_y\sigma_z - \sigma_x\tau_{yz}^2 - \sigma_y\tau_{xz}^2 - \sigma_z\tau_{xy}^2 + 2\tau_{xy}\tau_{yz}\tau_{zx}) = 0 \tag{2-17}$$

One can prove that there are always three real roots for Equation (2-17), i. e., there are always three real values σ_1, σ_2 and σ_3 that satisfy the equation. These are the principal stresses which we have alluded to earlier.

To get the direction cosines for any one of the computed principal stresses σ_i, we simply substitute σ_i into Equation (2-13) and solving any pair of this set simultaneously with Equation (2-15).

Not only does Equation (2-17) permit us to evaluate principal stresses at a point but it also permits us to make certain far-reaching conclusions about the state of stress at a point in general. We have pointed out that there is a unique quadratic surface for a given state of stress at a point. This means that the principal stresses depend only on the state of stress at a point and not on the coordinate system that one might choose to work with at the point. Thus, if we rotate xyz to some other reference $x''y''z''$ having the same original point, the Equation (2-17) for this new reference should yield the same set of roots σ_1, σ_2 and σ_3. For this to be the case for any general rotation of axes, we must conclude that the coefficient of the powers of σ in Equation (2-17) must be invariant under a rotation of reference, i. e.

$$\sigma_x + \sigma_y + \sigma_z = I_1 \tag{2-18a}$$

$$\sigma_x\sigma_y + \sigma_y\sigma_z + \sigma_z\sigma_x - \tau_{xy}^2 - \tau_{yz}^2 - \tau_{zx}^2 = I_2 \tag{2-18b}$$

$$\sigma_x\sigma_y\sigma_z - \sigma_x\tau_{yz}^2 - \sigma_y\tau_{zx}^2 - \sigma_z\tau_{xy}^2 + 2\tau_{xy}\tau_{yz}\tau_{zx} = I_3 \tag{2-18c}$$

where I_1, I_2 and I_3 are constants for a given set of values x, y, z. Of course, when we move to another point the set of quantities, I_1, I_2 and I_3 may change values. Thus, each of the aforementioned set of quantities forms a scalar field in a stressed body since the value of each quantity is only a function of position. These quantities are called the first, second, and third tensor invariants of stress respectively.

You will notice that the first-tensor invariant of stress is the sum of the main-diagonal terms in the stress tensor, i. e. , the trace of the stress tensor. One-third of this value is then the average normal stress at a point and we often call this the bulk stress σ_m, i. e.

$$\sigma_m = \frac{1}{3}(\sigma_x + \sigma_y + \sigma_z) \tag{2-19}$$

In the case of a fluid, minus the average normal stress, σ_m is simply called the pressure.

The second-tensor invariant of stress is easily shown to be the sum of three determinants formed from the matrix representation of the stress tensor.

$$\tag{2-20}$$

Determinant 1

$$\begin{vmatrix} \sigma_x & \tau_{xy} & \tau_{xz} \\ \tau_{yx} & \sigma_y & \tau_{yz} \\ \tau_{zx} & \tau_{zy} & \sigma_z \end{vmatrix}$$

Determinant 2

Two of the determinants are labeled whereas the third is that of the circled quantities. You may recognize that these submatrices form the minors of the terms in the principal diagonal. Thus I_2 is the sum of principal minors.

The third-tensor invariant of stress is simply the determinant of the entire matrix representation of the stress tensor.

2.6 Plane Stress

We now present a simplified stress distribution called plane stress which may be used to represent the state of stress in many practical problems. We define plane stress as a stress distribution where in all stress components in some one direction are zero. This direction is usually taken as the z direction. Hence, plane stress requires that

$$\tau_{xz} = \tau_{yz} = \sigma_z = 0 \tag{2-21}$$

Thin plates acted on by loads lying in the plane of symmetry of the plate, as shown in Figure 2-10, can often be considered to be in a state of plane stress with the z direction taken normal to the plate. Clearly with no loads normal to the plane of the plate, σ_z will be zero on both surfaces of the plate. Since the plate is thin we can also consider σ_z to be equal zero inside the plate. Also with forces oriented parallel to the plate surfaces we would expect τ_{xz} and τ_{yz} to be zero throughout.

We shall, in this section, reduce some of the general formulations developed earlier for this special but useful stress distribution. This will give us more feel for the more general equations and their implications and will also provide us with useful simplified working formulas.

Accordingly, consider an infinitesimal prismatic element under plane stress as is shown in Figure 2-11. The normal stress σ can be computed in terms of the stresses σ_x, σ_y and τ_{xy} by employing Equation (2-10) in the following manner

$$\sigma = \sigma_x l^2 + \sigma_y m^2 + 2\tau_{xy} lm \tag{2-22}$$

And the shear stress τ can be computed similarly by employing Equation (2-11), as follows

$$\tau = ll'\sigma_x + mm'\sigma_y + (lm' + ml')\tau_{xy} \tag{2-23}$$

For plane-stress problems it will be convenient to introduce the angle θ (Figure 2-11) to replace the direction cosines in the following manner

$$l = \cos\theta \quad l' = -\sin\theta$$
$$m = \sin\theta \quad m' = \cos\theta$$

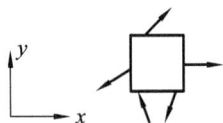

Figure 2-10　A state of plane stress

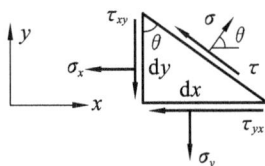

Figure 2-11　An infinitesimal prismatic element under plane stress

We then get the following pair of equations

$$\begin{cases} \sigma = \sigma_x \cos^2\theta + \sigma_y \sin^2\theta + 2\tau_{xy} \sin\theta\cos\theta \\ \tau = -\sigma_x \cos\theta\sin\theta + \sigma_y \sin\theta\cos\theta + (\cos^2\theta - \sin^2\theta)\tau_{xy} \end{cases} \tag{2-24}$$

Noting the following trigonometric identities

$$\begin{cases} \cos^2\theta = \dfrac{1}{2}(1 + \cos2\theta) \\ \sin^2\theta = \dfrac{1}{2}(1 - \cos2\theta) \\ 2\sin\theta\cos\theta = \sin2\theta \end{cases} \tag{2-25}$$

We may rewrite Equation (2-24) in the following manner

$$\sigma = \frac{\sigma_x + \sigma_y}{2} + \frac{\sigma_x - \sigma_y}{2}\cos2\theta + \tau_{xy}\sin2\theta \tag{2-26a}$$

$$\tau = \frac{\sigma_y - \sigma_x}{2}\sin2\theta + \tau_{xy}\cos2\theta \tag{2-26b}$$

Let us next turn to the computation of principal stress for the special case at hand. Equation (2-17) simplifies to the following form

$$\sigma^3 - (\sigma_x + \sigma_y)\sigma^2 + (\sigma_x\sigma_y - \tau_{xy}^2)\sigma = 0 \tag{2-27}$$

Canceling σ from the equation means that one of the roots is always zero. This principal stress must clearly correspond to the z direction. The other two roots are determined by solving the remaining quadratic equation. We then get

$$\sigma_1, \sigma_2 = \frac{\sigma_x + \sigma_y}{2} \pm \sqrt{\left(\frac{\sigma_x - \sigma_y}{2}\right)^2 + \tau_{xy}^2} \tag{2-28}$$

where σ_1 and σ_2 are the remaining principal stresses.

To get the orientation of the principal axes, we need only set $\tau = 0$ in Equation (2-26b) and solve for 2θ. Replacing θ by β for this result we get

$$\tan2\beta = \frac{2\tau_{xy}}{\sigma_x - \sigma_y} \tag{2-29}$$

There are two values of 2β which are $180°$ apart and which can satisfy Equation (2-29) for a given set of stresses. Therefore there will be two values of β which are 90° apart, that can be found from the preceding equation. These give the orientation of the principal planes at the point.

You can readily show that the maximum shear stress τ_{max} is so oriented that

$$\tan 2\beta' = -\frac{\sigma_x - \sigma_y}{2\tau_{xy}} \tag{2-30}$$

where β' now replaces θ. Again there are two values β' which are 90° apart, which can be found from Equation (2-30). Furthermore, we can conclude, by noting that the right sides of Equations (2-29) and (2-30) are negative reciprocals of each other, that the planes of maximum shear stress are oriented at 45° from the principal planes of stress, as is shown in Figure 2-12.

In the general three-dimensional state of stress, one can show that the plane of maximum shear is also oriented at 45° to a set of principal planes at a point. In the following sections, we set forth graphical aids which permit us to deduce these conditions easily.

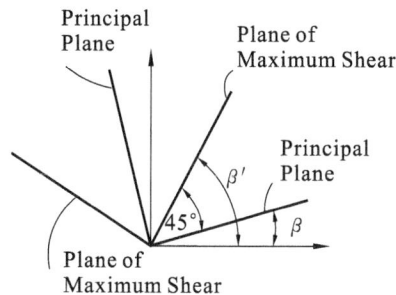

Figure 2-12　The orientation of principal axes

2.7　Mohr's Circle

A convenient graphical representation called the Mohr's circle, is in common use to represent the state of plane stress at a point. To employ Mohr's circle, we must set forth additional sign conventions. Consider the infinitesimal rectangular element shown in Figure 2-13 where we have shown positive shear and normal stresses without the first-order variations. In order to deal with Mohr's circle, we shall employ the following rule for shear stress: A shear stress will be taken as positive on a face of an element when it yields a clockwise moment about the center point O of the face. A shear stress yielding a counterclockwise moment about the center point O will then be taken as negative in reference to Mohr's circle. Thus τ_{xy} in Figure 2-13 will be considered negative and τ_{yx}

will be considered positive.

Let us now introduce the stress coordinates, as shown in Figure 2-14, where σ, i. e. , normal stress, is the abscissa and τ, i. e. , shear stress, is the ordinate. To draw Mohr's circle on this plane, we first plot the stresses for two orthogonal adjacent faces of an element, as for example faces a and b in Figure 2-13, using the sign convention for shear stress developed in the previous paragraph.

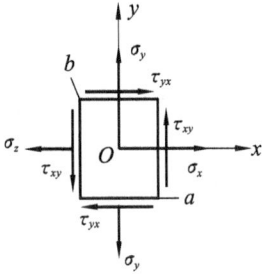

Figure 2-13 **An infinitesimal rectangular element**

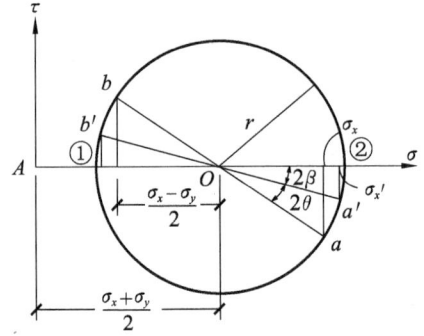

Figure 2-14 **The stress coordinates of Mohr's circle**

The mapped points of faces a and b are denoted in the stress diagram also as a and b, respectively. Now connect these points with a straight line so as to intersect the σ axis at O. With O as a center, draw a circle through points a and b as we have shown in Figure 2-14. This is the celebrated Mohr's circle.

How can we use the Mohr's circle? Suppose we wish to know the stress for reference $x'y'$ rotated through an angle θ from xy as shown in Figure 2-15. To find the point a' in the stress diagram corresponding to face a' in the physical diagram, we draw a line from point O in the stress diagram rotated from the Oa axis by an angle 2θ in the same direction as in the physical plane. The coordinates of the intersection a' of this axis with the Mohr's circle then give the shear and normal stresses corresponding to face a' in the physical diagram. Now by extending Oa', we may also form point b' on Mohr's circle. Since Ob' is rotated 180° from Oa', point b' corresponds to the state of stress on face b' which is rotated 90° from face a' in the physical diagram. We thus have the state of stress for faces a' and b' corresponding to the x' and y' axes, respectively.

Thus far we have shown how to construct Mohr's circle and how to employ it. Our next step is proving the validity of the method. To do this, we must show that stresses deduced from Mohr's circle satisfy Equation (2-26). Inspection of the Mohr's circle (Figure 2-14) leads one to the following conclusions

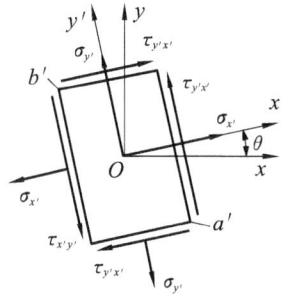

Figure 2-15 **The physical diagram for reference $x'y'$ rotated through an angle from xy**

$$OA = \frac{\sigma_x + \sigma_y}{2} \tag{2-31}$$

$$r = \sqrt{\left(\frac{\sigma_x - \sigma_y}{2}\right)^2 + \tau_{xy}^2} \tag{2-32}$$

These quantities have been shown in Figure 2-14. Using geometrical reasoning for Mohr's circle, we can say for the stress σ, which corresponds to point a' in the stress diagram

$$\sigma = OA + r\cos(2\beta - 2\theta) \tag{2-33}$$

Substituting Equation (2-31) into Equation (2-33) and expanding the cosine term, we get

$$\sigma = \frac{\sigma_x + \sigma_y}{2} + r(\cos2\beta\cos2\theta + \sin2\beta\sin2\theta) \tag{2-34}$$

But from the stress diagram it is clear on inspection that

$$\cos2\beta = \frac{\sigma_x - \sigma_y}{2r} \tag{2-35a}$$

$$\sin2\beta = \frac{\tau_{xy}}{r} \tag{2-35b}$$

Substituting these results into Equation (2-34), we get

$$\sigma = \frac{\sigma_x + \sigma_y}{2} + \frac{\sigma_x - \sigma_y}{2}\cos2\theta + \tau_{xy}\sin2\theta \tag{2-36}$$

which is the proper relation that we derived earlier.

In a similar manner, we first show for τ

$$\tau = r\sin(2\beta - 2\theta) = r(\sin2\beta\cos2\theta - \cos2\beta\sin2\theta) \tag{2-37}$$

Using Equation (2-35) we get

$$\tau = \tau_{xy}\cos2\theta - \frac{\sigma_x - \sigma_y}{2}\sin2\theta \tag{2-38}$$

Equation (2-38) checks with Equation (2-26b). Thus we have fully justified the construction and proposed the use of the Mohr's circle.

We may readily ascertain the principal stresses from Mohr's circle since they must be at points ① and ② in the stress diagram (Figure 2-14). The principal axes are rotated 2β in the stress diagram, and angle β counterclockwise from the xy axes in Figure 2-15.

Finally, we see from Mohr's circle that the maximum shear stress $\tau_{\max}$ has the value

$$\tau_{\max} = \sqrt{\left(\frac{\sigma_x - \sigma_y}{2}\right)^2 + \tau_{xy}^2} \tag{2-39}$$

and that it occurs on planes rotated $45°$ in the physical plane from the principal axes.

We shall not actually use Mohr's circle to get numerical values for particular stresses although one could employ the construction for this purpose. Rather we shall sketch Mohr's circle approximately to scale at times so as to get a visual picture of how the stress varies at a point for plane stress. For such purposes, the construction can be most valuable.

2.8   Three-dimensional Mohr's Circle

We shall now set forth the generalized Mohr's circle which is helpful in the study of three-dimensional stress at a point.

Consider a point O in a stressed body where the planes of the x, y, z axes coincide with the principal axes of stress at the point. An infinitesimal tetrahedron of the body at the aforementioned point is to be considered, as we have shown in Figure 2-16. The Face ABC of the tetrahedron has a unit normal $\boldsymbol{n}$ with direction cosines l, m and n. These direction cosines must of course satisfy the equation

$$1 = l^2 + m^2 + n^2 \tag{2-40}$$

And we can say for the normal stress, using Equation (2-4), that

$$\sigma = \sigma_1 l^2 + \sigma_2 m^2 + \sigma_3 n^2 \tag{2-41}$$

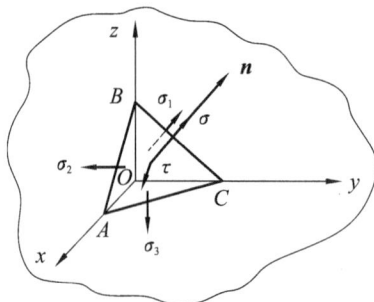

Figure 2-16   An infinitesimal tetrahedron of the body

Also, we have shown the maximum shear stress on Face ABC, and we denoted this shear stress as τ. According to τ, σ and the principal stresses, we utilize the equilibrium requirement that the resultant force on Face ABC must be equal and opposite to the resultant of the forces on the coordinate planes of the tetrahedron. Thus we can say for the squares of these forces

$$(\tau^2 + \sigma^2)(ABC)^2 = (\sigma_1 BOC)^2 + (\sigma_2 AOB)^2 + (\sigma_3 AOC)^2 \tag{2-42}$$

Dividing through by $(ABC)^2$ and making the substitution

$$\begin{cases} \left(\dfrac{BOC}{ABC}\right)^2 = l^2 \\[2mm] \left(\dfrac{AOB}{ABC}\right)^2 = m^2 \\[2mm] \left(\dfrac{AOC}{ABC}\right)^2 = n^2 \end{cases} \tag{2-43}$$

We get the desired equation

$$\tau^2 + \sigma^2 = \sigma_1^2 l^2 + \sigma_2^2 m^2 + \sigma_3^2 n^2 \tag{2-44}$$

Observing Equations (2-40), (2-41) and (2-44) we see that we have three equations involving the direction cosines l, m and n in terms of the principal stresses σ_i, as well as stresses σ and τ. We now solve for the direction cosines. Using Cramer's rule, we can state that

$$l^2 = \frac{\begin{vmatrix} 1 & 1 & 1 \\ \sigma & \sigma_2 & \sigma_3 \\ (\tau^2 + \sigma^2) & \sigma_2^2 & \sigma_3^2 \end{vmatrix}}{\begin{vmatrix} 1 & 1 & 1 \\ \sigma_1 & \sigma_2 & \sigma_3 \\ \sigma_1^2 & \sigma_2^2 & \sigma_3^2 \end{vmatrix}} \tag{2-45}$$

Assume that the principal stresses are all different, i. e., that there are no repeated roots (we shall examine the case of repeated roots later). Furthermore, let us say that $\sigma_1 > \sigma_2 > \sigma_3$. We then get the following result upon carrying out the determinants

$$l^2(\sigma_2\sigma_3^2 + \sigma_1\sigma_2^2 + \sigma_3\sigma_1^2 - \sigma_2\sigma_1^2 - \sigma_3\sigma_2^2 - \sigma_1\sigma_3^2) = \sigma_2\sigma_3(\sigma_3 - \sigma_2) + \tau^2(\sigma_3 - \sigma_2) + \sigma^2(\sigma_3 - \sigma_2) - \sigma(\sigma_3^2 - \sigma_2^2)$$

Dividing through by $(\sigma_3 - \sigma_2)$ and rearranging the equation we then get

$$\sigma^2 + \tau^2 - \sigma(\sigma_3 + \sigma_2) = -\sigma_2\sigma_3 + \frac{l^2}{\sigma_3 - \sigma_2}(\sigma_2\sigma_3^2 + \sigma_1\sigma_2^2 + \sigma_3\sigma_1^2 - \sigma_2\sigma_1^2 - \sigma_3\sigma_2^2 - \sigma_1\sigma_3^2) \tag{2-46}$$

You may readily demonstrate that the bracketed quantity on the right-hand side of Equation (2-46) may be reformulated as $(\sigma_1 - \sigma_2)(\sigma_1 - \sigma_3)(\sigma_3 - \sigma_2)$. We can then rewrite Equation (2-46) as

$$\sigma^2 + \tau^2 - \sigma(\sigma_3 + \sigma_2) = -\sigma_3\sigma_2 + l^2(\sigma_1 - \sigma_2)(\sigma_1 - \sigma_3) \tag{2-47}$$

To complete the square in Equation (2-47) we proceed as follows

$$\tau^2 + \left(\sigma - \frac{\sigma_3 + \sigma_2}{2}\right)^2 = \left(\frac{\sigma_3 + \sigma_2}{2}\right)^2 - \sigma_3\sigma_2 + l^2(\sigma_1 - \sigma_2)(\sigma_1 - \sigma_3) \tag{2-48}$$

But the expression $[(\sigma_3 + \sigma_2)/2]^2 - \sigma_3\sigma_2$ can be replaced by $[(\sigma_3 - \sigma_2)/2]^2$. We thus have

$$\tau^2 + \left(\sigma - \frac{\sigma_3 + \sigma_2}{2}\right)^2 = \left(\frac{\sigma_3 - \sigma_2}{2}\right)^2 + l^2(\sigma_1 - \sigma_2)(\sigma_1 - \sigma_3) \tag{2-49}$$

If we hold the direction cosine l fixed, then for a given set of principal stresses σ_i the preceding equation gives a circle on the $\tau\sigma$ plane. This circle represents all the possible normal stress and maximum shear-stress combinations found on ABC by rotating ABC about the x axis (this keeps l fixed). The center of this circle clearly is displaced along the σ axis by the value $(\sigma_3 + \sigma_2)/2$.

Let us consider the extreme values of zero and unity for l. We then have the following equations

For $l = 0$

$$\tau^2 + \left(\sigma - \frac{\sigma_2 + \sigma_3}{2}\right)^2 = \left(\frac{\sigma_2 - \sigma_3}{2}\right)^2 \tag{2-50}$$

For $l = 1$

$$\tau^2 + \left(\sigma - \frac{\sigma_2 + \sigma_3}{2}\right)^2 = \left(\frac{\sigma_2 - \sigma_3}{2}\right)^2 + (\sigma_1 - \sigma_2)(\sigma_1 - \sigma_3) = \left(\sigma_1 - \frac{\sigma_2 + \sigma_3}{2}\right)^2 \tag{2-51}$$

We now have two circles, the first of radius $|(\sigma_2 - \sigma_3)/2|$ and the second of radius $|\sigma_1 - (\sigma_2 + \sigma_3)/2|$. These are plotted in Figure 2-17. We may conclude that the stress pair τ, σ must lie between the two circles. This region has been cross-hatched in the diagram. We will not apply a sign convention for τ so that we need only consider the upper half of the stress plane.

By the same procedure we just employed, we can formulate two other sets of equations like Equations (2-50) and (2-51) involving, respectively, the direction cosines m and n. These sets of equations may be most simply reached by merely permuting the indices of Equations (2-50) and (2-51). Furthermore, pairs of circles for each set of equations can be drawn, representing the extreme values of the corresponding direction cosine. Figure 2-18 shows such a system of circles. Each pair has been drawn in a distinct way for each direction cosine and has been numbered ①,② and ③ where the particular direction cosine is zero, and ①',②' and ③' where the particular direction cosine is unity.

The stress pair σ, τ must always be between each pair of curves, as explained earlier in discussing the single set of circles in Figure 2-17. Thus in considering our three sets of

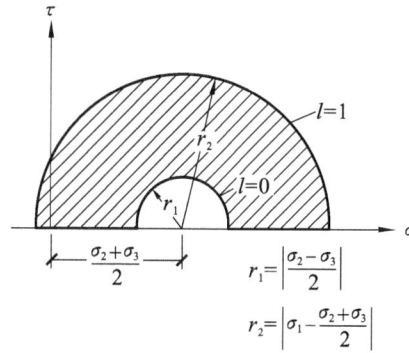

$$r_1 = \left| \frac{\sigma_2 - \sigma_3}{2} \right|$$

$$r_2 = \left| \sigma_1 - \frac{\sigma_2 + \sigma_3}{2} \right|$$

Figure 2-17 The stresses between $l=0$ and $l=1$

circles it is clear that the stress pair must be in the crosshatched region shown in Figure 2-18. This region is bounded by circles ①, ② and ③. Notice, furthermore, that these circles are tangent at the σ axis at positions σ_1, σ_2 and σ_3, and therefore can be drawn quite readily when these stresses are known. To simplify our discussion, we have drawn the circles in Figure 2-19. These are the so-called Mohr's circles of three-dimensional stress.

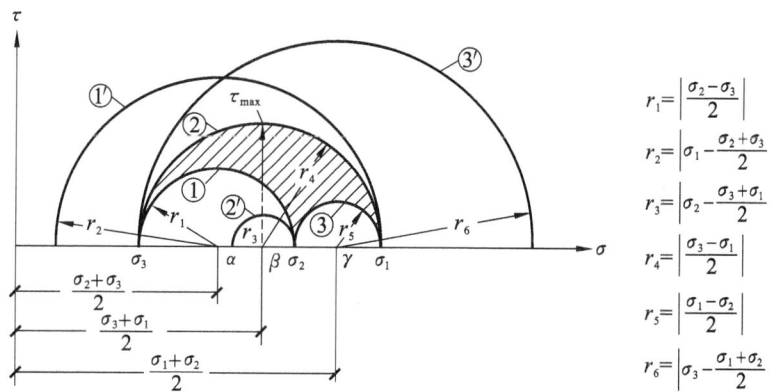

$$r_1 = \left| \frac{\sigma_2 - \sigma_3}{2} \right|$$

$$r_2 = \left| \sigma_1 - \frac{\sigma_2 + \sigma_3}{2} \right|$$

$$r_3 = \left| \sigma_2 - \frac{\sigma_3 + \sigma_1}{2} \right|$$

$$r_4 = \left| \frac{\sigma_3 - \sigma_1}{2} \right|$$

$$r_5 = \left| \frac{\sigma_1 - \sigma_2}{2} \right|$$

$$r_6 = \left| \sigma_3 - \frac{\sigma_1 + \sigma_2}{2} \right|$$

Figure 2-18 The Mohr's circles of three-dimensional stress

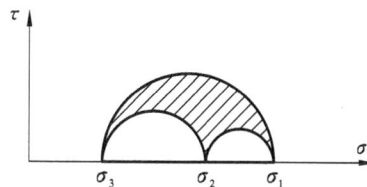

Figure 2-19 The simplified Mohr's circles of three-dimensional stress

How do we use Mohr's circles to determine a set of stresses σ and τ on any interface with direction cosines l, m, and n? We know that all circles corresponding to a given l must be drawn from a point on the σ axis that is a distance $(\sigma_2 + \sigma_3)/2$ from the origin. This is point α in Figure 2-18. The radius of the circle for the given l is given by Equation (2-52)

$$r=\left[\left(\frac{\sigma_3-\sigma_2}{2}\right)^2+l^2(\sigma_1-\sigma_2)(\sigma_1-\sigma_3)\right]^{\frac{1}{2}} \tag{2-52}$$

For the family of circles for m, we use β as the center. The radius for a particular m is computed using the preceding formula but permuting the symbols once. Finally, the circles for n have γ as a center and a radius found by Equation (2-52) with two permutations of the subscripts. The common point of intersection of the three circles then yields the proper stress pair τ,σ.

We have shown such a set of circles in Figure 2-20. Thus for a given set of principal stresses $\sigma_1,\sigma_2,\sigma_3$ and a given set of direction cosines relative to a set of axes oriented so that σ_1 is parallel to x,σ_2 parallel to y,etc. ,we are able to determine graphically a set of stresses σ,τ for a interface corresponding to the given set of direction cosines.

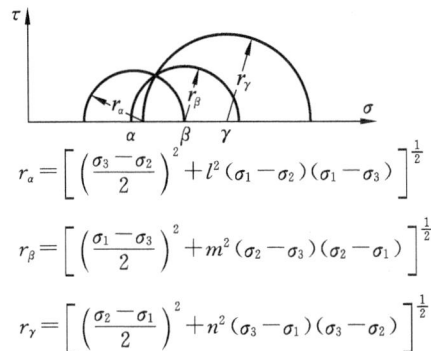

$$r_a=\left[\left(\frac{\sigma_3-\sigma_2}{2}\right)^2+l^2(\sigma_1-\sigma_2)(\sigma_1-\sigma_3)\right]^{\frac{1}{2}}$$

$$r_\beta=\left[\left(\frac{\sigma_1-\sigma_3}{2}\right)^2+m^2(\sigma_2-\sigma_3)(\sigma_2-\sigma_1)\right]^{\frac{1}{2}}$$

$$r_\gamma=\left[\left(\frac{\sigma_2-\sigma_1}{2}\right)^2+n^2(\sigma_3-\sigma_1)(\sigma_3-\sigma_2)\right]^{\frac{1}{2}}$$

Figure 2-20   A set of stress corresponding to the given set of direction cosines

As we pointed out in our discussion of Mohr's circle for plane stress, we do not generally propose that this graphic procedure be used for computing stress at a point. However, useful conclusions can readily be reached by a simple observation of the construction. For example, Figure 2-18 shows that for the case at hand the maximum shear stress at a point occurs at circle ②, where $m=0$. The value of the maximum shear stress is

$$\tau_{\max}=r_4=\frac{\sigma_1-\sigma_3}{2} \tag{2-53}$$

That is, the maximum shear stress equals one-half the difference between the maximum normal stress and the minimum normal stress. Thus,if $\sigma_1=100$ Pa and $\sigma_3=-50$ Pa, the maximum shear stress is 75 Pa. Since $m=0$ for this maximum shear condition,we can conclude that the plane of the maximum shear is tangent to the y axis. Substituting $(\sigma_1-\sigma_3)/2$ for τ in Equation (2-47) and replacing σ by $(\sigma_3+\sigma_1)/2$ in accordance with Figure 2-18 we can solve for l. You may readily show that

$$l=\pm\frac{1}{\sqrt{2}} \tag{2-54}$$

And we can by a similar procedure show that

$$n = \pm \frac{1}{\sqrt{2}} \tag{2-55}$$

Thus the plane of the maximum shear stress bisects the angle formed by the planes of maximum normal stress and minimum normal stress. We have shown this in Figure 2-21.

Thus far we have only considered the case where the principal stresses are different. What happens when two of the principal stresses are equal? Let us consider the case $\sigma_1 = \sigma_2 > \sigma_3$ (Figure 2-22). Examining Equation (2-52) involving l and the corresponding equation for m, we see that for all values of these direction cosines, r has the constant value.

$$r = \left| \frac{\sigma_3 - \sigma_2}{2} \right| = \left| \frac{\sigma_3 - \sigma_1}{2} \right| \tag{2-56}$$

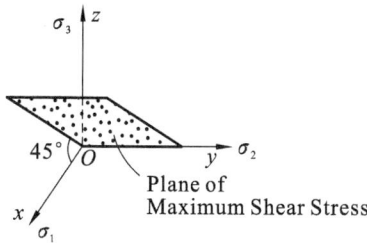

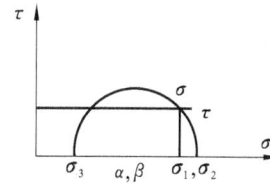

Figure 2-21   The plane of the maximum shear stress   Figure 2-22   The Mohr's circle on the case $\sigma_1 = \sigma_2 > \sigma_3$

As for direction cosine n we get for r

$$r = n \left[(\sigma_3 - \sigma_1)(\sigma_3 - \sigma_2) \right]^{\frac{1}{2}} = n(\sigma_1 - \sigma_3) \tag{2-57}$$

Thus we see that two of the Mohr's circles coincide reducing the cross-hatched region of Figure 2-19 into a line forming a semicircle. The semicircle corresponds to the Mohr's circles for l and m. The intersection of the Mohr's circle corresponding to n, shown only as an arc in the diagram, with the aforementioned semicircle gives us the proper point in the stress plane.

Note that the values of l and m have no effect on the result because the semicircle drawn in the diagram includes all possible values of these direction cosines. Thus for the case where stresses σ_1 and σ_2 are equal, the values of τ and σ are axially-symmetrically distributed about the z axis.

2.9   Closure

In this chapter, we examined the means by which we can communicate quantitatively how forces are transmitted through a solid. We introduced the concepts of normal and shear stress on an interface at a point and then showed that, knowing the stresses on three orthogonal interfaces at a point, we could determine stresses on any interface at that point. Although we thought in terms of a solid body in arriving at these results, we have pointed out that these conclusions apply for any continuous medium under static or dynamic conditions. Thus we are then able, using one reference, to describe in effect the distribution of force at any time in any continuous medium.

Next by examining the transformation equations for stress at a point, we were able to make a number of very valuable conclusions. Thus, we learned about principal axes and principal stresses as well as the three stress invariants at a point. Now there are many other quantities that have the same transformation equations for a rotation of axes at a point and consequently exhibit the aforementioned principal axes and invariants at a point as well as other common characteristics. Thus our discussion of tensors in this chapter is in no way limited to deformable solids although the idea of the tensor did come in naturally in our study of solids. The conclusions will extend to other engineering sciences, physics and mathematics.

In Chapter 3 we consider the geometry of deformation in solids. We show in the course of this discussion that strain is a second-order tensor. Once this has been done, we will immediately be able to apply all conclusions developed in Chapter 2 for stress from its transformation properties.

Problems

2-1   Label the stresses shown in the infinitesimal rectangular parallelepiped in Figure 2-23.

2-2   Label the stresses shown in Figure 2-24 and indicate the proper signs.

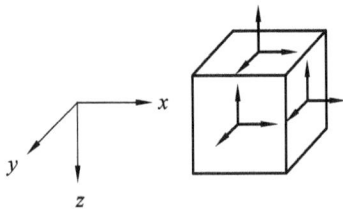

Figure 2-23   Figure of Problem 2-1   Figure 2-24   Figure of Problem 2-2

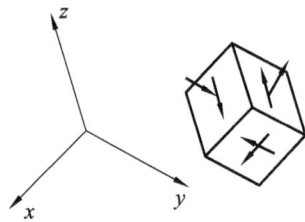

2-3   The stress components at orthogonal interfaces parallel to xyz at a point are known to be

$$\sigma_x = 1000 \text{ Pa} \qquad \tau_{xy} = 200 \text{ Pa}$$
$$\sigma_y = -600 \text{ Pa} \qquad \tau_{xz} = 0 \text{ Pa}$$
$$\sigma_z = 0 \text{ Pa} \qquad \tau_{yz} = -400 \text{ Pa}$$

Assuming that shear stresses with interchanged indices are equal $(\tau_{ij} = \tau_{ji})$, what is the normal stress in the direction $\boldsymbol{\varepsilon}$ such that

$$\boldsymbol{\varepsilon} = 0.11\boldsymbol{i} + 0.35\boldsymbol{j} + 0.93\boldsymbol{k}$$

2-4　Derive Equation (2-6).

2-5　The state of stress at a point for a given reference xyz is given by the following array of terms

$$\begin{pmatrix} 200 & 100 & 0 \\ 100 & 0 & 0 \\ 0 & 0 & 500 \end{pmatrix}$$

If a new set of axes x', y', z' is formed by rotating x, y, z 60° about the z axis, what is the array of stress terms for the x', y', z' axes at the point (Use tensor notation in setting up calculations)?

2-6　Going plane stress

$$\sigma_x = 500 \text{ Pa}$$
$$\sigma_y = 1000 \text{ Pa}$$
$$\tau_{xy} = 500 \text{ Pa}$$

What are the shear and normal stresses for a set of axes $x'y'$ rotated 45° from xy?

2-7　In Problem 2-6 compute the principal stresses and their directions.

2-8　Given the following state of plane stress

$$\sigma_x = -1000 \text{ Pa}$$
$$\sigma_y = 500 \text{ Pa}$$
$$\tau_{xy} = 1000 \text{ Pa}$$

What are the three tensor invariants at the point?

2-9　In Problem 2-8 determine the principal stresses and their directions.

2-10　Sketch Mohr's circle for the state of stress

$$\sigma_x = -1000 \text{ Pa}$$
$$\sigma_y = 2000 \text{ Pa}$$
$$\tau_{xy} = -500 \text{ Pa}$$

2-11　In Problem 2-10 show approximately the state of stress for an interface rotated 30° from x to y.

2-12　Show that another way of denoting plane stress is to have σ_z a principal stress and equal to zero.

2-13　Sketch Mohr's circle for the case where $\tau_{xy} = 500$ Pa, $\sigma_x = \sigma_y = 0$ Pa.

3　Strain

3.1　Introduction

We shall now develop methods by which we can express the displacement of a deformable solid. You will recall from previous courses in mechanics that a rigid-body movement may be described by the superposition of a translation equal to the actual motion of any point in the body, plus a rotation about an axis going through the selected point. This is the celebrated Chasle's theorem. The displacement we deal with in this chapter is more general in that, besides the possibility of rigid-body translation and rotation, there also may be deformation taking place that must be accounted for.

We shall now consider the undeformed geometry of some arbitrary solid, as shown in Figure 3-1. A stationary reference xyz has been shown. A position vector $\boldsymbol{r}$ locates any point P having coordinates x, y, z in the undeformed geometry. The deformed geometry is shown as a dotted line. Each point P in the undeformed geometry moves to a point P' having coordinates x', y', z'. We have indicated the displacement of point P to point P' by $\boldsymbol{u}$, called the displacement vector. In Cartesian components, the displacement vector is usually given as

$$\boldsymbol{u} = u_x \boldsymbol{i} + u_y \boldsymbol{j} + u_z \boldsymbol{k} \tag{3-1}$$

It should be clear that the displacement vector $\boldsymbol{u}$ will vary continuously from point to point and so it forms a vector field. We call this field the displacement field and we usually express it as a function of the coordinates of the undeformed geometry, i. e. , as $\boldsymbol{u}(x, y, z)$.

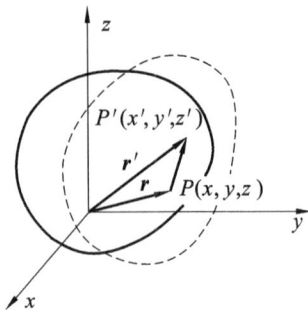

In using Cartesian tensor notation, it will be desirable to refer to the reference axes x, y and z as x_1, x_2 and x_3, respectively. Also, the unit vectors $\boldsymbol{i}, \boldsymbol{j}$ and $\boldsymbol{k}$ will be represented as $\boldsymbol{\varepsilon}_1, \boldsymbol{\varepsilon}_2$ and $\boldsymbol{\varepsilon}_3$, respectively, so that Equation (3-1) can be written as

$$\boldsymbol{u} = u_i \boldsymbol{\varepsilon}_i \tag{3-2}$$

where i goes from 1 to 3.

Figure 3-1　The undeformed geometry of some arbitrary solid

3.2 Small Domain Viewpoint

Consider a body undergoing deformation as shown in Figure 3-2. We have selected two arbitrary points P and Q in the body and have formed the vector $\boldsymbol{A}$ by connecting these points with a directed line segment. In the deformed state, points P and Q move to points P' and Q' and we have formed a second vector $\boldsymbol{A}'$ by connecting these points by a directed line segment as is shown in the diagram.

We are interested in computing $(\boldsymbol{A}' - \boldsymbol{A})$ which we shall denote as $\delta\boldsymbol{A}$. For this purpose, we have formed a vector polygon in Figure 3-2 by inserting the displacement vectors $\boldsymbol{u}_P$ and $\boldsymbol{u}_Q$ from points P and Q, respectively. We can then say

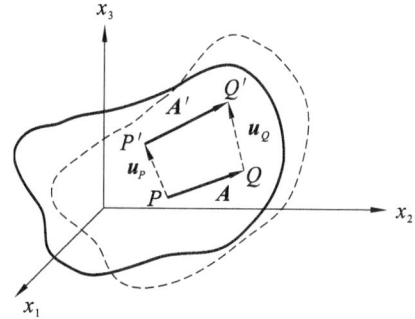

Figure 3-2   A body undergoing deformation

$$\boldsymbol{u}_P + \boldsymbol{A}' = \boldsymbol{A} + \boldsymbol{u}_Q \qquad (3\text{-}3)$$

Rearranging the equation, we get

$$\boldsymbol{A}' - \boldsymbol{A} = \delta\boldsymbol{A} = \boldsymbol{u}_Q - \boldsymbol{u}_P \qquad (3\text{-}4)$$

The displacement field $\boldsymbol{u}$ for the problem we shall discuss will always be an analytic function. For this reason, we can express $\boldsymbol{u}_Q$ in terms of $\boldsymbol{u}_P$ by employing a Taylor series expansion of the displacement $\boldsymbol{u}$ about point P. That is,

$$\boldsymbol{u}_Q = \boldsymbol{u}_P + \left(\frac{\partial \boldsymbol{u}}{\partial x_1}\right)_P \Delta x_1 + \left(\frac{\partial \boldsymbol{u}}{\partial x_2}\right)_P \Delta x_2 + \left(\frac{\partial \boldsymbol{u}}{\partial x_3}\right)_P \Delta x_3 + \cdots \qquad (3\text{-}5)$$

But $\Delta x_1, \Delta x_2$ and Δx_3 are simply the components A_1, A_2 and A_3, respectively. Therefore, incorporating the $\boldsymbol{A}$'s and employing summation indices, we may restate the preceding equation as

$$\boldsymbol{u}_Q = \boldsymbol{u}_P + \left(\frac{\partial \boldsymbol{u}}{\partial x_j}\right)_P A_j + \cdots \qquad (3\text{-}6)$$

If a vector $\boldsymbol{A}$ is very small, i. e., if we limit our consideration to a very small domain about point P, the higher order terms can be deleted in the preceding series giving us

$$\boldsymbol{u}_Q = \boldsymbol{u}_P + \left(\frac{\partial \boldsymbol{u}}{\partial x_j}\right)_P A_j \qquad (3\text{-}7)$$

Substituting Equation (3-7) into Equation (3-4), we then get

$$\delta\boldsymbol{A} = \left(\frac{\partial \boldsymbol{u}}{\partial x_j}\right)_P A_j \qquad (3\text{-}8)$$

Since point P is any arbitrary point in the body, the P subscript in the previous expression may be deleted. Thus, for the change of any vector $\boldsymbol{A}$ in a vanishingly small domain, we have

$$\delta A_i = \frac{\partial u_i}{\partial x_1} A_1 + \frac{\partial u_i}{\partial x_2} A_2 + \frac{\partial u_i}{\partial x_3} A_3 \tag{3-9}$$

or

$$\delta A_i = \frac{\partial u_i}{\partial x_j} A_j \tag{3-10}$$

We shall soon have much use for Equation (3-10) when we consider the deformation of vanishingly small elements of a body. In the next section, the small deformation restriction will be introduced. This restriction should not be confused with the small domain viewpoint which has been presented in this section.

The following examples illustrate the use of the formulations presented in this section.

Example 3-1 Given the following displacement field

$$\boldsymbol{u} = (xy\boldsymbol{i} + 3x^2 z\boldsymbol{j} + 4\boldsymbol{k}) \times 10^{-2}$$

and a very small segment $\Delta \boldsymbol{s}$ having the following direction cosines before deformation

$$l = 0.200, \quad m = 0.800, \quad n = 0.555.$$

This segment is directed away from point $(2,1,3)$. What is the new vector $\Delta \boldsymbol{s}'$ after the displacement field has been imposed?

Solution We first compute $\partial u_i / \partial x_j$. Thus

$$\frac{\partial u_1}{\partial x} = 0.01y \quad \frac{\partial u_1}{\partial y} = 0.01x \quad \frac{\partial u_1}{\partial z} = 0$$

$$\frac{\partial u_2}{\partial x} = 0.06xz \quad \frac{\partial u_2}{\partial y} = 0 \quad \frac{\partial u_2}{\partial z} = 0.03x^2$$

$$\frac{\partial u_3}{\partial x} = 0 \quad \frac{\partial u_3}{\partial y} = 0 \quad \frac{\partial u_3}{\partial z} = 0$$

We then have

$$\frac{\partial u_i}{\partial x_j} = \begin{bmatrix} 0.01y & 0.01x & 0 \\ 0.06xz & 0 & 0.03x^2 \\ 0 & 0 & 0 \end{bmatrix}$$

In a small domain around point $(2,1,3)$, we have

$$\left(\frac{\partial u_i}{\partial x_j}\right)_{(2,1,3)} = \begin{bmatrix} 0.01 & 0.02 & 0 \\ 0.36 & 0 & 0.12 \\ 0 & 0 & 0 \end{bmatrix}$$

Using Equation (3-10), we get

$$[\delta(\Delta s)]_1 = \left(\frac{\partial u_1}{\partial x_j}\right)_P (\Delta s)_j$$

$$= \left(\frac{\partial u_1}{\partial x}\right)_P (\Delta s)l + \left(\frac{\partial u_1}{\partial y}\right)_P (\Delta s)m + \left(\frac{\partial u_1}{\partial z}\right)_P (\Delta s)n$$

$$= \Delta s(0.002 + 0.016) = 0.018\Delta s$$

$$[\delta(\Delta s)]_2 = \left(\frac{\partial u_2}{\partial x_j}\right)_P (\Delta s)_j$$

$$= \left(\frac{\partial u_2}{\partial x}\right)_P (\Delta s)l + \left(\frac{\partial u_2}{\partial y}\right)_P (\Delta s)m + \left(\frac{\partial u_2}{\partial z}\right)_P (\Delta s)n$$

$$= \Delta s(0.072 + 0.0666) = 0.1386\Delta s$$

$$[\delta(\Delta s)]_3 = \left(\frac{\partial u_3}{\partial x_j}\right)_P (\Delta s)_j$$

$$= \left(\frac{\partial u_3}{\partial x}\right)_P (\Delta s)l + \left(\frac{\partial u_3}{\partial y}\right)_P (\Delta s)m + \left(\frac{\partial u_3}{\partial z}\right)_P (\Delta s)n$$

$$= \Delta s(0) = 0$$

The change in the vector (Δs) becomes

$$\delta(\Delta s) = (0.018\boldsymbol{i} + 0.1386\boldsymbol{j})\Delta s$$

The new vector $\Delta s'$ is then

$$\Delta s' = \Delta s + \delta(\Delta s)$$

$$= (0.200\boldsymbol{i} + 0.800\boldsymbol{j} + 0.555\boldsymbol{k})\Delta s + (0.018\boldsymbol{i} + 0.1386\boldsymbol{j})\Delta s$$

$$= (0.218\boldsymbol{i} + 0.9386\boldsymbol{j} + 0.555\boldsymbol{k})\Delta s$$

Example 3-2   Given the following displacement field

$$u_x = (x^2 + 2y^2 z + yz) \times 10^{-2}$$

$$u_y = [(y+z)x + 3x^2 z] \times 10^{-2}$$

$$u_z = (4y^3 + 2z^2) \times 10^{-2}$$

What is $\delta(\Delta s)$ for a segment Δs lying originally along the x axis at the specific position $(1,1,1)$?

Solution   The matrix $\partial u_i/\partial x_j$ then becomes

$$\frac{\partial u_i}{\partial x_j} = \begin{bmatrix} 2x & (4yz+z) & (2y^2+y) \\ (y+z+6xz) & x & (x+3x^2) \\ 0 & 12y^2 & 4z \end{bmatrix} \times 10^{-2}$$

To ascertain the displacement of a particle at $(1,1,1)$ we can use the preceding equation. Thus

$$(\boldsymbol{u})_{(1,1,1)} = [(1+2+1)\boldsymbol{i} + (2+3)\boldsymbol{j} + (4+2)\boldsymbol{k}] \times 10^{-2}$$

$$= (4\boldsymbol{i} + 5\boldsymbol{j} + 6\boldsymbol{k}) \times 10^{-2}$$

On the other hand, to find the new length of a segment Δs lying originally along the x axis at position $(1,1,1)$ we employ the relation

$$[\delta(\Delta s)]_i = \frac{\partial u_i}{\partial x}(\Delta x) + \frac{\partial u_i}{\partial y}(0) + \frac{\partial u_i}{\partial z}(0)$$

Thus

$$[\delta(\Delta s)]_1 = \frac{\partial u_x}{\partial x}(\Delta x) + \frac{\partial u_x}{\partial y}(0) + \frac{\partial u_x}{\partial z}(0) = 2x\Delta x \times 10^{-2}$$

$$[\delta(\Delta s)]_2 = \frac{\partial u_y}{\partial x}(\Delta x) + \frac{\partial u_y}{\partial y}(0) + \frac{\partial u_y}{\partial z}(0) = (y+z+6xz)\Delta x \times 10^{-2}$$

$$[\delta(\Delta s)]_3 = \frac{\partial u_z}{\partial x}(\Delta x) + \frac{\partial u_z}{\partial y}(0) + \frac{\partial u_z}{\partial z}(0) = 0$$

Thus we have for $\delta(\Delta s)$ the following result

$$[\delta(\Delta s)]_P = [2x\boldsymbol{i} + (y+z+6xz)\boldsymbol{j}](\Delta x) \times 10^{-2}$$

And for the specific position $(1,1,1)$ we have

$$[\delta(\Delta s)]_{(1,1,1)} = (2\boldsymbol{i}+8\boldsymbol{j})(\Delta x) \times 10^{-2}$$

3.3 Small Deformation Restriction

Consider two different displacement fields $\boldsymbol{u}^{(1)}$ and $\boldsymbol{u}^{(2)}$. Then, in any small domain the change of a vector $\boldsymbol{A}$ as a result of the first deformation $\boldsymbol{u}^{(1)}$ becomes

$$\delta A_i = \left(\frac{\partial u_i^{(1)}}{\partial x_j}\right)_a A_j \tag{3-11}$$

where $(\ )_a$ indicates that the enclosed term is evaluated in the undeformed geometry at a (Figure 3-3). It will be convenient to replace δA_i in the preceding equation by $(A'_i - A_i)$ where the prime indicates the deformed geometry associated with the first deformation. We then have, for A'_i

$$A'_i = A_i + \left(\frac{\partial u_i^{(1)}}{\partial x_j}\right)_a A_j \tag{3-12}$$

Now imagine a second deformation from the primed geometry to the double-primed geometry (Figure 3-3). That is, the deformed geometry from the first deformation is to be considered the initial geometry for the second deformation. The component A''_i then becomes

$$A''_i = A'_i + \left(\frac{\partial u_i^{(2)}}{\partial x_k}\right)_{a'} A'_k \tag{3-13}$$

where $(\ )_{a'}$ indicates the evaluation of the enclosed quantity at position a', i. e. , at the

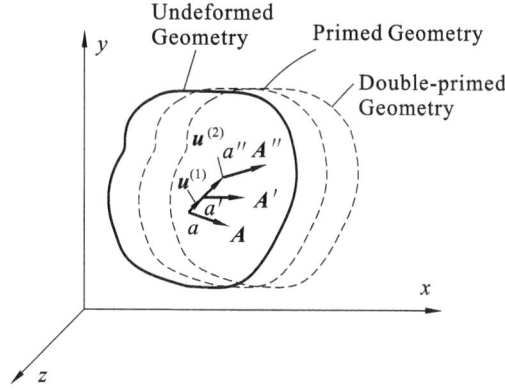

Figure 3-3 The diagram from undeformed to double primed geometry

position corresponding to the deformed geometry resulting from the first deformation. Note that we have used k as the dummy index rather than j. This will help us in later steps (It should be understood that there is no change in results by any change made for a dummy index). Now replace A'_i and A'_k in Equation (3-13), using Equation (3-12) for this purpose. Thus we get

$$A''_i = A_i + \left(\frac{\partial u_i^{(1)}}{\partial x_j}\right)_a A_j + \left(\frac{\partial u_i^{(2)}}{\partial x_k}\right)_{a'}\left[A_k + \left(\frac{\partial u_k^{(1)}}{\partial x_j}\right)_a A_j\right] \tag{3-14}$$

This equation may be expressed as

$$A''_i = A_i + \left(\frac{\partial u_i^{(1)}}{\partial x_j}\right)_a A_j + \left(\frac{\partial u_i^{(2)}}{\partial x_k}\right)_{a'} A_k + \left(\frac{\partial u_i^{(2)}}{\partial x_k}\right)_{a'}\left(\frac{\partial u_k^{(1)}}{\partial x_j}\right)_a A_j \tag{3-15}$$

Now change the dummy index in the third expression on the right-hand side of Equation (3-15) from k to j. We can then collect terms as follows

$$A''_i = A_i + \left[\left(\frac{\partial u_i^{(1)}}{\partial x_j}\right)_a + \left(\frac{\partial u_i^{(2)}}{\partial x_j}\right)_{a'}\right]A_j + \left(\frac{\partial u_i^{(2)}}{\partial x_k}\right)_{a'}\left(\frac{\partial u_k^{(1)}}{\partial x_j}\right)_a A_j \tag{3-16}$$

Let us next express the quantity $(\partial u_i^{(2)}/\partial x_j)_{a'}$, which is evaluated at Position a', as a Taylor series expansion about Position a.

$$\left(\frac{\partial u_i^{(2)}}{\partial x_j}\right)_{a'} = \left(\frac{\partial u_i^{(2)}}{\partial x_j}\right)_a + \left[\frac{\partial}{\partial x_k}\left(\frac{\partial u_i^{(2)}}{\partial x_j}\right)\right]_a u_k^{(1)} + \left[\frac{\partial^2}{\partial x_k \partial x_l}\left(\frac{\partial u_i^{(2)}}{\partial x_j}\right)\right]_a \frac{u_k^{(1)} u_l^{(1)}}{2} + \cdots \tag{3-17}$$

We now impose the small deformation restriction requiring $u_i^{(1)}$, $u_i^{(2)}$, $(\partial u_i^{(2)}/\partial x_j)_a$ and $(\partial u_i^{(2)}/\partial x_j)_{a'}$ to be very small. This means in Equation (3-17) that only the first term need be retained on the right-hand side. In other words, for our purposes

$$\left(\frac{\partial u_i^{(2)}}{\partial x_j}\right)_{a'} = \left(\frac{\partial u_i^{(2)}}{\partial x_j}\right)_a \tag{3-18}$$

so that we can use the undeformed geometry for computing the effects of successive deformations, making the subscripts a and a' in the preceding equations unnecessary. Furthermore, in Equation (3-16), we can neglect the products of the derivatives as negligible compared to the derivatives themselves. In short, we can rewrite Equation (3-16) for small deformations in the following way

$$A''_i - A_i = [\delta A_i]_{\text{total}} = \left(\frac{\partial u_i^{(1)}}{\partial x_j} + \frac{\partial u_i^{(2)}}{\partial x_j} \right)_a A_j \qquad (3-19)$$

By having made the small deformation restriction, the following basic simplifications occur:

(1) The total deformation for a series of so-called infinitesimal displacements can be found by merely superposing the individual displacements each computed separately from the original geometry. This should be evident from Equation (3-19) and is the superposition principle.

(2) The order of imposing infinitesimal displacements does not have an effect on the total deformation. This is the commutative law—also evident from Equation (3-19).

We shall restrict ourselves entirely to the small deformation case. Large deformation theory lacking the simplifications above is exceedingly difficult and is beyond the scope of this book. Fortunately, most engineering problems can be handled by using small deformation theory.

We can say in summary that we shall generally use the small domain viewpoint as presented in Section 3.2 so as to permit us the use of Equation (3-10). Understand that this viewpoint has nothing whatsoever to do with large or small deformations—it can be used for both. In addition, we shall impose the small deformation restriction. Thus the use of small domain viewpoint and the small deformation restriction means that we shall be considering the deformation of small elements of a body undergoing small deformations.

3.4　Rigid-body Rotation and Pure Deformation of an Element

In section 3.2 we developed formulations—Equations (3-9) and (3-10)—by which we can determine changes of length and orientation of line segments in vanishingly small elements of a body undergoing deformation. Clearly, knowing how such line segments in an element behave is equivalent to knowing how this element deforms. Accordingly, the terms $\partial u_i / \partial x_j$ will be the key quantities for such studies. In this section, we show how these terms can give us rigid-body rotation of an element as well as the so-called pure deformation.

As a first step, we express $\partial u_i / \partial x_j$ thus

$$\frac{\partial u_i}{\partial x_j} = \frac{1}{2}\left(\frac{\partial u_i}{\partial x_j} + \frac{\partial u_j}{\partial x_i}\right) + \frac{1}{2}\left(\frac{\partial u_i}{\partial x_j} - \frac{\partial u_j}{\partial x_i}\right) \tag{3-20}$$

We can think of this equation as forming two matrices ε_{ij} and ω_{ij} which combine to give the matrix $\partial u_i / \partial x_j$.

Thus

$$\frac{\partial u_i}{\partial x_j} = \varepsilon_{ij} + \omega_{ij} \tag{3-21}$$

where

$$\varepsilon_{ij} = \frac{1}{2}\left(\frac{\partial u_i}{\partial x_j} + \frac{\partial u_j}{\partial x_i}\right) \tag{3-22}$$

$$\omega_{ij} = \frac{1}{2}\left(\frac{\partial u_i}{\partial x_j} - \frac{\partial u_j}{\partial x_i}\right) \tag{3-23}$$

We can now express Equation (3-10) as follows

$$\delta A_i = (\varepsilon_{ij} + \omega_{ij})A_j \tag{3-24}$$

We shall carefully investigate the quantities ε_{ij} and ω_{ij} in this and succeeding sections.

Consider the matrix ω_{ij} first. Let us evaluate all the terms for this matrix using Equation (3-23). Thus

$$\omega_{11} = \frac{1}{2}\left(\frac{\partial u_1}{\partial x_1} - \frac{\partial u_1}{\partial x_1}\right) = 0 \quad \omega_{12} = \frac{1}{2}\left(\frac{\partial u_1}{\partial x_2} - \frac{\partial u_2}{\partial x_1}\right) \quad \omega_{13} = \frac{1}{2}\left(\frac{\partial u_1}{\partial x_3} - \frac{\partial u_3}{\partial x_1}\right)$$

$$\omega_{21} = \frac{1}{2}\left(\frac{\partial u_2}{\partial x_1} - \frac{\partial u_1}{\partial x_2}\right) \quad \omega_{22} = \frac{1}{2}\left(\frac{\partial u_2}{\partial x_2} - \frac{\partial u_2}{\partial x_2}\right) = 0 \quad \omega_{23} = \frac{1}{2}\left(\frac{\partial u_2}{\partial x_3} - \frac{\partial u_3}{\partial x_2}\right) \tag{3-25}$$

$$\omega_{31} = \frac{1}{2}\left(\frac{\partial u_3}{\partial x_1} - \frac{\partial u_1}{\partial x_3}\right) \quad \omega_{32} = \frac{1}{2}\left(\frac{\partial u_3}{\partial x_2} - \frac{\partial u_2}{\partial x_3}\right) \quad \omega_{33} = \frac{1}{2}\left(\frac{\partial u_3}{\partial x_3} - \frac{\partial u_3}{\partial x_3}\right) = 0$$

We have deliberately placed the equations to suggest a matrix array of terms. It should be clear by inspection that the main diagonal terms of ω_{ij} are zero and that the corresponding off-diagonal terms are negatives of each other. We can thus say

$$\omega_{ij} = -\omega_{ji} \tag{3-26}$$

Clearly when $i = j$ the ω_{ij} must be zero to satisfy the equation. Such a matrix is called a skew-symmetric or an antisymmetric matrix.

We shall first show that the off-diagonal terms of ω_{ij} give the rigid-body rotation components of an element. Accordingly, consider an element of a body at point P undergoing infinitesimal deformation as shown in Figure 3-4. As a result of deformation the element at P moves to P'. Thus there is a translation of the element given as $\boldsymbol{u}(P)$. In addition, we can expect a rigid-body rotation of this element as well as a change in shape. Let us assume now that there is only a translation and rigid-body rotation of the element as we go from the unprimed to the primed geometry. Examine the line segment Δy as shown in Figure 3-4. The determination of the amount of rotation of this segment about the x axis (or any axis parallel to the x axis) clearly will give us the component of the rotation of the element about the x axis. In Figure 3-5, we have shown an enlarged view of this segment in the deformed geometry having denoted it as $\Delta y'$. The displacement components in the z direction of the end points of the segment Δy have also been shown in the diagram. Using these displacement components, we can express the rotation of Δy about the x axis in the following manner.

$$(\delta\boldsymbol{\phi})_x = \frac{\left(u_z + \dfrac{\partial u_z}{\partial y}\Delta y\right) - u_z}{(\Delta y')_{y'}} \tag{3-27}$$

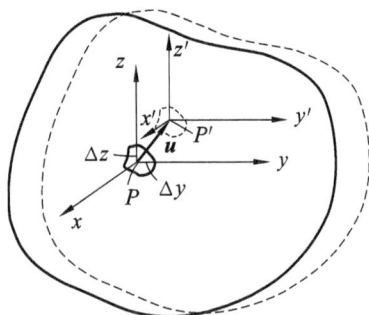

Figure 3-4　An element of a body at point P
undergoing infinitesimal deformation

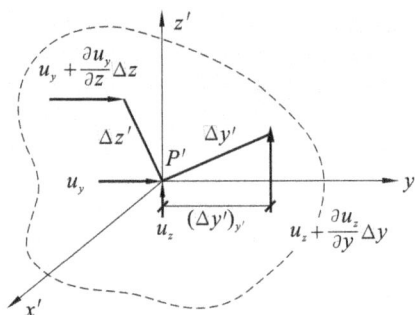

Figure 3-5　An enlarged view of the segment
in the deformed geometry

where $(\Delta y')_{y'}$ is the projection of $\Delta y'$ along the y' axis as indicated in Figure 3-5. For small deformation, we can assume in Equation (3-27) that $(\Delta y')_{y'} = \Delta y$ so that $(\delta\boldsymbol{\phi})_x$ becomes

$$(\delta\boldsymbol{\phi})_x = \frac{\partial u_z}{\partial y} \tag{3-28}$$

We also could have used the segment Δz shown in Figure 3-4 for a similar study. Observing Figure 3-5, we can accordingly say

$$(\delta\boldsymbol{\phi})_x = \frac{-\left(u_y + \dfrac{\partial u_y}{\partial z}\Delta z - u_y\right)}{\Delta z} \tag{3-29}$$

Note the signs of the terms are determined so as to maintain the right-hand rule for rotation. Equation (3-29) becomes

$$(\delta\boldsymbol{\phi})_x = -\frac{\partial u_y}{\partial z} \tag{3-30}$$

Adding Equations (3-28) and (3-30), we can express $(\delta\boldsymbol{\phi})_x$ in the following manner

$$(\delta\boldsymbol{\phi})_x = \frac{1}{2}\left(\frac{\partial u_z}{\partial y} - \frac{\partial u_y}{\partial z}\right) \tag{3-31}$$

But the term on the right side of this equation is simplified into ω_{zy}. By similar arguments we can show that $(\delta\boldsymbol{\phi})_y$ and $(\delta\boldsymbol{\phi})_z$ equal ω_{xz} and ω_{yx} respectively. We thus have the result

$$(\delta\boldsymbol{\phi})_x = \omega_{zy}$$
$$(\delta\boldsymbol{\phi})_y = \omega_{xz} \tag{3-32}$$
$$(\delta\boldsymbol{\phi})_z = \omega_{yx}$$

Thus the off-diagonal terms of ω_{ij} represent rotation components of a rigid-body rotation. The rotation vector $\delta\boldsymbol{\phi}$ can be then given as

$$\delta\boldsymbol{\phi} = \omega_{zy}\boldsymbol{i} + \omega_{xz}\boldsymbol{j} + \omega_{yx}\boldsymbol{k} \tag{3-33}$$

We can now readily show that the contribution to δA_i from ω_{ij} in Equation (3-24) is that which results from rigid-body rotation. You will recall from rigid-body mechanics that for a vector $\boldsymbol{A}$ fixed in a rigid-body, the time rate of change of $\boldsymbol{A}$ is

$$\frac{\mathrm{d}\boldsymbol{A}}{\mathrm{d}t} = \boldsymbol{\omega} \times \boldsymbol{A} \tag{3-34}$$

where $\boldsymbol{\omega}$ is the angular velocity of the rigid-body.

This equation may be written as

$$\mathrm{d}\boldsymbol{A} = \boldsymbol{\omega}\mathrm{d}t \times \boldsymbol{A} = \mathrm{d}\boldsymbol{\phi} \times \boldsymbol{A} \tag{3-35}$$

where $\mathrm{d}\boldsymbol{A}$ represents the change in vector $\boldsymbol{A}$ due to the rotation $\mathrm{d}\boldsymbol{\phi}$. Taking the i th component of Equation (3-35) and replacing d by δ, we may write

$$(\delta\boldsymbol{A})_i = (\delta\boldsymbol{\phi} \times \boldsymbol{A})_i \tag{3-36}$$

Examine for simplicity the case where $i = 1$. We get, carrying out the cross product

$$(\delta\boldsymbol{A})_1 = (\delta\boldsymbol{\phi})_2 A_3 - (\delta\boldsymbol{\phi})_3 A_2 \tag{3-37}$$

Replacing $(\delta\boldsymbol{\phi})_2$ and $(\delta\boldsymbol{\phi})_3$ using Equation (3-32), we have

$$(\delta\boldsymbol{A})_1 = \omega_{13} A_3 - \omega_{21} A_2 \tag{3-38}$$

Since $\omega_{21} = -\omega_{12}$ and $\omega_{11} = 0$, we can write Equation (3-37) as follows

$$(\delta\boldsymbol{A})_1 = \omega_{11} A_1 + \omega_{12} A_2 + \omega_{13} A_3 = \omega_{1j} A_j \tag{3-39}$$

Similarly considering other components of $\delta\boldsymbol{A}$ in Equation (3-36), we can conclude that

$$(\delta\boldsymbol{A})_i = \delta A_i = \omega_{ij} A_j \tag{3-40}$$

But the preceding equation is identical to Equation (3-24) with only the matrix ω_{ij} present. We can thus conclude that ω_{ij} in Equation (3-40) gives the rigid-body rotation contribution to the deformation of an element of the body undergoing infinitesimal deformation. We have now established physical interpretations of the terms of ω_{ij} as well as the nature of the contributions of ω_{ij} to the deformation of an element. As is to be expected ω_{ij} is called the rotation matrix.

Let us next examine the terms in ε_{ij} as we did the terms in ω_{ij}. Thus

$$\varepsilon_{11}=\frac{\partial u_1}{\partial x_1} \qquad \varepsilon_{12}=\frac{1}{2}\left(\frac{\partial u_1}{\partial x_2}+\frac{\partial u_2}{\partial x_1}\right) \qquad \varepsilon_{13}=\frac{1}{2}\left(\frac{\partial u_1}{\partial x_3}+\frac{\partial u_3}{\partial x_1}\right)$$

$$\varepsilon_{21}=\frac{1}{2}\left(\frac{\partial u_2}{\partial x_1}+\frac{\partial u_1}{\partial x_2}\right) \qquad \varepsilon_{22}=\frac{\partial u_2}{\partial x_2} \qquad \varepsilon_{23}=\frac{1}{2}\left(\frac{\partial u_2}{\partial x_3}+\frac{\partial u_3}{\partial x_2}\right) \qquad (3\text{-}41)$$

$$\varepsilon_{31}=\frac{1}{2}\left(\frac{\partial u_3}{\partial x_1}+\frac{\partial u_1}{\partial x_2}\right) \qquad \varepsilon_{32}=\frac{1}{2}\left(\frac{\partial u_3}{\partial x_2}+\frac{\partial u_2}{\partial x_3}\right) \qquad \varepsilon_{33}=\frac{\partial u_3}{\partial x_3}$$

An inspection of this matrix reveals that it is asymmetric matrix. We have thus formed a skew-symmetric matrix ω_{ij} in Equation (3-25) and a symmetric matrix ε_{ij} in Equation (3-41). Since ω_{ij} represents rigid-body rotation of the element we must then conclude that the matrix ε_{ij} represents pure deformation of the element. For this reason we call ε_{ij} the strain matrix. Later, after we have shown the tensorial nature of ε_{ij}, we shall call it the strain tensor. In the following section we will carefully investigate the nature of the terms of ε_{ij}.

Example 3-3 A body has deformed so as to have the following deformation field:

$$u_1=0.003x_1+0.002x_2$$
$$u_2=-0.001x_1+0.0005x_3$$
$$u_3=0.0006x_1+0.003x_2-0.003x_3$$

What are ω_{ij} and ε_{ij}?

Solution The matrix $\partial u_i/\partial x_j$ is easily determined by inspection from the preceding equation

$$\frac{\partial u_i}{\partial x_j}=\begin{bmatrix} 0.003 & 0.002 & 0 \\ -0.001 & 0 & 0.0005 \\ 0.0006 & 0.003 & -0.003 \end{bmatrix}$$

Hence,

$$\omega_{11}=0 \quad \omega_{12}=\frac{1}{2}\times(0.002+0.001)=0.0015 \quad \omega_{13}=\frac{1}{2}\times(0-0.0006)=-0.0003$$

$$\omega_{21}=\frac{1}{2}\times(-0.001-0.002)=-0.0015 \quad \omega_{22}=0 \quad \omega_{23}=\frac{1}{2}\times(0.0005-0.003)=-0.00125$$

$$\omega_{31}=\frac{1}{2}\times(0.0006-0)=0.0003 \quad \omega_{32}=\frac{1}{2}\times(0.003-0.0005)=0.00125 \quad \omega_{33}=0$$

The rotation components of $\delta\boldsymbol{\phi}$ are given by Equation (3-32). Thus

$$(\delta\boldsymbol{\phi})_1=\omega_{32}=0.00125\text{rad}$$
$$(\delta\boldsymbol{\phi})_2=\omega_{13}=-0.0003\text{rad}$$
$$(\delta\boldsymbol{\phi})_3=\omega_{21}=-0.0015\text{rad}$$

Finally, the strain matrix for this displacement becomes

$\varepsilon_{11}=0.003$ $\qquad$ $\varepsilon_{12}=\dfrac{1}{2}\times(0.002-0.001)=0.0005$ $\quad$ $\varepsilon_{13}=\dfrac{1}{2}\times(0+0.0006)=0.0003$

$\varepsilon_{21}=\dfrac{1}{2}\times(-0.001+0.002)=0.0005$ $\quad$ $\varepsilon_{22}=0$ $\qquad$ $\varepsilon_{23}=\dfrac{1}{2}\times(0.0005+0.003)=0.00175$

$\varepsilon_{31}=\dfrac{1}{2}\times(0.0006+0)=0.0003$ $\qquad$ $\varepsilon_{32}=\dfrac{1}{2}\times(0.003+0.0005)=0.00175$ $\quad$ $\varepsilon_{33}=-0.003$

Example 3-3 is a case of affine deformation. Note that the strain matrix and the rotation matrix are composed of constants. This means that each small element of the body has the same rotation and the same pure deformation as every other element. Such deformation is called homogeneous deformation.

Example 3-4 Given the following displacement field

$$u_x=(3x^2y+6)\times10^{-2}$$
$$u_y=(y^2+6xz)\times10^{-2}$$
$$u_z=(6z^2+2yz+10)\times10^{-2}$$

What is the rotation of an element at position $x=1, y=0, z=2$?

Solution We must first ascertain the matrix $\partial u_i/\partial x_j$ for this displacement field. Thus

$$\frac{\partial u_x}{\partial x}=0.06xy \qquad \frac{\partial u_x}{\partial y}=0.03x^2 \qquad \frac{\partial u_x}{\partial z}=0$$

$$\frac{\partial u_y}{\partial x}=0.06z \qquad \frac{\partial u_y}{\partial y}=0.02y \qquad \frac{\partial u_y}{\partial z}=0.06x$$

$$\frac{\partial u_z}{\partial x}=0 \qquad \frac{\partial u_z}{\partial y}=0.02z \qquad \frac{\partial u_z}{\partial z}=0.12z+0.02y$$

Hence

$$\frac{\partial u_i}{\partial x_j}=\begin{pmatrix} 0.06xy & 0.03x^2 & 0 \\ 0.06z & 0.02y & 0.06x \\ 0 & 0.02z & 0.12z+0.02y \end{pmatrix}$$

We next get the rotation matrix ω_{ij} from this field. By inspection we get

$$\omega_{ij}=\begin{vmatrix} 0 & \dfrac{1}{2}(3x^2-6z) & 0 \\ -\dfrac{1}{2}(3x^2-6z) & 0 & \dfrac{1}{2}(6x-2z) \\ 0 & -\dfrac{1}{2}(6x-2z) & 0 \end{vmatrix}\times10^{-2}$$

For the element, we have

$$\omega_{ij} = \begin{bmatrix} 0 & -0.045 & 0 \\ 0.045 & 0 & 0.01 \\ 0 & -0.01 & 0 \end{bmatrix}$$

And hence we have, for the rotation components

$$(\delta\boldsymbol{\phi})_1 = \omega_{32} = -0.01\text{rad}$$

$$(\delta\boldsymbol{\phi})_2 = \omega_{13} = 0\text{rad}$$

$$(\delta\boldsymbol{\phi})_3 = \omega_{21} = 0.045\text{rad}$$

3.5 Physical Interpretation of Strain Terms

We have thus far shown how, knowing the displacement field $\boldsymbol{u}$ for a general infinitesimal deformation, we can ascertain the matrices ω_{ij} and ε_{ij} which, when evaluated at a point, give us the rigid-body rotation and the pure deformation of a vanishingly small element of the body at the point. We have related the terms of ω_{ij} with the rotation vector $\delta\boldsymbol{\phi}$. Our next step will be to associate the terms of ε_{ij}, evaluated at a point, with certain physically meaningful geometric interpretations that will prove very useful to us as we proceed further.

As a first step, consider a line segment Δx along the x axis connecting points P and Q as shown in Figure 3-6. In the deformed state P goes to P' and Q goes to Q'. The projection of $\overline{P'Q'}$ in the x direction, which we denote as $(\overline{P'Q'})_x$, is computed in terms of the original length Δx and the displacement in the x direction of the points P and Q in the following way

$$(\overline{P'Q'})_x = \Delta x + (u_x)_Q - (u_x)_P \tag{3-42}$$

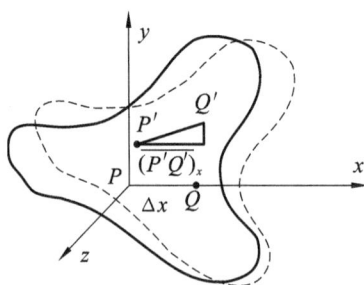

Figure 3-6　A line segment Δx along the x axis connecting points P and Q

We next express $(u_x)_Q$ in Equation (3-42) as a Taylor series expansion about point P as was done in Section 3.2. We then get

$$(\overline{P'Q'})_x = \Delta x + \left[(u_x)_P + \left(\frac{\partial u_x}{\partial x}\right)_P \Delta x + \cdots \right] - (u_x)_P \tag{3-43}$$

The net x component of the elongation of the segment Δx can then be given as

$$(\overline{P'Q'})_x - \Delta x = \left(\frac{\partial u_x}{\partial x}\right)_P \Delta x + \cdots$$

Dividing through by Δx, we next take the limit of each term as $\Delta x \to 0$. Thus

$$\lim_{\Delta x \to 0}\frac{(\overline{P'Q'})_x - \Delta x}{\Delta x} = \left(\frac{\partial u_x}{\partial x}\right)_P \tag{3-44}$$

Note that the right side of the preceding equation is the strain term ε_{xx} at point P. We thus see, considering the left side of the equation, that ε_{xx} is the elongation in the x direction of an infinitesimal line segment originally in the x direction, per unit of original length. We can in this way give similar interpretations for ε_{yy} and ε_{zz}. These are the diagonal terms of the strain matrix and we call these strains normal strains. In general, we can say that ε_{pp} at a point is the elongation in the pth coordinate direction of a vanishing small line segment originally in the pth coordinate direction per unit of the original length of the segment. Since we are limiting the discussion to small deformations, we can use $\overline{P'Q'}$ in Equation (3-44) instead of its component $(\overline{P'Q'})_x$ with negligible change in the result. Thus, ε_{pp} can also be interpreted as merely the change in length of a segment originally in the pth coordinate direction per unit of original length.

Now let us consider line segments $\overline{PQ}$ of length Δx along the x axis and $\overline{PR}$ of length Δy along the y axis as has been shown in Figure 3-7. In the deformed state, P, Q and R move to P', Q' and R', respectively. We shall be interested in the projection of $\overline{P'Q'}$ and $\overline{P'R'}$ on the xy plane and so we have shown this view in Figure 3-8. The angle α is the angle between the projection of $\overline{P'R'}$ and the y direction; β is the angle between the projection of $\overline{P'Q'}$ and the x direction. The displacement of point P in the x direction has been shown simply as u_x and the displacement of point R in the x direction has been evaluated using a Taylor series expansion in the form $[u_x + (\partial u_x/\partial y)\Delta y + \cdots]$ (Note for this expansion $\Delta x = \Delta z = 0$). Finally note that the component of the projected length of $\overline{P'R'}$ in the y direction has been indicated in the diagram as

$$(\overline{P'R'})_y = \Delta y + [\delta(\Delta y \boldsymbol{j})]_y \tag{3-45}$$

Consulting Figure 3-8, it is a simple matter now to evaluate $\tan\alpha$. Thus

$$\tan\alpha = \frac{\dfrac{\partial u_x}{\partial y}\Delta y + \cdots}{\Delta y + [\delta(\Delta y \boldsymbol{j})]_y} \tag{3-46}$$

Now take the limit as $\Delta y \to 0$. The higher-order

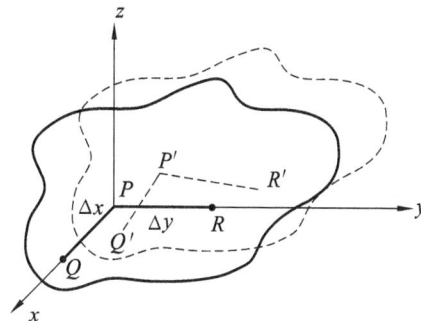

Figure 3-7 The deformed state of line segments $\overline{PQ}$ and $\overline{PR}$

terms all vanish in the limit and we end with the simple relation

$$\tan\alpha = \frac{\partial u_x}{\partial y} \tag{3-47}$$

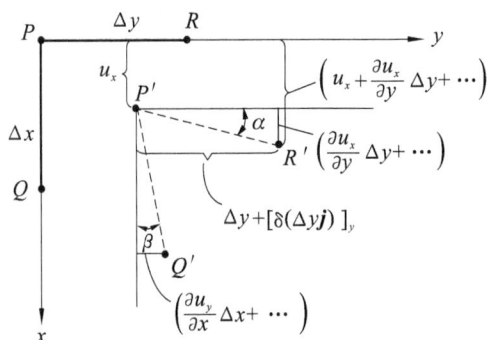

Figure 3-8 **The projection of $\overline{P'Q'}$ and $\overline{P'R'}$ on the xy plane**

For small deformation, $\tan\alpha = \alpha$, and so we have

$$\alpha = \frac{\partial u_x}{\partial y} \tag{3-48}$$

By similar formulations, we can say for the angle β

$$\beta = \frac{\partial u_y}{\partial x} \tag{3-49}$$

The sum of the angles $(\alpha + \beta)$ is the decrease in right angles of the pair of infinitesimal line segments at P when we project the deformed geometry onto the plane formed by the line segments in the undeformed geometry. Because of the small deformation requirement, however, the change of right angle between the infinitesimal segments in the deformed geometry can be used in place of the angle found by projecting the deformed geometry back onto the xy plane. Using Equations (3-48) and (3-49), we can say

$$\alpha + \beta = \frac{\partial u_x}{\partial y} + \frac{\partial u_y}{\partial x} = 2\varepsilon_{xy} \tag{3-50}$$

Thus we see that $2\varepsilon_{ij}$ is the decrease of right angle between infinitesimal orthogonal line segments $\mathrm{d}x_i$ and $\mathrm{d}x_j$ at a point as a result of deformation. We thus have reached a physical interpretation of the off-diagonal terms of the strain matrix. Such strains are called shear strains. It is customary to use the shear angle γ_{ij} to represent the total decrease of right angle between $\mathrm{d}x_i$ and $\mathrm{d}x_j$. Thus

$$\gamma_{ij} = 2\varepsilon_{ij} \tag{3-51}$$

We shall use the shear angle γ_{ij} freely in subsequent chapters.

Let us now make further physical interpretations for the strain elements as they pertain to an infinitesimal three-dimensional rectangular parallelepiped such as is shown in Figure 3-9. Let us consider first that the strain matrix ε_{ij} for this element has only normal strains which are nonzero. From our discussion, we can conclude that the rectangular parallelepiped remains a rectangular parallelepiped during deformation. It should be pointed out, however, that the element may also have rigid-body rotation; and as a result, the sides of the rectangular parallelepiped may not be parallel to the coordinate axes after deformation (This possibility has been shown in the diagram).

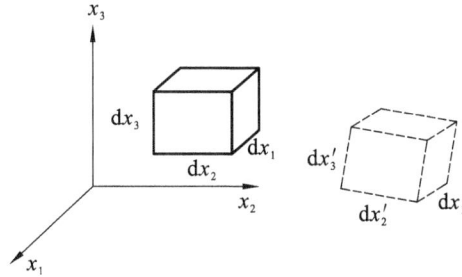

Figure 3-9   An infinitesimal three-dimensional rectangular parallelepiped

Using our geometrical interpretation of normal strain, the new lengths of the sides of the rectangular parallelepiped then can be given as

$$\begin{cases} dx_1' = dx_1(1+\varepsilon_{11}) \\ dx_2' = dx_2(1+\varepsilon_{22}) \\ dx_3' = dx_3(1+\varepsilon_{33}) \end{cases} \tag{3-52}$$

It will be of interest to compute the change in volume of this element. Accordingly, we now express the change in volume of the element as follows

$$dx_1'dx_2'dx_3' - dx_1dx_2dx_3 = dx_1dx_2dx_3(1+\varepsilon_{11})(1+\varepsilon_{22})(1+\varepsilon_{33}) - dx_1dx_2dx_3 \tag{3-53}$$

For small deformations, we may make the following approximation

$$(1+\varepsilon_{11})(1+\varepsilon_{22})(1+\varepsilon_{33}) = 1 + (\varepsilon_{11}+\varepsilon_{22}+\varepsilon_{33}) \tag{3-54}$$

where we have neglected the products of the strains. Now substituting the foregoing expression into Equation (3-53) we get the result

$$dx_1'dx_2'dx_3' - dx_1dx_2dx_3 = (\varepsilon_{11}+\varepsilon_{22}+\varepsilon_{33})dx_1dx_2dx_3 \tag{3-55}$$

We then may say

$$\frac{dx_1'dx_2'dx_3' - dx_1dx_2dx_3}{dx_1dx_2dx_3} = \varepsilon_{11}+\varepsilon_{22}+\varepsilon_{33} \tag{3-56}$$

We can immediately interpret Equation (3-56) as equating the change of volume per unit volume at a point, which we term the cubical dilatation, with the sum of the normal strains.

Next consider a state of strain at a point where the only nonzero strains are the shear strains. It is clear that the rectangular parallelepiped of the previous discussion loses its rectangular shape.

Thus we can conclude, considering a vanishing small rectangular parallelepiped of material in the undeformed geometry, that normal strains in the direction of the edges may cause dilatation, i. e. , change in volume per unit volume, without affecting the mutual orthogonality of the edges. On the other hand, shear strain destroys the orthogonality of the edges, and thus affects the basic shape of the element, but does not affect the volume.

We now have interpretations of the diagonal terms, i. e. , the normal strains, and the off-diagonal terms, i. e. , the shear strains, of the strain matrix. The cubical dilatation concept furthermore gives us an interpretation of the trace of the strain matrix.

Our next step is showing that the strain terms ε_{ij} form a second-order tensor. When this is done, much further information as to the behavior of strain at a point will be available to us.

3.6　Transformation Equations for Strain

Making use of the geometrical interpretations of the strain terms presented in Section 3. 5, we shall now show that strain terms form a second-order tensor field. To do this, you will recall, we must show that at each point the strain terms transform according to an equation of the form given by Equation (2-10) and Equation (2-11) when there is a rotation of axes at the point.

As a first step let us evaluate the normal strain at point P in the direction indicated by the unit vector $\boldsymbol{n}$ as has been shown in Figure 3-10. We have shown a line segment $\overline{PQ}$ of length Δn and we have identified the deformed geometry with primed letters. We have the following formulation for the displacement of point P in the $\boldsymbol{n}$ direction where we have projected the displacements in the coordinate directions onto the direction $\boldsymbol{n}$.

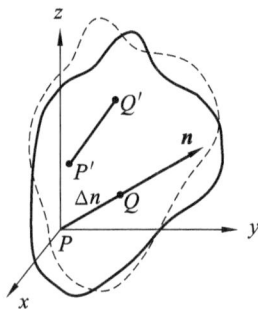

$$(u_n)_P = (u_x)_P l + (u_y)_P m + (u_z)_P n \qquad (3\text{-}57)$$

Furthermore, the displacement of point Q in the direction $\boldsymbol{n}$ can be expressed as a Taylor series expansion about point P in the following manner

$$(u_n)_Q = (u_n)_P + \left(\frac{\partial u_n}{\partial x}\right)_P \Delta x + \left(\frac{\partial u_n}{\partial y}\right)_P \Delta y + \left(\frac{\partial u_n}{\partial z}\right)_P \Delta z + \cdots$$

$$(3\text{-}58)$$

Figure 3-10　The deformed geometry of a line segment $\overline{PQ}$

Solving for $(u_n)_Q - (u_n)_P$ in Equation (3-58) and substituting on the right-hand side for u_n from Equation (3-57), we get

$$(u_n)_Q - (u_n)_P = \left(\frac{\partial u_x}{\partial x}\right)_P \Delta x l + \left(\frac{\partial u_y}{\partial x}\right)_P \Delta x m + \left(\frac{\partial u_z}{\partial x}\right)_P \Delta x n + \left(\frac{\partial u_x}{\partial y}\right)_P \Delta y l +$$

$$\left(\frac{\partial u_y}{\partial y}\right)_P \Delta y m + \left(\frac{\partial u_z}{\partial y}\right)_P \Delta y n + \left(\frac{\partial u_x}{\partial z}\right)_P \Delta z l + \left(\frac{\partial u_y}{\partial z}\right)_P \Delta z m + \left(\frac{\partial u_z}{\partial z}\right)_P \Delta z n + \cdots$$

(3-59)

Now we divide both sides by Δn, that is, the original length of $\overline{PQ}$ and take the limit as $\Delta n \to 0$. The higher-order terms vanish and we get

$$\lim_{\Delta n \to 0} \frac{(u_n)_Q - (u_n)_P}{\Delta n} = \left(\frac{\partial u_x}{\partial x}\right)_P \frac{\Delta x}{\Delta n} l + \left(\frac{\partial u_y}{\partial x}\right)_P \frac{\Delta x}{\Delta n} m + \left(\frac{\partial u_z}{\partial x}\right)_P \frac{\Delta x}{\Delta n} n + \left(\frac{\partial u_x}{\partial y}\right)_P \frac{\Delta y}{\Delta n} l +$$

$$\left(\frac{\partial u_y}{\partial y}\right)_P \frac{\Delta y}{\Delta n} m + \left(\frac{\partial u_z}{\partial y}\right)_P \frac{\Delta y}{\Delta n} n + \left(\frac{\partial u_x}{\partial z}\right)_P \frac{\Delta z}{\Delta n} l + \left(\frac{\partial u_y}{\partial z}\right)_P \frac{\Delta z}{\Delta n} m + \left(\frac{\partial u_z}{\partial z}\right)_P \frac{\Delta z}{\Delta n} n$$

(3-60)

The left side clearly is the normal strain ε_{nn} according to our geometrical interpretation. On the right side of the equation, we can replace $\Delta x/\Delta n, \Delta y/\Delta n, \Delta z/\Delta n$ by l, m, n. Since the foregoing equation is valid at any point P, we can drop the subscript P. We may then write

$$\varepsilon_{nn} = \frac{\partial u_x}{\partial x} l^2 + \frac{\partial u_y}{\partial y} m^2 + \frac{\partial u_z}{\partial z} n^2 + \left(\frac{\partial u_x}{\partial y} + \frac{\partial u_y}{\partial x}\right) lm +$$

$$\left(\frac{\partial u_x}{\partial z} + \frac{\partial u_z}{\partial x}\right) ln + \left(\frac{\partial u_y}{\partial z} + \frac{\partial u_z}{\partial y}\right) mn$$

(3-61)

Using the definition of the strain terms, we may express the preceding equation as

$$\varepsilon_{nn} = \varepsilon_{xx} l^2 + \varepsilon_{yy} m^2 + \varepsilon_{zz} n^2 + 2(\varepsilon_{xy} lm + \varepsilon_{xz} ln + \varepsilon_{yz} nm)$$

(3-62)

We now consider the shear-strain terms. In Figure 3-11, we have drawn segments $\overline{PQ}$ of length Δn in the $\boldsymbol{n}$ direction and $\overline{PR}$ of length Δs in the $\boldsymbol{s}$ direction. The deformed geometry is indicated by points P', Q' and R'. The shear strain ε_{ns} may be given by the following formulation

$$\varepsilon_{ns} = \frac{1}{2}\left(\frac{\partial u_n}{\partial s} + \frac{\partial u_s}{\partial n}\right)$$

(3-63)

We can express u_n and u_s in terms of displacements along the coordinate directions, as we did earlier, in the following manner

$$\begin{cases} u_n = u_x l + u_y m + u_z n \\ u_s = u_x l' + u_y m' + u_z n' \end{cases}$$

(3-64)

We then have for Equation (3-63), employing the preceding relations

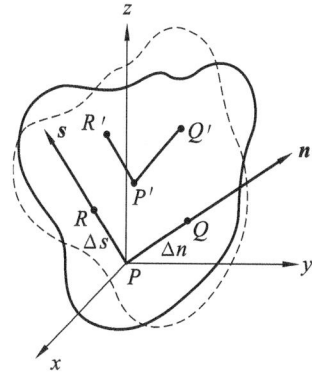

Figure 3-11　The deformed geometry of segments $\overline{PQ}$ and $\overline{PR}$

$$\varepsilon_{ns} = \frac{1}{2} \left(\frac{\partial u_x}{\partial s} l + \frac{\partial u_y}{\partial s} m + \frac{\partial u_z}{\partial s} n + \frac{\partial u_x}{\partial n} l' + \frac{\partial u_y}{\partial n} m' + \frac{\partial u_z}{\partial n} n' \right) \tag{3-65}$$

We can express the derivatives in the preceding formulation as follows

$$\begin{cases} \frac{\partial u_x}{\partial s} = \left(\frac{\partial u_x}{\partial x} l' + \frac{\partial u_x}{\partial y} m' + \frac{\partial u_x}{\partial z} n' \right), \frac{\partial u_y}{\partial s} = \left(\frac{\partial u_y}{\partial x} l' + \frac{\partial u_y}{\partial y} m' + \frac{\partial u_y}{\partial z} n' \right), \frac{\partial u_z}{\partial s} = \left(\frac{\partial u_z}{\partial x} l' + \frac{\partial u_z}{\partial y} m' + \frac{\partial u_z}{\partial z} n' \right) \\ \frac{\partial u_x}{\partial n} = \left(\frac{\partial u_x}{\partial x} l + \frac{\partial u_x}{\partial y} m + \frac{\partial u_x}{\partial z} n \right), \frac{\partial u_y}{\partial n} = \left(\frac{\partial u_y}{\partial x} l + \frac{\partial u_y}{\partial y} m + \frac{\partial u_y}{\partial z} n \right), \frac{\partial u_z}{\partial n} = \left(\frac{\partial u_z}{\partial x} l + \frac{\partial u_z}{\partial y} m + \frac{\partial u_z}{\partial z} n \right) \end{cases} \tag{3-66}$$

Hence, we have

$$\begin{aligned} \varepsilon_{ns} = \frac{1}{2} \Big[& 2 \times \frac{\partial u_x}{\partial x} ll' + 2 \times \frac{\partial u_y}{\partial y} mm' + 2 \times \frac{\partial u_z}{\partial z} nn' + \left(\frac{\partial u_x}{\partial y} + \frac{\partial u_y}{\partial x} \right) (lm' + l'm) + \\ & \left(\frac{\partial u_x}{\partial z} + \frac{\partial u_z}{\partial x} \right) (ln' + l'n) + \left(\frac{\partial u_y}{\partial z} + \frac{\partial u_z}{\partial y} \right) (mn' + nm') \Big] \\ = & \varepsilon_{xx} ll' + \varepsilon_{yy} mm' + \varepsilon_{zz} nn' + \varepsilon_{xy} (lm' + l'm) + \varepsilon_{xz} (ln' + l'n) + \varepsilon_{yz} (mn' + nm') \end{aligned} \tag{3-67}$$

By letting s in the preceding formulation be n, we can consider the preceding equation to represent the general transformation equation for strain. We can thus conclude that strain is a second-order tensor.

Since we have shown that strain is a second-order tensor, we can apply all the conclusions reached in Chapter 2 for the case of the stress tensor to the case of the strain tensor. That is,

(1) We can associate a second-order tensor called a strain quadric for each state of strain.

(2) There are three orthogonal directions where, for two of these directions, the normal strains take on the extreme values for the particular state of strain. Also the shear strains are zero for these axes. These are the principal axes of strain.

(3) The three tensor invariants for strain become

$$\varepsilon_{11} + \varepsilon_{22} + \varepsilon_{33} = I_\varepsilon \tag{3-68}$$

$$\begin{vmatrix} \varepsilon_{22} & \varepsilon_{23} \\ \varepsilon_{32} & \varepsilon_{33} \end{vmatrix} + \begin{vmatrix} \varepsilon_{11} & \varepsilon_{13} \\ \varepsilon_{31} & \varepsilon_{33} \end{vmatrix} + \begin{vmatrix} \varepsilon_{11} & \varepsilon_{12} \\ \varepsilon_{21} & \varepsilon_{22} \end{vmatrix} = II_\varepsilon \tag{3-69}$$

$$\begin{vmatrix} \varepsilon_{11} & \varepsilon_{12} & \varepsilon_{13} \\ \varepsilon_{21} & \varepsilon_{22} & \varepsilon_{23} \\ \varepsilon_{31} & \varepsilon_{32} & \varepsilon_{33} \end{vmatrix} = III_\varepsilon \tag{3-70}$$

(4) We can solve for the principal strains by employing the same cubic equation, in Chapter 2, that was used for stress. Thus

$$\varepsilon^3 - (I_\varepsilon) \varepsilon^2 + (II_\varepsilon) \varepsilon - (III_\varepsilon) = 0 \tag{3-71}$$

where (I_ε), (II_ε) and (III_ε) are the three tensor invariants previously given.

(5) We can formulate the case of plane strain just as we did plane stress where now $\varepsilon_{xz} = \varepsilon_{yz} = \varepsilon_{zz} = 0$. This type of strain corresponds to a prismatic body with the longitudinal axis in the z direction and with an external loading that does not change with z. Finally, the ends of this prism must be rigidly constrained. An example of this would be a dam, as shown in Figure 3-12. All the formulations given for plane stress in Chapter 2 can be carried over to plane strain. This includes the Mohr's circle.

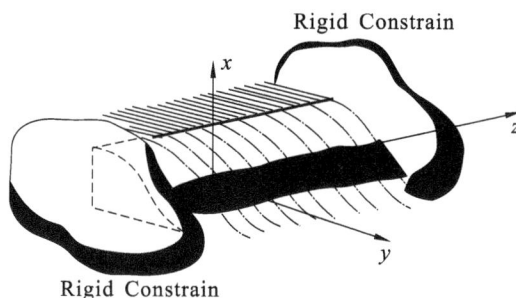

Figure 3-12 A dam where its ends must be rigidly constrained

We present a number of problems at the end of this chapter which require you to employ the general characteristics of second-order tensors.

3.7 Strain Terms in Two-dimensional Polar Coordinates

In solving plane problems such as circular objects, it is generally convenient to use polar coordinates. In polar coordinates, ε_r represents the normal strain of radial segment, ε_θ represents the normal strain of circumferential segment, $\gamma_{r\theta}$ represents the shear angle which is the change of right angle between radial and circumferential segment, u_r represents the radial displacement and u_θ represents the circumferential displacement.

It is assumed that there is only radial displacement with no circumferential displacement, as shown in Figure 3-13. Due to the radial displacement, the radial segment PA moves to $P'A'$, the circumferential segment PB moves to $P'B'$, and the displacements of three points P, A, B are shown as

$$PP' = u_r \qquad AA' = u_r + \frac{\partial u_r}{\partial r}\mathrm{d}r \qquad BB' = u_r + \frac{\partial u_r}{\partial \theta}\mathrm{d}\theta$$

Therefore, the normal strain of the radial segment PA is

$$\varepsilon_r = \frac{P'A' - PA}{PA} = \frac{AA' - PP'}{PA} = \frac{\left[u_r + \dfrac{\partial u_r}{\partial r}\mathrm{d}r\right] - u_r}{\mathrm{d}r} = \frac{\partial u_r}{\partial r} \qquad (3\text{-}72)$$

Normal strain of circumferential segment PB:

$$\varepsilon_\theta = \frac{P'B' - PB}{PB} = \frac{(r + u_r)\mathrm{d}\theta - r\mathrm{d}\theta}{r\mathrm{d}\theta} = \frac{u_r}{r} \tag{3-73}$$

The rotation angle of radial segment PA:

$$\alpha = 0 \tag{3-74}$$

The rotation angle of circular segment PB:

$$\beta = \frac{BB' - PP'}{PB} = \frac{\left(u_r + \dfrac{\partial u_r}{\partial \theta}\mathrm{d}\theta\right) - u_r}{r\mathrm{d}\theta} = \frac{1}{r}\frac{\partial u_r}{\partial \theta} \tag{3-75}$$

Then the shear angle is

$$\gamma_{r\theta} = \alpha + \beta = \frac{1}{r}\frac{\partial u_r}{\partial \theta} \tag{3-76}$$

It is assumed that there is only circumferential displacement, as shown in Figure 3-14. Due to the circumferential displacement, the radial line PA moves to $P''A''$, the circumferential line PB moves to $P''B''$, and the displacements of the three point P,A,B are shown as

$$PP'' = u_\theta \quad AA'' = u_\theta + \frac{\partial u_\theta}{\partial r}\mathrm{d}r \quad BB'' = u_\theta + \frac{\partial u_\theta}{\partial \theta}\mathrm{d}\theta$$

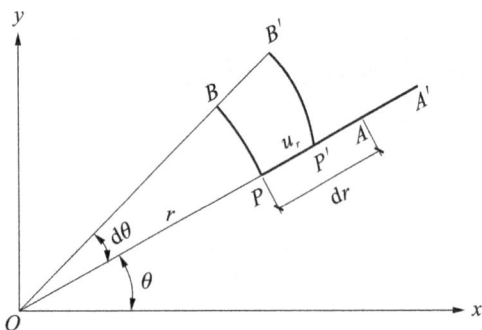

Figure 3-13 Geometric diagram with radial displacement

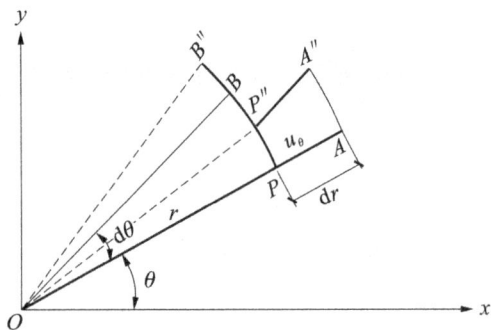

Figure 3-14 Geometric diagram with circumferential displacement

Therefore, the normal strain of the radial segment PA:

$$\varepsilon_r = 0 \tag{3-77}$$

The normal strain of the circumferential segment PB:

$$\varepsilon_\theta = \frac{P''B'' - PB}{PB} = \frac{BB'' - PP''}{PB} = \frac{\left(u_\theta + \dfrac{\partial u_\theta}{\partial \theta}\mathrm{d}\theta\right) - u_\theta}{r\mathrm{d}\theta} = \frac{1}{r}\frac{\partial u_\theta}{\partial \theta} \tag{3-78}$$

The rotation angle of the radial segment PA:

$$\alpha = \frac{AA'' - PP''}{PA} = \frac{\left(u_\theta + \dfrac{\partial u_\theta}{\partial r}\mathrm{d}r\right) - u_\theta}{\mathrm{d}r} = \frac{\partial u_\theta}{\partial r} \tag{3-79}$$

The rotation angle of the annular segment PB:

$$\beta = -\angle POP'' = -\frac{PP''}{OP} = -\frac{u_\theta}{r} \tag{3-80}$$

Then the shear angle is

$$\gamma_{r\theta} = \alpha + \beta = \frac{\partial u_\theta}{\partial r} - \frac{u_\theta}{r} \tag{3-81}$$

Therefore, if there is displacement along both radial and circumferential directions, it is available to be superimposed separately

$$\begin{cases} \varepsilon_r = \dfrac{\partial u_r}{\partial r} \\[2mm] \varepsilon_\theta = \dfrac{u_r}{r} + \dfrac{1}{r}\dfrac{\partial u_\theta}{\partial \theta} \\[2mm] \gamma_{r\theta} = \dfrac{1}{r}\dfrac{\partial u_r}{\partial \theta} + \dfrac{\partial u_\theta}{\partial r} - \dfrac{u_\theta}{r} \end{cases} \tag{3-82}$$

These are the strain terms in polar coordinates.

3.8   Equations of Compatibility

Let us consider further the strain displacement relations

$$\varepsilon_{ij} = \frac{1}{2}\left(\frac{\partial u_i}{\partial x_j} + \frac{\partial u_j}{\partial x_i}\right) \tag{3-83}$$

If the displacement field is given, we can compute the strain tensor field easily by carrying out partial derivatives in accordance with the foregoing equations. The inverse problem of finding the displacement field from a strain field is not so simple. Here the displacement field, composed of three functions u_i, must be determined by integration of the six partial differential equations represented by Equation (3-83). In order to insure a single-valued, continuous solution for u_i, we must impose certain restrictions on the strain functions ε_{ij}. That is, we cannot take any tensor field ε_{ij} and expect it automatically to be associated with a single-valued, continuous displacement field. But actual deformations are single-valued. Furthermore, the deformations of interest to us are continuous. Hence the restrictions on ε_{ij} stemming from these considerations will apply to all our formulations. The resulting equations are termed the compatibility equations. The compatibility equations relate the strains properly to satisfy the aforestated conditions.

A complete development of the compatibility equations is beyond the level of this book. We shall proceed by formulating the compatibility equations as necessary conditions required for the proper relation between a single-valued, continuous displacement field and the strain field. Later, we shall point out without proof when these equations satisfy the sufficiency requirements. As a first step, we rewrite Equation (3-83) using unabridged notation

$$\varepsilon_{xx} = \frac{\partial u_x}{\partial x} \tag{3-84a}$$

$$\varepsilon_{yy} = \frac{\partial u_y}{\partial y} \tag{3-84b}$$

$$\varepsilon_{zz} = \frac{\partial u_z}{\partial z} \tag{3-84c}$$

$$\gamma_{xy} = \frac{\partial u_y}{\partial x} + \frac{\partial u_x}{\partial y} \tag{3-84d}$$

$$\gamma_{xz} = \frac{\partial u_z}{\partial x} + \frac{\partial u_x}{\partial z} \tag{3-84e}$$

$$\gamma_{yz} = \frac{\partial u_z}{\partial y} + \frac{\partial u_y}{\partial z} \tag{3-84f}$$

Our procedure will be to eliminate the displacements from these equations. First, differentiate Equation (3-84a) with respect to y twice and Equation (3-84b) with respect to x twice. Thus

$$\frac{\partial^2 \varepsilon_{xx}}{\partial y^2} = \frac{\partial^3 u_x}{\partial y^2 \partial x} \tag{3-85a}$$

$$\frac{\partial^2 \varepsilon_{yy}}{\partial x^2} = \frac{\partial^3 u_y}{\partial x^2 \partial y} \tag{3-85b}$$

Adding the preceding equations, we get

$$\frac{\partial^2 \varepsilon_{xx}}{\partial y^2} + \frac{\partial^2 \varepsilon_{yy}}{\partial x^2} = \frac{\partial^3 u_x}{\partial y^2 \partial x} + \frac{\partial^3 u_y}{\partial x^2 \partial y} \tag{3-86}$$

Next take the mixed derivative of Equation (3-84d) with respect to x and y. We get

$$\frac{\partial^2 \gamma_{xy}}{\partial x \partial y} = \frac{\partial^3 u_x}{\partial x \partial y^2} + \frac{\partial^3 u_y}{\partial y \partial x^2} \tag{3-87}$$

Assuming that all the derivatives are continuous, we can interchange the order of partial differentiation. We then see that the right sides of Equation (3-86) and Equation (3-87) are equal, thus permitting us to equate the left sides of these equations in the following way

$$\frac{\partial^2 \varepsilon_{xx}}{\partial y^2} + \frac{\partial^2 \varepsilon_{yy}}{\partial x^2} = \frac{\partial^2 \gamma_{xy}}{\partial x \partial y} \tag{3-88}$$

We may form two more equations in a similar way, i. e.

$$\frac{\partial^2 \varepsilon_{yy}}{\partial z^2} + \frac{\partial^2 \varepsilon_{zz}}{\partial y^2} = \frac{\partial^2 \gamma_{yz}}{\partial y \partial z} \tag{3-89a}$$

$$\frac{\partial^2 \varepsilon_{zz}}{\partial x^2} + \frac{\partial^2 \varepsilon_{xx}}{\partial z^2} = \frac{\partial^2 \gamma_{zx}}{\partial z \partial x} \tag{3-89b}$$

Equation (3-88) and Equation (3-89) form three of the six compatibility equations to be formulated. To get the remaining equations we take the mixed derivative of Equation (3-84a) with respect to z and y. Thus

$$\frac{\partial^2 \varepsilon_{xx}}{\partial y \partial z} = \frac{\partial^3 u_x}{\partial y \partial z \partial x} \tag{3-90}$$

Now take the partial derivative of Equation (3-84d) with respect to z and x. We get

$$\frac{\partial^2 \gamma_{xy}}{\partial x \partial z} = \frac{\partial^3 u_x}{\partial x \partial z \partial y} + \frac{\partial^3 u_y}{\partial z \partial^2 x} \tag{3-91}$$

Also take the partial derivative of Equation (3-84e) with respect to y and x. Thus

$$\frac{\partial^2 \gamma_{xz}}{\partial x \partial y} = \frac{\partial^3 u_x}{\partial x \partial y \partial z} + \frac{\partial^3 u_z}{\partial^2 x \partial y} \tag{3-92}$$

Finally, take the partial derivative of Equation (3-84f) with respect to x twice. Thus

$$\frac{\partial^2 \gamma_{yz}}{\partial x^2} = \frac{\partial^3 u_z}{\partial x^2 \partial y} + \frac{\partial^3 u_y}{\partial x^2 \partial z} \tag{3-93}$$

Now add Equation (3-91) and Equation (3-92) and subtract Equation (3-93) from the result. By further changes of the order of partial derivatives, we arrive at the following equation

$$-\frac{\partial^2 \gamma_{yz}}{\partial x^2} + \frac{\partial^2 \gamma_{xz}}{\partial x \partial y} + \frac{\partial^2 \gamma_{xy}}{\partial x \partial z} = 2 \frac{\partial^3 u_x}{\partial x \partial y \partial z} \tag{3-94}$$

We may now equate twice the left side of Equation (3-90) with the left side of Equation (3-94) to form another of the compatibility equations, i. e.

$$2\frac{\partial^2 \varepsilon_{xx}}{\partial y \partial z} = \frac{\partial}{\partial x}\left(-\frac{\partial \gamma_{yz}}{\partial x} + \frac{\partial \gamma_{xz}}{\partial y} + \frac{\partial \gamma_{xy}}{\partial z}\right) \tag{3-95}$$

In a similar manner, the following additional equations can be formed

$$\begin{cases} 2\dfrac{\partial^2 \varepsilon_{yy}}{\partial z \partial x} = \dfrac{\partial}{\partial y}\left(-\dfrac{\partial \gamma_{zx}}{\partial y} + \dfrac{\partial \gamma_{yx}}{\partial z} + \dfrac{\partial \gamma_{yz}}{\partial x}\right) \\[3mm] 2\dfrac{\partial^2 \varepsilon_{zz}}{\partial x \partial y} = \dfrac{\partial}{\partial z}\left(-\dfrac{\partial \gamma_{xy}}{\partial z} + \dfrac{\partial \gamma_{zy}}{\partial x} + \dfrac{\partial \gamma_{zx}}{\partial y}\right) \end{cases} \tag{3-96}$$

We shall now write the entire set of compatibility equations

$$\frac{\partial^2 \varepsilon_{xx}}{\partial y^2} + \frac{\partial^2 \varepsilon_{yy}}{\partial x^2} = \frac{\partial^2 \gamma_{xy}}{\partial x \partial y} \tag{3-97a}$$

$$\frac{\partial^2 \varepsilon_{yy}}{\partial z^2} + \frac{\partial^2 \varepsilon_{zz}}{\partial y^2} = \frac{\partial^2 \gamma_{yz}}{\partial y \partial z} \tag{3-97b}$$

$$\frac{\partial^2 \varepsilon_{zz}}{\partial x^2} + \frac{\partial^2 \varepsilon_{xx}}{\partial z^2} = \frac{\partial^2 \gamma_{zx}}{\partial z \partial x} \tag{3-97c}$$

$$2\frac{\partial^2 \varepsilon_{xx}}{\partial y \partial z} = \frac{\partial}{\partial x}\left(-\frac{\partial \gamma_{yz}}{\partial x} + \frac{\partial \gamma_{zx}}{\partial y} + \frac{\partial \gamma_{xy}}{\partial z}\right) \tag{3-97d}$$

$$2\frac{\partial^2 \varepsilon_{yy}}{\partial z \partial x} = \frac{\partial}{\partial y}\left(-\frac{\partial \gamma_{zx}}{\partial y} + \frac{\partial \gamma_{yx}}{\partial z} + \frac{\partial \gamma_{yz}}{\partial x}\right) \tag{3-97e}$$

$$2\frac{\partial^2 \varepsilon_{zz}}{\partial x \partial y} = \frac{\partial}{\partial z}\left(-\frac{\partial \gamma_{xy}}{\partial z} + \frac{\partial \gamma_{zy}}{\partial x} + \frac{\partial \gamma_{zx}}{\partial y}\right) \tag{3-97f}$$

The strain tensor field must satisfy the preceding equations if the strain field is to correspond to a single-valued, continuous, and thus physically meaningful deformation.

3.9 A Note on Simply and Multiply Connected Bodies

In order to make statements concerning the sufficiency conditions of the compatibility equations, we need to distinguish between two classes of bodies, namely, simply connected bodies and multiply connected bodies.

We define a simply connected body as one where every closed path in the body can be continuously shrunk to a point without cutting a boundary. The definition permits us to move the path in any way in the process of shrinking it to a point. Thus the body shown in Figure 3-15 is simply connected because any path a can be shrunk to a point without cutting the outside boundary surface S_1 or the closed internal boundary surface S_2 which encloses a cavity inside the material.

A multiply connected body is one where there exists one or more paths which cannot be shrunk to a point in the manner described in the previous paragraph. An example of multiply connected region is the ring shown in Figure 3-16 where clearly path a cannot possibly be shrunk to a point without cutting the boundary.

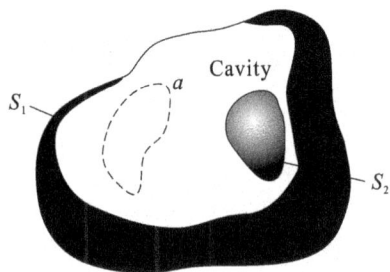

Figure 3-15   A simply connected body

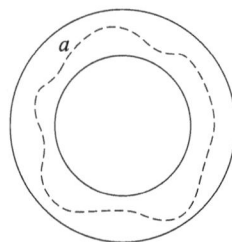

Figure 3-16   A multiply connected body

We state now without proof that the satisfaction of the compatibility equations comprises, for a body with one point fixed, both the necessary and sufficient conditions for a single-valued, continuous displacement field when the body is simply connected. If the body is multiply connected, the compatibility equations give only the necessary conditions for a single-valued, continuous displacement field. Other conditions, which we shall not discuss here, must be imposed for the sufficiency requirement.

3.10 Closure

In this chapter, we have developed the rotation matrix and the strain matrix. These formulations permit us to describe the rigid-body rotation and the pure deformation of a vanishingly small element of the body for the case of small or so-called infinitesimal deformation. The formulations of this chapter were purely geometric and thus valid for any continuous medium that undergoes a small deformation under load. In fluids and viscoelastic media, it is the time rate of deformation that is of significance in analysis and so we are then interested in strain rates, etc. In Chapter 2 on the other hand, we formulated stress at a point using Newton's law. These formulations again are valid for all continuous media.

In relating stress and deformation, we find that the nature of the continuous medium becomes important. In Chapter 4, we shall examine experimental data showing relations between the stress and the strain tensor for certain media. In addition to presenting certain experimental results, we shall also present useful extrapolations and idealizations of these results. This will set the stage for solution of stress and strain distributions beginning in Chapter 5.

Problems

3-1   Given the following displacement field

$$\boldsymbol{u}=(x^2+y)\boldsymbol{i}+(3+z)\boldsymbol{j}+(x^2+2y)\boldsymbol{k}$$

What is the deformed position of a point originally at $(3,1,-2)$?

3-2   Two points in the undeformed geometry are originally at $(0,0,1)$ and $(2,0,-1)$. What is the distance between these points after deformation (Assuming the displacement field given in Problem 3-1 is imposed on the body)?

3-3   A displacement field is given as

$$\boldsymbol{u}=(0.16x^2+\sin y)\boldsymbol{i}+\left(0.1z+\frac{x}{y^3}\right)\boldsymbol{j}+0.004\boldsymbol{k}$$

As a result of deformation what is the increase in distance between two points, which in the undeformed geometry have position vectors

$$r_1 = 10i + 3j$$
$$r_2 = 4k + 3j$$

3-4　A displacement field is given as

$$u_i = \lambda_{ij} x_j$$

where λ_{ij} form a set of constants, is called an affine deformation. If

$$\lambda_{ij} = \begin{pmatrix} +0.2 & -0.05 & -0.1 \\ +0.03 & +0.1 & -0.02 \\ +0.003 & -0.2 & +0.03 \end{pmatrix}$$

what is the displacement of a point whose position vector from the fixed point in the undeformed geometry is

$$r = i - j + 3k$$

3-5　Show that for affine deformations (Problem 3-4).

(1) Plane sections remain plane during deformation.

(2) Straight lines remain straight lines during deformation.

3-6　In Problem 3-4 what is the equation of the surface which was the xy plane in the undeformed geometry [$Hint$: A simple procedure is to consider two position vectors in the xy plane. Find normal n to these vectors in the deformed geometry. Now the plane desired must have in it position vectors $r' = (x'i + y'j + z'k)$ such that $n \cdot r' = 0$]?

3-7　Given the following displacement field

$$u = [y^2 i + 3yz j + (4 + 6x^2)k] \times 10^{-2}$$

What is the matrix $\partial u_i / \partial x_j$ for a small domain about any point xyz? For the point $(0,1,3)$ what is this matrix?

3-8　Using the displacement field given in Problem 3-7, what is the increase in length per unit length in the y direction of an infinitesimal segment at the $(1,0,2)$ along the y axis?

3-9　The following displacement field is given

$$u = \frac{x}{100} i + \frac{y^2}{200} j + \left(0.02 + \frac{y^2 z}{500}\right) k$$

What is the displacement of a point at position

$$r = 6i + 2j$$

What is the matrix for giving the deformation in a small domain at the origin?

3-10　Explain why only infinitesimal rotations of rigid bodies could be considered representable by vectors.

3-11　Given the following displacement field

$$u=[x^2i+(2y^2+3z)j+10k]\times10^{-3}$$

(1) What is the increase in distance between two points $(0,1,0)$ and $(2,-1,3)$ in the undeformed geometry as a result of deformation?

(2) What is the elongation per unit length in the y direction of a line segment originally in the y direction in a small domain about $(1,3,2)$?

3-12　In Problem 3-11, what is the change of a small vector

$$\Delta s=0.002i+0.003j+0.0004k$$

at position $(1,0,1)$ as a result of deformation? What is the new vector representation in the deformed geometry?

3-13　Suppose we have two successive small deformations represented by the deformation matrices at point P

$$\left(\frac{\partial u_i^{(1)}}{\partial x_j}\right)_P=\begin{bmatrix}0.02 & 0.01 & 0 \\ 0 & 0.01 & -0.02 \\ 0 & 0 & -0.02\end{bmatrix}$$

and

$$\left(\frac{\partial u_i^{(2)}}{\partial x_j}\right)_P=\begin{bmatrix}0.01 & 0.015 & -0.02 \\ 0 & 0 & -0.01 \\ 0 & -0.03 & 0.04\end{bmatrix}$$

What is the total change at point P of a vector Δs given by

$$\Delta s=(6i+10j+2k)\times10^{-3}$$

3-14　A body is pinned at point O. If it rotates an angle φ about the z axis at that point, what is the matrix $\partial u_i/\partial x_j$ for this movement (*Hint*: Consider successively line segments in the x, y and z directions)?

3-15　A body element undergoes a small rotation $\delta\phi$ given as follows

$$\delta\phi=0.0002i+0.0005j-0.0002k \quad \text{(rad)}$$

What is the matrix $\partial u_i/\partial x_j$ at this point?

3-16　Given the following displacement field

$$u=[(6y+5z)i+(-6x+3z)j+(-5x-3y)k]\times10^{-3}$$

Show that this displacement field is that of rigid-body rotation. What is the rotation vector $\delta\phi$ for the body?

3-17　An element at P is subject to a deformation given by the deformation matrix

$$\left(\frac{\partial u_i}{\partial x_j}\right)_P=\begin{bmatrix}0.01 & 0 & 0 \\ -0.02 & 0.03 & 0 \\ 0 & -0.02 & -0.01\end{bmatrix}$$

What are the rotation matrix, the strain matrix, and the angle of rotation for this element?

3-18　Given the displacement field

$$\boldsymbol{u}=[(x^3+10)\boldsymbol{i}+3yz\boldsymbol{j}+(z^2-yx)\boldsymbol{k}]\times10^{-2}$$

What is the rigid-body translation of the body? What is the rotation of an element at position $(2,1,0)$?

3-19　Given the following displacement field

$$\boldsymbol{u}=[(y^2+3z)\boldsymbol{i}+(x+3yz)\boldsymbol{j}+(\sin z)\boldsymbol{k}]\times10^{-2}$$

Formulate the strain matrix and the rotation matrix for an element at $(1,2,3)$. What is the rigid-body rotation of this element? What is the displacement of this element?

3-20　Given the following displacement field

$$u_x=0.06x+0.05y-0.01z$$
$$u_y=0.01x-0.03z$$
$$u_z=-0.02x+0.01z$$

What is the normal strain ε_{xx} at all points of the body? If there is a line segment 10^{-3} m long parallel to the x axis in the undeformed geometry what will be the new length of this line segment?

3-21　Using the geometrical definition of normal strain determine ε_{kk} in Problem 3-20 for a direction in the xy plane $30°$ from the x axis.

3-22　In Problem 3-17 what are the shear strain ε_{xy} and the shear angle γ_{xy} at the point P?

3-23　Given the displacement field

$$\boldsymbol{u}=[xy\boldsymbol{i}+3y\boldsymbol{j}+(6+10z)\boldsymbol{k}]\times10^{-2}$$

What are the normal strains ε_{xx}, ε_{yy} and the shear strain ε_{yz} at position $(2,1,3)$?

3-24　Using the displacement field given in Problem 3-18, what is the change in right angle of two infinitesimal line segments parallel to the y and z axes at some position xyz in the body? What is the rotation of this pair of line segments?

3-25　In Problem 3-23 determine the change of right angle between segments parallel to dx and dy at position $(3,0,-2)$. Also what is the change in volume per unit volume at this point?

3-26　Do Problem 3-21 using the transformation formula for strain.

3-27　The following state of strain exists at a point in a body

$$\boldsymbol{\varepsilon}_{ij}=\begin{bmatrix}0.01 & -0.02 & 0\\-0.02 & 0.03 & -0.01\\0 & -0.01 & 0\end{bmatrix}$$

In a direction p having the direction cosines $l=0.6, m=0, n=0.8$, what is ε_{pp}?

3-28　In Problem 3-27, a set of axes x', y', z' is chosen as is shown in Figure 3-17. What is the strain tensor at the point of interest for this new reference?

3-29　The principal strains at a point in a body are $\varepsilon_{xx}=0.002, \varepsilon_{yy}=0.001, \varepsilon_{zz}=0$. What is the shear angle $\gamma_{x'y'}$ using figure of Problem 3-28 to get the direction of the coordinate axes?

3-30　In Problem 3-27, what are the three strain-tensor invariants at the point of interest?

3-31　Using the Figure 3-18, show for plane strain that

$$\varepsilon_{x'x'}=\frac{\varepsilon_{xx}+\varepsilon_{yy}}{2}+\frac{\varepsilon_{xx}-\varepsilon_{yy}}{2}\cos2\theta+\frac{\gamma_{xy}}{2}\sin2\theta$$

and

$$\gamma_{x'y'}=\gamma_{xy}\cos2\theta-(\varepsilon_{xx}-\varepsilon_{yy})\sin2\theta$$

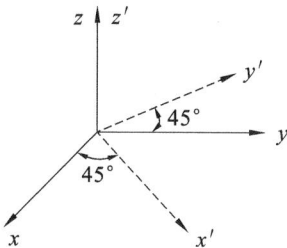

Figure 3-17　Figure of Problem 3-28　　　　Figure 3-18　Figure of Problem 3-31

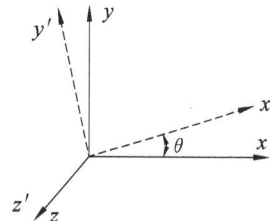

3-32　Using the formulation in Problem 3-31 determine the strains at a point for axes x', y', z' rotated $30°$ about the z axis from reference xyz. We are given the following information at the point

$$\varepsilon_{xx}=0.01$$
$$\varepsilon_{yy}=-0.02$$
$$\varepsilon_{xy}=0.02$$

What are the principal strains at the point and the direction of the principal axes?

3-33　Develop a formulation for maximum shear strain at a point for the case of plane strain. Also determine the direction of the axes for maximum shear strain.

3-34　Show for the simple case of plane strain that the axis for maximum shear-strain angle $(\gamma_{x'y'})_{max}$ must be rotated $45°$ from the principal axes x, y. Also show the maximum shear-strain angle $(\gamma_{x'y'})_{max}$ equals the difference between the principal normal strains.

3-35　What is the maximum shear-strain angle $(\gamma_{x'y'})_{max}$ for Problem 3-23?

3-36　What is the maximum shear-strain angle $(\gamma_{x'y'})_{max}$ for Problem 3-32?

3-37　How would you construct a Mohr's circle for the case of plane strain?

3-38　Construct Mohr's circle for the state of strain in Problem 3-32. Find the strain for the axes x', y' given by Figure 3-18, where $\theta = 30°$.

3-39　Express the compatibility equations for plane strain.

3-40　Given the following plane-strain distribution:

$$\varepsilon_{xx} = 3x^2 y$$

$$\varepsilon_{yy} = 4y^2 x$$

$$\varepsilon_{xy} = yx + x^3$$

Are the compatibility equations satisfied?

4　Stress-strain Relations

4.1　Introduction

In previous chapters we have separately considered stress and strain. It is intuitively clear that these quantities are related and in this chapter we shall make an introductory inquiry into such relations.

Almost all the working knowledge we now possess stems from macroscopic testing of materials and it is the results of such tests as well as the macroscopic theories stemming from such tests that concern us in this chapter. For some time, however, solid-state physicists and engineers have been intensively studying the microscopic bases for mechanical properties, i. e. , actions at the atomic and molecular level. Much progress has been made along this avenue of approach, although a thorough understanding of the mechanisms involved has as yet not been reached. It is expected that in the future we shall turn more and more to this fundamental approach. Modern technology is putting our structures into more complex environments and under more complex conditions for which macroscopic laboratory tests, such as the ones we shall describe in this chapter, are becoming less meaningful. Needed for a better understanding of how a material is to behave under a combination of conditions, such as high temperature, dynamic loads, radiation, temperature gradients, vibration, etc. , is a comprehension of how mechanical action relates to atomic and molecular structure. We cannot presently employ these results in a quantitative manner to a great extent, but certainly such material may be quite helpful qualitatively.

In this chapter, we shall first consider the simple uniaxial loading case and then we shall endeavor to extrapolate our uniaxial results and concepts to the general state of stress.

4.2　The Tensile Test

The most basic test in the study of stress-strain relations is the simple tensile test where a cylindrical specimen of the type shown in Figure 4-1 is subjected by a tensile test machine to a force F along the centerline of the specimen. The distance L between two points on the specimen is measured at all times by Gauge 1 as seen in the diagram. Another Gauge 2 mean-while measures the diameter D of the cylinder. As the force F is varied, we measure L and D for each setting of F. Hence, at any setting, we have the following information.

(1) Actual stress $(\sigma_z)_{\text{act}}$— computed as F/A_{act}, where A_{act} is the cross-sectional area of the cylinder found by employing the actual diameter D given by Gauge 2.

(2) Engineering stress $(\sigma_z)_{\text{eng}}$— computed as F/A_0, where A_0 is the initial unstrained cross-sectional area of the cylinder.

(3) Strain ε_{zz}— computed by the ratio $\Delta L/L_0$, where ΔL is found using Gauge 1 and L_0 is the unstrained length.

It is customary to plot the engineering stress $(\sigma_z)_{\text{eng}}$ versus the strain ε_{zz} for such a test. Because the volume of the specimen will change only slightly, there will take place a contraction of the cross-sectional area A_{act} as the tensile load is increased so that the engineering stress will be less than the actual stress at all times. For small loads, A_{act} will not be appreciably smaller than A_0 so that little difficulty is encountered by using the simpler engineering stress. However, for large loads there will be a significant difference between A_{act} and A_0 with the result that the curve $(\sigma_z)_{\text{eng}}$ versus ε_{zz} will appear to be "unnatural". Nevertheless, because $(\sigma_z)_{\text{eng}}$ is related simply by a direct proportionality to F through the constant $1/A_0$, and because the lateral contractions are not easily measured accurately, engineers are often motivated to use the engineering stress $(\sigma_z)_{\text{eng}}$ rather than the actual stress $(\sigma_z)_{\text{act}}$.

We have shown a stress-strain diagram from a simple tensile test in Figure 4-2. This is a typical curve for a low-carbon steel specimen. Although stress-strain curves may be quite different for other materials, we shall consider this curve in some detail so as to set forth most easily general definitions. Notice that the curve is a straight line at the early stages of the loading; that is, the stress is proportional to strain and we may then state

$$\sigma_z = E\varepsilon_{zz} \tag{4-1}$$

where the proportionality constant E is called Young's modulus having units (MPa) as you may easily verify yourself. Essentially this result was reached about 300 years ago by Robert Hooke, who as a result of his experiments with metallic rods under axially app-lied tensile loads, concluded that the extension is proportional to the force, a relation known by every high school student as Hooke's law. The stress at which the linear relationship between stress and strain ceases is called the proportional limit. Its value, however, is not easily measured.

Not all materials have a finite straight-line portion at the outset of the stress-strain diagram. For instance, rubber is a material that generally does not have this, and a stress-strain curve for a particular specimen of this material is shown in Figure 4-3. Despite the apparent difference in appearance between the curves for low-carbon steel and rubber in Figure 4-2 and Figure 4-3, there is an important similarity to be pointed out. That is, if that particular rubber specimen is unloaded to zero load, it will return to its original geometry along the loading curve as will the steel specimen provided that the load on the latter develops a stress σ_z sufficiently below the proportional limit (Figure 4-4). Thus both materials may act in a perfectly elastic manner.

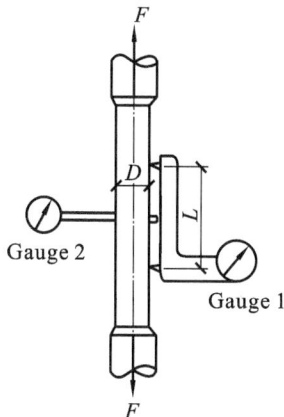

Figure 4-1   **The simple tensile test**

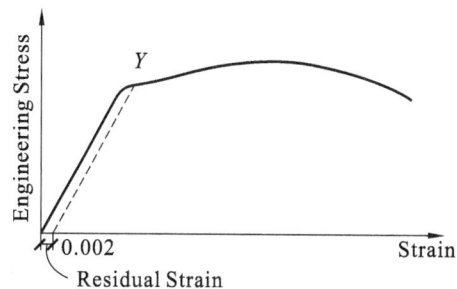

Figure 4-2   **The stress-strain diagram from a simple tensile test**

For the low-carbon steel specimen there is a stress level close to the proportional limit so that when the specimen is unloaded from a stress level above the aforementioned level the original geometry is no longer reached. We call this stress level, which gives the limit of elastic behavior and the onset of inelastic behavior, the elastic limit. For steels, the elastic limit is very close to the proportional limit, so close that one doesn't always make a distinction between these two stresses. A material having a proportional limit close to the elastic limit, such as steel, is termed a linear elastic materials.

Returning to the stress-strain diagram for steel, it is to be pointed out that the elastic limit like the proportional limit is difficult to measure accurately. Hence, engineers employ as a more useful definition of the beginning of nonelastic behavior the yield stress or the yield point, which is the value of stress resulting in a small specified residual strain (usually 0. 002) upon unloading. The point Y on the stress-strain diagram in Figure 4-2 corresponds to the yield point.

Our discussion of the stress-strain diagram has taken us thus far only to the yield point. In the domain up to the yield point, the actual cross-sectional area and the original area of the specimen differ by a very small amount and so it does not matter which area one uses for computations. We have been using A_0 for reasons set forth earlier. However, as pointed out earlier, at all times during the tensile test, a continual decrease in the cross-sectional area of the specimen takes place as the load is applied(In the case of a compression test, there is clearly a corresponding increase in the cross-sectional area as the load is applied). This lateral effect is called the Poisson effect. After the yield point, there may be a rapid increase in the strain ε_{xx} and simultaneously there will then be a rapid change in the cross-sectional area owing to the Poisson effect. This will cause the values of engineering stress and actual stress to diverge appreciably from each other. To illustrate this, Figure 4-5 shows sample stress-strain curves using both the actual stress and the engineering stress for a tensile test (Also shown dotted is the corresponding compression test using engineering stress). You will notice that the actual stress continually increases until the specimen breaks.

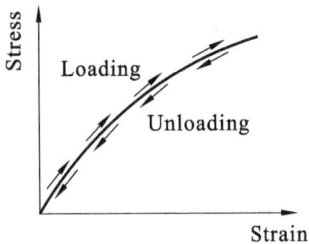

Figure 4-3　The stress-strain
curve for a rubber specimen

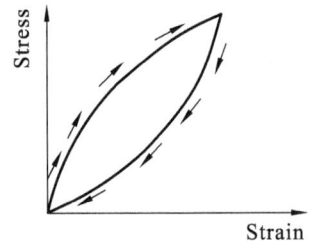

Figure 4-4　The stress-strain curve
for a particular specimen

As pointed out earlier, the engineering stress is proportional to the force F and so the maximum load capability of the specimen is developed at the maximum elevation of the engineering stress-strain curve. The engineering stress at this point, denoted as U in the diagram, is termed the ultimate stress. We may express the approach to the ultimate stress condition mathematically by considering the relation $F = \sigma_z A$. Using the symbol δ for the elongation ΔL we can then say, differentiating the preceding expression with respect to δ

$$\frac{\mathrm{d}F}{\mathrm{d}\delta} = A\,\frac{\mathrm{d}\sigma_z}{\mathrm{d}\delta} + \sigma_z\,\frac{\mathrm{d}A}{\mathrm{d}\delta} \tag{4-2}$$

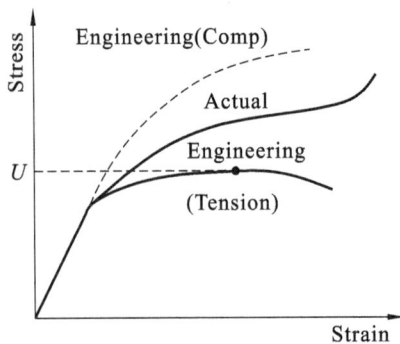

Figure 4-5　A stress-strain curve using both the
actual stress and the engineering stress

For small values of force, i. e. , for conditions below the yield point, the quantity $\mathrm{d}A/\mathrm{d}\delta$, which clearly is always negative for a tensile test, is extremely small in magnitude compared to the term $A(\mathrm{d}\sigma_z/\mathrm{d}\delta)$, which is always positive for a tensile test. However, increasing F beyond the yield point results in a significant rate of increase in the magnitude of the quantity $\mathrm{d}A/\mathrm{d}\delta$ until, at the ultimate stress condition, the right side of Equation (4-2) becomes zero. What happens after this point has been reached? Our stress-strain curve tells us that the load-carrying capacity of the specimen then falls off as the strain increases further. Actually when the ultimate stress has just been passed in a test, the specimen cannot maintain the applied load and the test proceeds precipitously to the destruction of the specimen. So quickly does this happen that it is difficult to record data for the portion of the stress-strain curve beyond the ultimate-stress point.

The loss in load-carrying capacity beyond the ultimate-stress point just described does not occur as a result of rapid area decrease of the entire specimen. Rather it occurs as a result of a rapid area decrease at some localized portion of the specimen. We call this action "necking" of the specimen. The position in the specimen where necking takes place depends primarily on local imperfections of the material. We have shown a diagram of a specimen loaded to destruction in Figure 4-6. The necking action, can be easily seen by observing the broken portion of the specimen. When there is large inelastic deformation occurring rapidly in a small domain, as in the necked region of the tensile specimen, we say there is plastic flow in this domain.

We thus have examined one of the most important of structural materials. What about other materials? We have already considered some rubber materials in this section and we see that there can be great departures from the case exemplified by mild steel. However we can use mild steel as a basis of comparison in our discussion of other materials so as to make our communication more meaningful. Further-

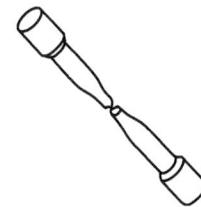

Figure 4-6 The diagram of a specimen loaded to destruction

more, the definitions that we set up while discussing the mild steel case hold for general discussions. Figure 4-7 shows stress-strain diagram for various steel and aluminum alloys. In Table 4-1 we have listed some of the parameters we have been discussing in this section for certain important materials(For more precise, detailed information of this type you are

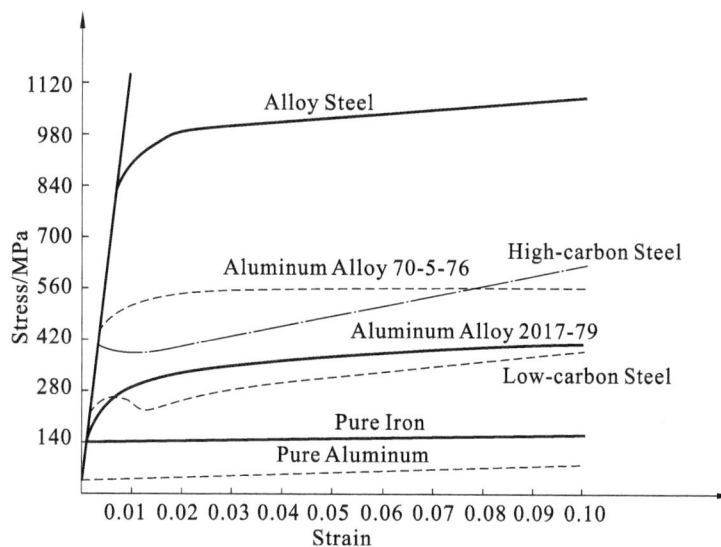

Figure 4-7 The stress-strain diagram for various steel and aluminum alloys

urged to consult structural or materials handbooks).

Table 4-1   **Some Mechanical Properties of Common Engineering Materials**

Material	E	U	Y	G
	Modulus of Elasticity/MPa*	Ultimate Stress/MPa	Yield Stress (0.002)/MPa	Modulus of Shear/MPa
Aluminum	6.90×10^4	4.14×10^2	3.11×10^2	2.76×10^4
Brass (cast)	8.97×10^4	3.11×10^2	1.38×10^2	3.45×10^4
Copper (hard drawn)	1.17×10^5	3.80×10^2	2.76×10^2	4.14×10^4
Cast Iron	9.66×10^4	1.38×10^2	—	3.86×10^4
Magnesium	4.49×10^4	2.42×10^2	1.59×10^2	1.66×10^4
Structural Steel	2.00×10^5	4.14×10^2	2.42×10^2	8.28×10^4
Stainless Steel	1.93×10^5	8.28×10^2	5.52×10^2	6.90×10^4

    *: Modulus of elasticity is about the same in tension and compression for all materials listed except cast iron, where only the tensile modulus has been given.

4.3   Strain-hardening

In most ductile materials, such as mild steel, aluminum and copper, it is observed that ever-increasing actual stress is required for continued deformation beyond the yield point. This is the case for the stress-strain diagram shown in Figure 4-5. We call this effect strain-hardening. Strain-hardening can be explained qualitatively for metals by the use of dislocation theory.

There is another important phenomenon in the plastic range that is referred to as strain-hardening. It has to do with the unloading of a specimen, having a linear elastic range, from the plastic range. In Section 4.2 we discussed the unloading of a linear elastic material when the load was in the elastic range as well as the unloading of a non-linear elastic material. In those cases, complete removal of load results in restoration of the original geometry, that is, no permanent set. Furthermore, the unloading path must retrace the loading path in the stress-strain diagram. In unloading a material with a linear elastic range from a load in the plastic range, we do not retrace the loading path but instead move along a new path which is essentially parallel to the linear elastic portion of the original loading path. This has been shown in Figure 4-8 where the initial loading has been stopped at A and the first unloading is shown to take place along a straight line to point B on the abscissa. Thus we have introduced a permanent set given by OB on the abscissa. The elastic recovery, on the other hand, clearly is *BE*. Now on a second loading we move along the path *BA*. A second unloading from a stress below that corresponding to point *A* will essentially move along path *BA* back to *B* and so we have for practical purposes a linear elastic range from *B* to *A*. An inspection of Figure 4-8 will indicate that the yield point has been raised for the second loading as a result of the first loading into the plastic range. The raising of the yield point by this action is the second phenomenon referred to as strain-hardening. Beyond the new yield point, the second loading proceeds along *AC* which is along the stress curve that would be followed by an uninterrupted first loading. At *C*, a second unloading is shown and the same process is repeated.

It is to be pointed out that unloading and reloading curves do not exactly overlap. Instead they form a small hysteresis loop as shown in Figure 4-9 in an exaggerated manner. There is an energy loss during a cycle represented by the area of the hysteresis loop. This energy loss is very small.

Finally it should be pointed out that the change in yield point by strain-hardening is observable only in the direction of initial loading. That is, there is no increase in yield stress in the material at right angles to the direction of the initial loading.

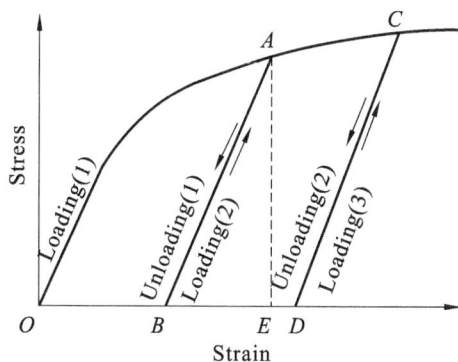

Figure 4-8   The diagram of strain-hardening phenomenon

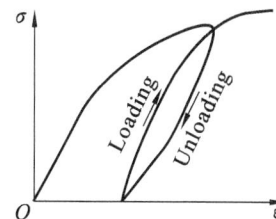

Figure 4-9   A small hysteresis loop

4.4   Other Properties Pertaining to the Tensile Test

On the basis of the simple tensile test described earlier and the simple compression test, which is essentially the same except for direction, we can make additional useful classifications which are meaningful in describing mechanical behavior of materials.

First we can form two classes of materials on the basis of the behavior of a specimen in a tensile test carried out to the point where the specimen fractures. Materials exhibiting little or no plastic deformation up to fracture, such as glass, are called brittle materials. Materials exhibiting substantial plastic deformation up to the point of fracture, such as mild steel, are called ductile materials. For a brittle material, a stress-strain curve carried out in a tensile test will differ from the corresponding curve carried out as a compression test. Furthermore, brittle materials exhibit a considerable scatter in breaking points found by a series of many tests. Ductile materials, on the other hand, give essentially the same stress-strain curve for tensile or compression test and have yield points, breaking points, etc. , which are considerably more reproducible in a series of tests.

It was pointed out earlier that a tensile load on the specimen causes a lateral contraction which we called the Poisson effect. By the same mechanism, a compression test induces a lateral extension. In the linear elastic range, we find from these tests that the lateral strain is proportional to the longitudinal strain and may be expressed as follows

$$\varepsilon_{lat} = -\nu\varepsilon_{long} \tag{4-3}$$

where the constant of proportionality ν is called Poisson's ratio and ranges for engineering materials from 0. 2 to 0. 5. In Section 4. 6, we shall see that Poisson's ratio can be considered as one of the fundamental constants characterizing the general mechanical behavior of linear, elastic, homogeneous, isotropic materials. Meanwhile, it is clear that simple tensile tests permit the evaluation of this constant.

In the tensile tests discussed up to this time, we have assumed that the temperature of the specimen was uniform and near room temperature. Also we pointed out that tests were conducted slowly enough to avoid dynamic effects and fast enough to avoid long-time effects. Knowledge of the behavior of materials under conditions where temperature is not close to room temperature and where short- or long-time effects become significant is becoming increasingly more essential in our progressing technology. We therefore shall discuss these effects briefly in the future.

Time effects. A rapid rate of loading will result in a different stress-strain curve other than a slow rate of loading. This will particularly be so for a soft material or for structural materials at elevated temperatures. In such cases, information as to the rate of loading must be included as pertinent information for the stress-strain curve. Essentially there will be a raise of the yield stress as the rate of loading is increased.

At the other extreme, we now examine long-time effects. We shall consider the case of a specimen on which a constant tensile load is maintained over a long period of time. For some materials, including structural materials at elevated temperature, there will take place small but continued increase in strain with time. This phenomenon is called creep. In such structures as boilers, reactor shells, and gas turbines this effect can be of great significance. If the load is maintained over a long enough time, the specimen may fail even though the stress is initially considerably lower than the ultimate stress. For testing purposes, creep strain is defined as the strain developed after the specimen has been brought up to the desired load. Hence, creep strain occurs under the action of a constant stress and is usually plotted as a function of time for different values of stress, such as in the plot shown in Figure 4-10. Note that the greater the stress, the greater the creep strain at any given time.

Temperature effects. We have already pointed out how elevated temperatures give rise to the phenomenon of creep. We can also state that decreasing the temperature results in an increase in the slope of the linear-elastic portion of a stress-strain diagram. This, of course, means that the modulus of elasticity increase with decreasing temperature. And for increasing temperature, the slope of the linear part of the stress-strain diagram decreases, indicating a decreasing modulus of elasticity with increasing temperature.

Another important thermal effect results from the fact that most materials expand as a result of an increase in temperature. If the temperature field is nonuniform there will result in a stress field which is called thermal stress.

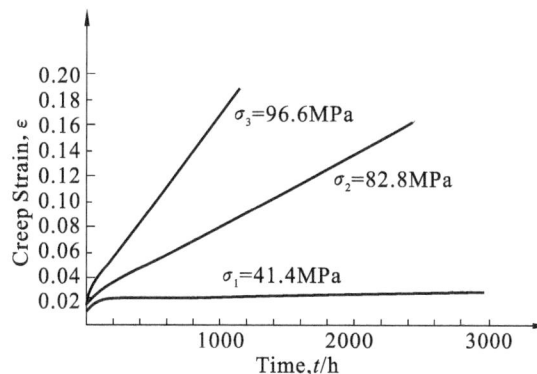

Figure 4-10 **A function of time for different values of stress**

4.5 Idealized One-dimensional Stress-strain Laws

It should be apparent by now that stress-strain relations in general are of great complexity, with the possibility of many ramifications. We introduced some of these complexities in the previous sections. To permit analytical treatment of material behavior under certain conditions, we employ at times idealizations of stress-strain relations.

The most simple stress-strain idealization is of course the rigid-body idealization shown in Figure 4-11. We have used such a model in rigid-body mechanics courses and you will recall that in Chapter 1 we were able to employ rigid-body models in ascertaining supporting forces for structures supported in a statically determinate manner.

In Figure 4-12, we have shown the stress-strain curve for a linear perfectly elastic material. This model is the one we shall employ in the major portion of the book. We must not forget that the stress-strain diagram is taken from a simple one-dimensional state of stress and accordingly in Section 4.6 we shall generalize this model for a general state of stress. The resulting formulation is called the generalized Hooke's law. We shall be able to use these results for the analysis of bodies composed of the usual structural materials, such as steel and aluminum, in cases where the stress has not exceeded the yield stress.

There are situations where there may be plastic deformations involved which far exceed the elastic deformations present and it may be profitable to formulate the idealization of a stress-strain diagram shown in Figure 4-13 which embodies rigid-body behavior up to a certain stress and then exhibits perfectly plastic behavior. If we include strain-hardening in the plastic range our model becomes more accurate albeit more complex. We have shown in Figure 4-14 the idealization rigid plastic behavior with strain-hardening to illustrate this case.

There may be times when the elastic deformations cannot be deleted from considerations and where there is little strain-hardening. For such cases, one may be able to employ the idealization shown in Figure 4-15 called the elastic perfectly-plastic stress-strain curve. Finally, allowing for strain-hardening, we get the curve shown in Figure 4-16 which is reasonably close to certain actual stress-strain diagrams.

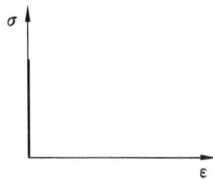

Figure 4-11　Rigid-body behavior

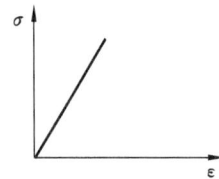

Figure 4-12　Perfectly elastic behavior

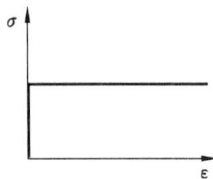

Figure 4-13　Rigid perfectly-plastic behavior

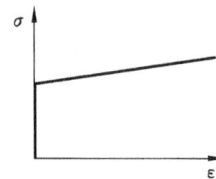

Figure 4-14　Idealization rigid plastic behavior with strain-hardening

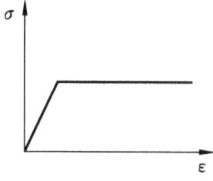

Figure 4-15 Elastic perfectly-plastic behavior

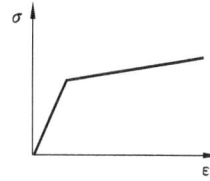

Figure 4-16 Elastic-plastic behavior with strain-hardening

In succeeding the part of this chapter, we shall make use of the generalized form of perfectly elastic behavior. Also we shall have the opportunity of using some of the other idealizations for problems involving simple states of stress in the plastic range.

4.6 Generalized Hooke's Law

In the one-dimensional experiments, we found that many structural materials exhibit a linear-relationship between stress and strain for stresses sufficiently below the yield stress as shown in Equation (4-1).

We shall now postulate that, in a more general state of stress, materials behaving according to Equation (4-1) for the unidimensional stress will have the following relations between stress and strain

$$\begin{cases} \sigma_x = C_{11}\varepsilon_{xx} + C_{12}\varepsilon_{yy} + C_{13}\varepsilon_{zz} + C_{14}\gamma_{xy} + C_{15}\gamma_{yz} + C_{16}\gamma_{xz} \\ \sigma_y = C_{21}\varepsilon_{xx} + C_{22}\varepsilon_{yy} + C_{23}\varepsilon_{zz} + C_{24}\gamma_{xy} + C_{25}\gamma_{yz} + C_{26}\gamma_{xz} \\ \sigma_z = C_{31}\varepsilon_{xx} + C_{32}\varepsilon_{yy} + C_{33}\varepsilon_{zz} + C_{34}\gamma_{xy} + C_{35}\gamma_{yz} + C_{36}\gamma_{xz} \\ \tau_{xy} = C_{41}\varepsilon_{xx} + C_{42}\varepsilon_{yy} + C_{43}\varepsilon_{zz} + C_{44}\gamma_{xy} + C_{45}\gamma_{yz} + C_{46}\gamma_{xz} \\ \tau_{yz} = C_{51}\varepsilon_{xx} + C_{52}\varepsilon_{yy} + C_{53}\varepsilon_{zz} + C_{54}\gamma_{xy} + C_{55}\gamma_{yz} + C_{56}\gamma_{xz} \\ \tau_{xz} = C_{61}\varepsilon_{xx} + C_{62}\varepsilon_{yy} + C_{63}\varepsilon_{zz} + C_{64}\gamma_{xy} + C_{65}\gamma_{yz} + C_{66}\gamma_{xz} \end{cases} \qquad (4\text{-}4)$$

where the terms C_{ij} form a 6×6 matrix of constants whose values depend on the material. We are saying in this formulation that each stress at a point is linearly related to all the strains at the point. Clearly such a relation will degenerate to that of Equation (4-1) for a simple uniaxial stress in the direction of one of the coordinates xyz. Using the z direction and noting that the shear strains for our axes are zero, we can say

$$\sigma_z = C_{31}\varepsilon_{xx} + C_{32}\varepsilon_{yy} + C_{33}\varepsilon_{zz} \qquad (4\text{-}5)$$

But we have shown that ε_{yy} and ε_{xx} are each proportionally related to ε_{zz}. Thus

$$\sigma_z = C_{31}(-\nu\varepsilon_{zz}) + C_{32}(-\nu\varepsilon_{zz}) + C_{33}\varepsilon_{zz} \qquad (4\text{-}6)$$

Hence

$$\sigma_z = (-\nu C_{31} - \nu C_{32} + C_{33})\varepsilon_{zz} \tag{4-7}$$

And so we see that Equation (4-7) is of the same form as Equation (4-1). We can thus justify the generalized formulation partially by the fact that it includes in the simpler stress-strain relations which we deduced from our tensile tests. The primary justification for the use of the linear stress-strain law, however, is an indirect one. We find by using this relation in the handling of linear elastic materials, the analytical results that we reach are in agreement with experimental results.

If a material has the same composition throughout, we say that it is homogeneous. It would then appear that the linear elastic homogeneous body has 36 constants or elastic moduli which are needed to relate stress and strain below the proportional limit. It would at first seem that we have a frightfully complicated situation on our hands for the handling of problems. However, most of the materials that we deal with in engineering applications have mechanical properties which are not dependent on any particular directions. That is, the generalized Hooke's law for some new reference $x'y'z'$ would retain the same constants C_{ij} when the law is expressed for the xyz reference as it is done in Equation (4-4). Thus the equation for $\sigma_{x'}$ would be given as

$$\sigma_{x'} = C_{11}\varepsilon_{x'x'} + C_{12}\varepsilon_{y'y'} + C_{13}\varepsilon_{z'z'} + C_{14}\gamma_{x'y'} + C_{15}\gamma_{y'z'} + C_{16}\gamma_{x'z'} \tag{4-8}$$

where the terms C_{ij} have the same values as in Equation (4-4). When a material behaves mechanically in the same manner for all directions, we say that the material is mechanically isotropic. We shall now show that for isotropic materials the 36 elastic moduli degenerate to only two elastic moduli.

In order to do this, we impose a certain sequence of rotations of the axes x, y, z requiring each time that the stress-strain relations for the new orientation have the same form as for the original set of axes.

First, consider a $180°$ rotation about the z axis as shown in Figure 4-17. The direction cosines between the various axes are given in the following tabular form

$$\begin{array}{c|ccc} & x & y & z \\ \hline x' & -1 & 0 & 0 \\ y' & 0 & -1 & 0 \\ z' & 0 & 0 & 1 \end{array} \tag{4-9}$$

The stresses $\sigma_{x'}, \sigma_{y'}, \sigma_{z'}, \tau_{x'y'}, \tau_{y'z'}, \tau_{x'z'}$ in the new reference are related to the stress $\sigma_x, \sigma_y, \sigma_z, \tau_{xy}, \tau_{yz}, \tau_{zx}$ in the old reference and the following results can then be gotten when we employ the direction cosines from Equation (4-9)

$$\sigma_{x'} = \sigma_x \qquad \tau_{x'y'} = \tau_{xy}$$

$$\sigma_{y'} = \sigma_y \qquad \tau_{y'z'} = -\tau_{yz} \tag{4-10}$$

$$\sigma_{z'} = \sigma_z \qquad \tau_{x'z'} = -\tau_{xz}$$

By using the transformation relation for strain with the direction cosines given by Equation (4-9), we reach, in a similar manner, the following strain relations

$$\varepsilon_{x'x'} = \varepsilon_{xx} \qquad \gamma_{x'y'} = \gamma_{xy}$$

$$\varepsilon_{y'y'} = \varepsilon_{yy} \qquad \gamma_{y'z'} = -\gamma_{yz} \tag{4-11}$$

$$\varepsilon_{z'z'} = \varepsilon_{zz} \qquad \gamma_{x'z'} = -\gamma_{xz}$$

Next replace the stresses and strains in Equation (4-8) according to the transformation results given by Equations (4-10) and (4-11). We get

$$\sigma_x = C_{11}\varepsilon_{xx} + C_{12}\varepsilon_{yy} + C_{13}\varepsilon_{zz} + C_{14}\gamma_{xy} - C_{15}\gamma_{yz} - C_{16}\gamma_{xz} \tag{4-12}$$

Comparing Equation (4-12) with the first equation in the system of Equations (4-4) we see that a necessary requirement for isotropy is that

$$C_{15} = C_{16} = 0 \tag{4-13}$$

Similarly by examining the other stresses in this manner we conclude additionally that for isotropy the following conditions must hold

$$C_{25} = 0 \quad C_{36} = 0 \quad C_{51} = 0 \quad C_{54} = 0 \quad C_{63} = 0$$

$$C_{26} = 0 \quad C_{45} = 0 \quad C_{52} = 0 \quad C_{61} = 0 \quad C_{64} = 0 \tag{4-14}$$

$$C_{35} = 0 \quad C_{46} = 0 \quad C_{53} = 0 \quad C_{62} = 0$$

The matrix of elastic moduli then simplifies to the form

$$\begin{bmatrix} C_{11} & C_{12} & C_{13} & C_{14} & 0 & 0 \\ C_{21} & C_{22} & C_{23} & C_{24} & 0 & 0 \\ C_{31} & C_{32} & C_{33} & C_{34} & 0 & 0 \\ C_{41} & C_{42} & C_{43} & C_{44} & 0 & 0 \\ 0 & 0 & 0 & 0 & C_{55} & C_{56} \\ 0 & 0 & 0 & 0 & C_{65} & C_{66} \end{bmatrix} \tag{4-15}$$

Observing Figure 4-17, it should be clear that, for the foregoing matrix, mechanical behavior in the x direction is the same as corresponding mechanical behavior in the x' direction and that this is also true for the y and y' directions. Thus we say that there is elastic symmetry about the plane yz and xz. To achieve symmetry about plane xy we impose the condition of isotropy for a 180° rotation of axes about the x axis as has been shown in Figure 4-18. We may show that matrix is reduced to the following form.

$$\begin{bmatrix} C_{11} & C_{12} & C_{13} & 0 & 0 & 0 \\ C_{21} & C_{22} & C_{23} & 0 & 0 & 0 \\ C_{31} & C_{32} & C_{33} & 0 & 0 & 0 \\ 0 & 0 & 0 & C_{44} & 0 & 0 \\ 0 & 0 & 0 & 0 & C_{55} & 0 \\ 0 & 0 & 0 & 0 & 0 & C_{66} \end{bmatrix} \tag{4-16}$$

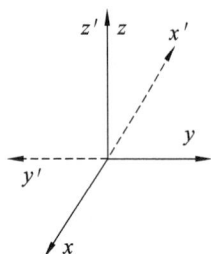

Figure 4-17　180° rotation about z axis

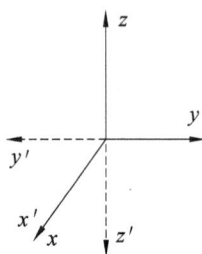

Figure 4-18　180° rotation about x axis

This matrix of moduli then represents the case of elastic symmetry three orthogonal planes. This means that there is no change in mechanics behavior when the x, y, z directions are reversed. A material behaving in the way is called orthotropic. These cases of limited isotropy are very important for certain materials. And for crystals this is an important consideration. In this book, however, we shall be concerned only with completely isotropic behavior.

By a series of additional rotations of axes, we can reduce the number of independent moduli to two by proceeding as in the first rotation of axes. We have shown that the generalized Hooke's law reduces to the following form

$$\begin{aligned} \sigma_x &= (2G+\lambda)\varepsilon_{xx} + \lambda(\varepsilon_{yy} + \varepsilon_{zz}) \\ \sigma_y &= (2G+\lambda)\varepsilon_{yy} + \lambda(\varepsilon_{xx} + \varepsilon_{zz}) \\ \sigma_z &= (2G+\lambda)\varepsilon_{zz} + \lambda(\varepsilon_{xx} + \varepsilon_{yy}) \\ \tau_{xy} &= G\gamma_{xy} \\ \tau_{yz} &= G\gamma_{yz} \\ \tau_{xz} &= G\gamma_{xz} \end{aligned} \tag{4-17}$$

where λ is called Lame's constant and G is called the shear modulus of elasticity.

How is Young's modulus E related to the constants λ and G? To answer this solve for the strains in terms of stresses in the foregoing equations. We get, by purely algebraic means

$$\varepsilon_{xx} = \frac{\lambda+G}{G(3\lambda+2G)}\sigma_x - \frac{\lambda}{2G(3\lambda+2G)}(\sigma_y + \sigma_z) \tag{4-18a}$$

$$\varepsilon_{yy} = \frac{\lambda + G}{G(3\lambda + 2G)}\sigma_y - \frac{\lambda}{2G(3\lambda + 2G)}(\sigma_x + \sigma_z) \qquad (4\text{-}18b)$$

$$\varepsilon_{zz} = \frac{\lambda + G}{G(3\lambda + 2G)}\sigma_z - \frac{\lambda}{2G(3\lambda + 2G)}(\sigma_x + \sigma_y) \qquad (4\text{-}18c)$$

$$\gamma_{xy} = \frac{1}{G}\tau_{xy} \qquad (4\text{-}18d)$$

$$\gamma_{xz} = \frac{1}{G}\tau_{xz} \qquad (4\text{-}18e)$$

$$\gamma_{yz} = \frac{1}{G}\tau_{yz} \qquad (4\text{-}18f)$$

In the simple tensile test, it is clear that $\sigma_y = \sigma_x = 0$, so that we would have from Equation (4-18c)

$$\varepsilon_{zz} = \frac{\lambda + G}{G(3\lambda + 2G)}\sigma_z \qquad (4\text{-}19)$$

Comparing Equation (4-19) with Equation (4-1), we may conclude that

$$E = \frac{G(3\lambda + 2G)}{\lambda + G} \qquad (4\text{-}20)$$

To set forth another form of Hooke's law involving Poisson's ratio note, for a simple tensile test, note that

$$\varepsilon_{xx} = -\nu\varepsilon_{zz}$$

Now employing Equation (4-18a) to replace ε_{xx} and Equation (4-18c) to replace ε_{zz} in the foregoing equation and employing E in accordance with Equation (4-20), we get

$$\frac{\sigma_x}{E} - \frac{\lambda}{2G(3\lambda + 2G)}(\sigma_y + \sigma_z) = -\nu\left[\frac{\sigma_z}{E} - \frac{\lambda}{2G(3\lambda + 2G)}(\sigma_x + \sigma_y)\right] \qquad (4\text{-}21)$$

Now setting $\sigma_x = \sigma_y = 0$ for the one-dimensional state of stress, we have on canceling σ_z

$$\frac{\lambda}{2G(3\lambda + 2G)} = \frac{\nu}{E} \qquad (4\text{-}22)$$

Substituting Equation (4-20) and Equation (4-22) into Equation (4-18), we can express Hooke's law in the following form

$$\varepsilon_{xx} = \frac{1}{E}[\sigma_x - \nu(\sigma_y + \sigma_z)] \qquad (4\text{-}23a)$$

$$\varepsilon_{yy} = \frac{1}{E}[\sigma_y - \nu(\sigma_x + \sigma_z)] \qquad (4\text{-}23b)$$

$$\varepsilon_{zz} = \frac{1}{E}[\sigma_z - \nu(\sigma_x + \sigma_y)] \qquad (4\text{-}23c)$$

$$\gamma_{xy} = \frac{1}{G}\tau_{xy} \qquad (4\text{-}23d)$$

$$\gamma_{xz} = \frac{1}{G}\tau_{xz} \qquad\qquad (4\text{-}23\text{e})$$

$$\gamma_{yz} = \frac{1}{G}\tau_{yz} \qquad\qquad (4\text{-}23\text{f})$$

where we have used constant ν, E and G.

We shall generally use the foregoing forms of the generalized Hooke's law in this book. However, it should not be forgotten that only two of the constants are independent. It will therefore be helpful to relate the constants given in the preceding form of the generalized Hooke's law. To do this, solve for λ from Equation (4-20) to get

$$\lambda = \frac{2G^2 - EG}{E - 3G} \qquad\qquad (4\text{-}24)$$

Now solve for λ using Equation (4-22). Thus

$$\lambda = \frac{4G^2\nu}{E - 6G\nu} \qquad\qquad (4\text{-}25)$$

Finally, equating the right-hand sides of Equations (4-24), (4-25) and solving for G we get the desired relation

$$G = \frac{E}{2(1+\nu)} \qquad\qquad (4\text{-}26)$$

Since E and ν can be measured easily in a simple tensile test, they are often considered the fundamental constants. Equation (4-26) then permits us to evaluate G in terms of these constant.

You may readily demonstrate that Equation (4-23) in tensor notation are given as

$$\varepsilon_{ij} = \frac{1+\nu}{E}\tau_{ij} - \frac{\nu}{E}\sigma_k\delta_{ij} \qquad\qquad (4\text{-}27)$$

where δ_{ij} called the Kronecker delta function, is unity when $i=j$ and is zero when $i\neq j$, and $\tau_{ij}=\sigma_i$ when $i=j$.

4.7   Closure

In this chapter, we shall first consider the simple uniaxial loading case and then we shall endeavor to extrapolate our uniaxial results and concepts to the general state of stress. We introduced the most basic test in the study of stress-strain relations which is called as the simple tensile test, and illustrated some concepts such as actual stress, engineering stress, strain, proportional limit, elastic limit and yield point. To permit analytical treatment of material behavior under certain conditions, we employed idealizations of stress-strain relations. We considered the simple uniaxial loading case and then we extrapolated those uniaxial results and concepts to the general state of stress. We have shown that the generalized Hooke's law.

Problems

4-1   A tensile specimen has a diameter of 15 mm . The increase in length recorded by the longitudinal gauge for a load of 4500 N is 0. 025 mm for an original length of 50 mm. The decrease in diameter as measured by the lateral gauge is 0. 00125 mm. What is the actual stress and the engineering stress? Compute the modulus of elasticity.

4-2   Figure 4-19 is shown a hypothetical stress-strain curve. What is the proportional limit, the ultimate stress and the modulus of elasticity for this material?

4-3   In Problem 4-1 determine Poisson's ratio.

4-4   Explain when a nonlinear elastic material can be used as a vibration absorber.

4-5   If in the tensile test corresponding to the graph in Figure 4-19, the material is unloaded from a stress of 380 MPa, what is the elastic recovery of strain? What is the permanent set? What is the proportional limit of the material when it is reloaded?

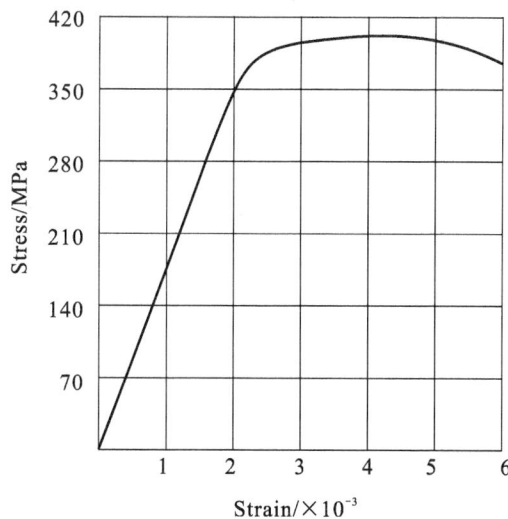

Figure 4-19   Figure of Problem 4-2

4-6   Is the perfectly plastic behavior of a solid material the same as that of a fluid? Explain it.

4-7   Justify Equation (4-13).

4-8   What are the nonzero coefficients of the matrix C_{ij} for elastic symmetry about one plane?

4-9   We have for steel the following data

$$E=2.1\times10^5 \text{ MPa}$$
$$G=1.1\times10^5 \text{ MPa}$$

Compute the Lame constant. For a state of strain at a point in this material given as

$$\varepsilon_{ij} = \begin{bmatrix} 0.001 & 0 & -0.002 \\ 0 & -0.003 & 0.0005 \\ -0.002 & 0.0005 & 0 \end{bmatrix}$$

determine the stress tensor at this point.

4-10   For a shear modulus of 6.9×10^4 MPa and an elastic modulus of 1.7×10^5 MPa compute the strain tensor for the following state of stress

$$\sigma_{ij} = \begin{bmatrix} 10 & -50 & 0 \\ -50 & 5 & 5 \\ 0 & 5 & -20 \end{bmatrix} \text{MPa}$$

4-11   Show that for an isotropic, linear, elastic material, principal axes of stress correspond to principal axes of strain(This is not true for nonisotropic materials although it is true for isotropic nonlinear materials).

4-12   For a Poisson's ratio of 0.30 and a Young's modulus of 2.1×10^5 MPa, determine the strain tensor for the following state of stress

$$\sigma_{ij} = \begin{bmatrix} 0 & 10 & -20 \\ 10 & 5 & -30 \\ -20 & -30 & 10 \end{bmatrix} \text{MPa}$$

5 Basic Equations of Elasticity

5.1 Introduction

In previous chapters we have studied stress and strain as separate distinct entities with little regard to the requirements for ascertaining the stress or strain distribution in a particular body. We have, however, shown in Chapter 4 that, for linear elastic materials, we could determine a strain distribution once the stress distribution was known and vice versa. We shall now set forth the means by which for given forces or displacements we can determine stress and strain field in certain bodies. Accordingly, our next task will be to set forth the basic laws that we must satisfy to reach physically meaningful solutions to problems that we shall very carefully pose.

In so doing, we shall not rely as heavily on tensor notation as in earlier chapters. The main advantage of this notation appears in the basic developments of the earlier chapters. By using this notion, we could then proceed at a rather high level of generality and see important mathematical relations that would otherwise be obscure. However, when we start working problems we shall often tend toward the unabridged notion, as you will note in progressing through the book.

5.2 Equations of the Theory of Elasticity

Any continuous medium must satisfy two basic laws: Newton's law and Conservation of energy.

In addition to these basic laws, certain subsidiary or constitutive laws which apply to specific materials, must also be satisfied. For linear elastic materials, we have already presented the generalized Hooke's law.

We now examine the first basic law as it pertains to our theory. Since this book is primarily concerned with bodies in equilibrium relative to inertial space, it is then necessary that Newton's law must be satisfied for the equilibrium condition by each element of the body. This requires the total force on any element to be zero. Then consider Figure 5-1, where an element of the body in the form of a rectangular parallelepiped has been shown. The stresses on the faces and the body-force components per unit volume have been shown. On the outer faces, we have expressed the stresses contributing force in the x direction as Taylor series expansions limited to two terms. By summing forces in the x direction we get

$$-\sigma_x \mathrm{d}y\mathrm{d}z + \left(\sigma_x + \frac{\partial \sigma_x}{\partial x}\mathrm{d}x\right)\mathrm{d}y\mathrm{d}z - \tau_{yx}\mathrm{d}x\mathrm{d}z + \left(\tau_{yx} + \frac{\partial \tau_{yx}}{\partial y}\mathrm{d}y\right)\mathrm{d}x\mathrm{d}z - \tau_{zx}\mathrm{d}x\mathrm{d}y +$$

$$\left(\tau_{zx} + \frac{\partial \tau_{zx}}{\partial z}\mathrm{d}z\right)\mathrm{d}x\mathrm{d}y + B_x\mathrm{d}x\mathrm{d}y\mathrm{d}z = 0 \tag{5-1}$$

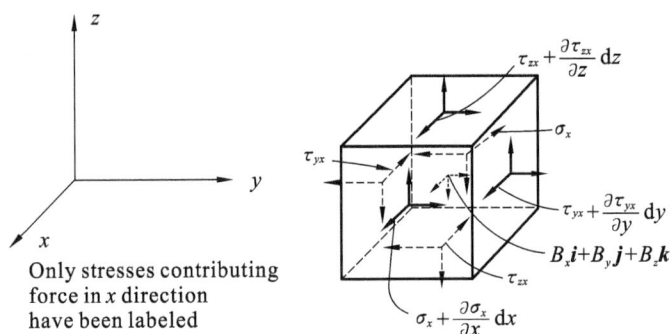

Only stresses contributing force in x direction have been labeled

Figure 5-1　An element of the body in the form of a rectangular parallelepiped

Simplifying the equation by canceling terms and dividing through by $(\mathrm{d}x\mathrm{d}y\mathrm{d}z)$ we get

$$\frac{\partial \sigma_x}{\partial x} + \frac{\partial \tau_{yx}}{\partial y} + \frac{\partial \tau_{zx}}{\partial z} + B_x = 0 \tag{5-2}$$

By considering forces in the remaining coordinate directions, we can formulate two other similar equations. Using the symmetry property of the stress tensor, we can give Newton's law in the following form

$$\begin{cases} \dfrac{\partial \sigma_x}{\partial x} + \dfrac{\partial \tau_{xy}}{\partial y} + \dfrac{\partial \tau_{xz}}{\partial z} + B_x = 0 \\[2mm] \dfrac{\partial \tau_{yx}}{\partial x} + \dfrac{\partial \sigma_y}{\partial y} + \dfrac{\partial \tau_{yz}}{\partial z} + B_y = 0 \\[2mm] \dfrac{\partial \tau_{zx}}{\partial x} + \dfrac{\partial \tau_{zy}}{\partial y} + \dfrac{\partial \sigma_z}{\partial z} + B_z = 0 \end{cases} \tag{5-3}$$

In Cartesian tensor notion, the Equation (5-3) becomes

$$\frac{\partial \tau_{ij}}{\partial x_j} + B_i = 0 \tag{5-4}$$

where $\tau_{ij} = \sigma_i$ when $i = j$. For the "quasi-static" problems described earlier, we can still use the preceding equations, provided that we include the inertial forces as part of the body force distribution ***B***.

For the linear elastic medium which we shall consider in our basic theory, there will neither be appreciable heat transfer nor internal friction in the problems to be examined. For this reason, the satisfaction of Newton's law insures the satisfaction of the conservation of energy. And, just as in the case of rigid-body mechanics, Newton's law and energy methods provide alternate avenues of approach.

The constitutive law that will play a key role in the theory is, as already noted, the generalized Hooke's law, which we present again as

$$
\begin{cases}
\varepsilon_{xx} = \dfrac{1}{E}[\sigma_x - \nu(\sigma_y + \sigma_z)] \\[2mm]
\varepsilon_{yy} = \dfrac{1}{E}[\sigma_y - \nu(\sigma_x + \sigma_z)] \\[2mm]
\varepsilon_{zz} = \dfrac{1}{E}[\sigma_z - \nu(\sigma_x + \sigma_y)] \\[2mm]
\gamma_{xy} = \dfrac{1}{G}\tau_{xy} \\[2mm]
\gamma_{xz} = \dfrac{1}{G}\tau_{xz} \\[2mm]
\gamma_{yz} = \dfrac{1}{G}\tau_{yz}
\end{cases}
\tag{5-5}
$$

The Cartesian tensor representation of the preceding equation is

$$
\varepsilon_{ij} = \frac{1+\nu}{E}\tau_{ij} - \frac{\nu}{E}(\sigma_k)\delta_{ij}
\tag{5-6}
$$

We showed in Chapter 3 as a result of geometrical considerations that the strain tensor is related to the displacement field in the following way

$$
\begin{cases}
\varepsilon_{xx} = \dfrac{\partial u_x}{\partial x} \quad \gamma_{xy} = \dfrac{\partial u_x}{\partial y} + \dfrac{\partial u_y}{\partial x} \\[2mm]
\varepsilon_{yy} = \dfrac{\partial u_y}{\partial y} \quad \gamma_{yz} = \dfrac{\partial u_y}{\partial z} + \dfrac{\partial u_z}{\partial y} \\[2mm]
\varepsilon_{zz} = \dfrac{\partial u_z}{\partial z} \quad \gamma_{xz} = \dfrac{\partial u_x}{\partial z} + \dfrac{\partial u_z}{\partial x}
\end{cases}
\tag{5-7}
$$

In tensor notation we then have

$$
\varepsilon_{ij} = \frac{1}{2}\left(\frac{\partial u_i}{\partial x_j} + \frac{\partial u_j}{\partial x_i}\right)
\tag{5-8}
$$

We thus have a total of 15 equations, namely, three for equilibrium, six for Hooke's law, and six for strain displacement relations. From these equations, we can solve for six stress components, six strain components, and three displacement components—a total of 15 unknowns.

We showed in Chapter 3 that if displacements were not considered explicitly the strains had to satisfy certain equations called compatibility equations as a necessary requirement in representing single-valued, continuous displacement fields, i. e.

$$
\begin{cases}
\dfrac{\partial^2 \varepsilon_{xx}}{\partial y^2} + \dfrac{\partial^2 \varepsilon_{yy}}{\partial x^2} = \dfrac{\partial^2 \gamma_{xy}}{\partial x \partial y} \\[2mm]
\dfrac{\partial^2 \varepsilon_{yy}}{\partial z^2} + \dfrac{\partial^2 \varepsilon_{zz}}{\partial y^2} = \dfrac{\partial^2 \gamma_{yz}}{\partial y \partial z} \\[2mm]
\dfrac{\partial^2 \varepsilon_{zz}}{\partial x^2} + \dfrac{\partial^2 \varepsilon_{xx}}{\partial z^2} = \dfrac{\partial^2 \gamma_{zx}}{\partial z \partial x} \\[2mm]
2\dfrac{\partial^2 \varepsilon_{xx}}{\partial y \partial z} = \dfrac{\partial}{\partial x}\left(-\dfrac{\partial \gamma_{yz}}{\partial x} + \dfrac{\partial \gamma_{zx}}{\partial y} + \dfrac{\partial \gamma_{xy}}{\partial z} \right) \\[2mm]
2\dfrac{\partial^2 \varepsilon_{yy}}{\partial z \partial x} = \dfrac{\partial}{\partial y}\left(-\dfrac{\partial \gamma_{zx}}{\partial y} + \dfrac{\partial \gamma_{yx}}{\partial z} + \dfrac{\partial \gamma_{yz}}{\partial x} \right) \\[2mm]
2\dfrac{\partial^2 \varepsilon_{zz}}{\partial x \partial y} = \dfrac{\partial}{\partial z}\left(-\dfrac{\partial \gamma_{xy}}{\partial z} + \dfrac{\partial \gamma_{zy}}{\partial x} + \dfrac{\partial \gamma_{zx}}{\partial y} \right)
\end{cases}
\tag{5-9}
$$

And in tensor notation, the preceding equations are contained in the following equation

$$
\frac{\partial^2 \varepsilon_{ij}}{\partial x_k \partial x_l} + \frac{\partial^2 \varepsilon_{kl}}{\partial x_i \partial x_j} = \frac{\partial^2 \varepsilon_{ik}}{\partial x_j \partial x_l} + \frac{\partial^2 \varepsilon_{jl}}{\partial x_i \partial x_k}
\tag{5-10}
$$

The equations presented in this section comprise the fundamental equations of elasticity.

5.3　Equilibrium Differential Equations in Two-dimensional Polar Coordinates

When solving plane problems, polar coordinates are more convenient than rectangular coordinates for circular objects. In polar coordinates, the position of P at any point in the plane is represented by the radial coordinate r and the circumferential coordinate θ, as shown in Figure 5-2.

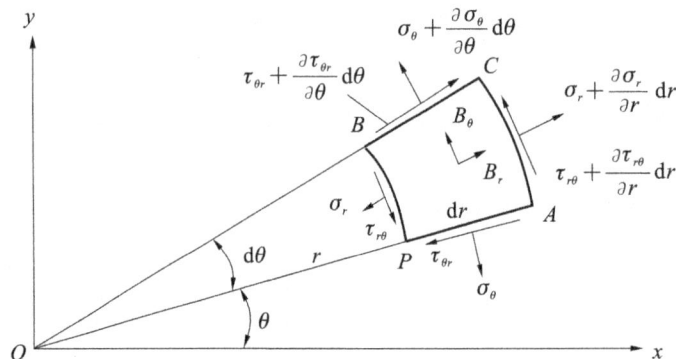

Figure 5-2　Force analysis of micro element in polar coordinates

In order to indicate the stress component in polar coordinates, take out the unit *PACB*, as shown in Figure 5-2. σ_r indicates that the normal stress along r direction is called radial normal stress, and σ_θ indicates that the normal stress along θ direction is called circumferential normal stress; $\tau_{r\theta}$ and $\tau_{\theta r}$ represent shear stress. According to the reciprocal theorem of shear stress, $\tau_{r\theta} = \tau_{\theta r}$. The sign of each stress component is consistent with the rectangular coordinate system. The stress components shown in the figure are positive. The radial and circumferential body force components are represented by B_r and B_θ respectively.

Similar to the rectangular coordinate system, since the stress changes with r, if the radial normal stress on plane *PB* is σ_r, the stress on plane *AC* is $\sigma_r + \frac{\partial \sigma_r}{\partial r} dr$; Similarly, the shear stresses on these two planes are $\tau_{r\theta}$ and $\tau_{r\theta} + \frac{\partial \tau_{r\theta}}{\partial r} dr$ respectively. The circumferential normal stresses on the two surfaces of *PA* and *BC* are σ_θ and $\sigma_\theta + \frac{\partial \sigma_\theta}{\partial \theta} d\theta$ respectively, and the shear stresses are $\tau_{\theta r}$ and $\tau_{\theta r} + \frac{\partial \tau_{\theta r}}{\partial \theta} d\theta$ respectively. There are two differences from the rectangular coordinate system. Firstly, the *PB* and *AC* arc surface areas of the element are different, and secondly, the two circumferential planes *PA* and *CB* are not parallel.

If the thickness of the unit is equal to 1, then the *PB* and *AC* areas are equal to $r d\theta$ and $(r+dr) d\theta$, respectively; the areas of both *PA* and *BC* are equal to dr, and the volume of the unit is equal to $r d\theta dr$. Since $d\theta$ is small, $\sin \frac{d\theta}{2}$ can be taken as $\frac{d\theta}{2}$ and $\cos \frac{d\theta}{2}$ is taken as 1.

The forces on the unit are projected onto the radial axis through the unit center, and the radial equilibrium equation is

$$\left[\sigma_r + \frac{\partial \sigma_r}{\partial r} dr \right] (r + dr) d\theta - \sigma_r r d\theta - \left[\sigma_\theta + \frac{\partial \sigma_\theta}{\partial \theta} d\theta \right] dr \frac{d\theta}{2} - \sigma_\theta dr \frac{d\theta}{2} +$$

$$\left[\tau_{\theta r} + \frac{\partial \tau_{\theta r}}{\partial \theta} d\theta \right] dr - \tau_{\theta r} dr + B_r r d\theta dr = 0 \tag{5-11}$$

Replacing $\tau_{\theta r}$ with $\tau_{r\theta}$. After simplification, dividing by $r d\theta dr$, and then omitting high order terms, we can get

$$\frac{\partial \sigma_r}{\partial r} + \frac{1}{r} \frac{\partial \tau_{\theta r}}{\partial \theta} + \frac{\sigma_r - \sigma_\theta}{r} + B_r = 0 \tag{5-12}$$

All forces are projected onto the circumferential axis through the unit center, and the circumferential equilibrium equation is

$$\left[\sigma_\theta + \frac{\partial \sigma_\theta}{\partial \theta}d\theta\right]dr - \sigma_\theta dr + \left[\tau_{r\theta} + \frac{\partial \tau_{r\theta}}{\partial r}dr\right](r + dr)d\theta - \tau_{r\theta}r d\theta +$$

$$\left[\tau_{\theta r} + \frac{\partial \tau_{\theta r}}{\partial \theta}d\theta\right]dr\frac{d\theta}{2} - \tau_{\theta r}dr\frac{d\theta}{2} + B_r r d\theta dr = 0 \tag{5-13}$$

Replacing $\tau_{\theta r}$ with $\tau_{r\theta}$. After simplification, dividing by $r d\theta dr$, and then omitting high order terms, we can get

$$\frac{1}{r}\frac{\partial \sigma_\theta}{\partial \theta} + \frac{\partial \tau_{r\theta}}{\partial r} + \frac{2\tau_{r\theta}}{r} + B_\theta = 0 \tag{5-14}$$

Therefore, the equilibrium equation in two-dimensional polar coordinate system is presented in Equation (5-15).

$$\begin{cases} \dfrac{\partial \sigma_r}{\partial r} + \dfrac{1}{r}\dfrac{\partial \tau_{r\theta}}{\partial \theta} + \dfrac{\sigma_r - \sigma_\theta}{r} + B_r = 0 \\[3mm] \dfrac{1}{r}\dfrac{\partial \sigma_\theta}{\partial \theta} + \dfrac{\partial \tau_{r\theta}}{\partial r} + \dfrac{2\tau_{r\theta}}{r} + B_\theta = 0 \end{cases} \tag{5-15}$$

5.4 Boundary-value Problems

It is clear that, in subjecting the interior elements of a body to a given body-force distribution while simultaneously subjecting the boundary to given surface tractions, one imposes a definite unique state of stress and strain in the body. A problem posed to us with these data, i. e., body-force distributions and surface tractions, is termed a boundary-value problem of the first kind.

Similarly, we may also conclude that, in subjecting the interior elements of a body to a given body-force distribution while simultaneously subjecting the boundary to a given displacement, one imposes a definite unique state of stress and strain in the body. A problem posed to us with these data, i. e., body-force distributions and surface displacements is called a boundary-value problem of the second kind.

In this book, we shall consider only the first boundary-value problem.

Let us now examine the formulation of the surface traction. For this purpose we employ the vector $\boldsymbol{T}$, which gives the force per unit area at a point on the boundary. This is shown in Figure 5-3, where the total surface force on the infinitesimal boundary surface element ABC is

$$\boldsymbol{F} = \boldsymbol{T}d\boldsymbol{A} \tag{5-16}$$

It should be clearly understood that $\boldsymbol{T}$ is not necessarily collinear with the vector $d\boldsymbol{A}$

since it includes friction force as well as normal force intensity on the surface element. We shall next relate T, which will be a known function in the first boundary-value problem, with the stress inside the body as one approaches the boundary. To do this we employ Newton's law for the element of Figure 5-3 which has now shown the enlarged one in Figure 5-4 with the stresses indicated on the faces parallel to the xyz coordinates. Summing forces in the x direction we get

$$T_xABC = \sigma_xCOB + \tau_{xy}COA + \tau_{xz}BOA \tag{5-17}$$

Using l, m and n to represent the direction cosines of the normal to the area ABC, we can replace COB, COA and BOA by $lABC, mABC$ and $nABC$, respectively. Dividing through by ABC, we then get

$$T_x = \sigma_xl + \tau_{xy}m + \tau_{xz}n \tag{5-18}$$

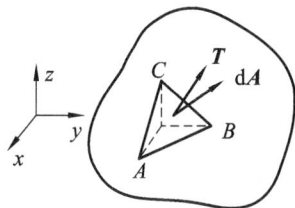

Figure 5-3　The element of the body
approaching the boundary

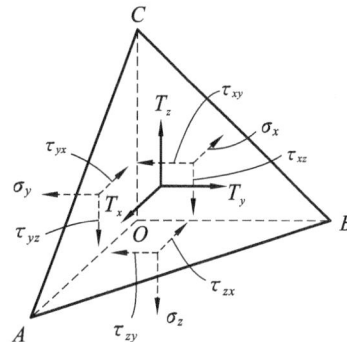

Figure 5-4　The enlarged element

By processing in a similar manner for the other coordinates and using the complementary property of shear, we may reach the following equations

$$T_x = \sigma_xl + \tau_{xy}m + \tau_{xz}n \tag{5-19a}$$

$$T_y = \tau_{yx}l + \sigma_ym + \tau_{yz}n \tag{5-19b}$$

$$T_z = \tau_{zx}l + \tau_{zy}m + \sigma_zn \tag{5-19c}$$

The preceding equations are so-called boundary conditions that must be satisfied for a given problem. Furthermore, we must point out that T and B cannot be specified arbitrarily but must be so related that they satisfy the equilibrium requirements for the entire body taken as a free body (This takes us back to the statics of a rigid-body).

5.5　Method of Analysis

We have considered the basic laws of continuum mechanics and have decided, for the frictionless,isothermal,elastic body which we are presently considering,that only Newton's law is required. Furthermore,we have the equations stemming from geometrical considerations of small deformation. Finally,we have a constitutive law for a linear elastic material. We posed two kinds of boundary-value problems which could be solved for a material having all the properties and geometrical limitations just enumerated. We shall next discuss briefly how we can go about solving these problems. At the present stage of our study,three methods of attack may be outlined.

(1) Carry out a direct integration of the pertinent differential equations for the particular boundary-values problem. This "head-on" approach is successful only in relatively simple situations because of the complexity of the equations involved.

(2) Select certain possible functions for stress or strain distributions which satisfy the pertinent differential equations. Then,examining the general boundary conditions, formulate boundary-value problems that the functions might represent. This is the so-called inverse method . When several simple solutions have been so established,we may sometimes superpose these solutions to form solutions of problems of practical interest.

(3) Combine methods (1) and (2). This is the so-called semi-inverse method where one selects certain portions of the stress or strain formulations while other portions are developed by direct integration procedures.

There are times when analytical solutions can be established easily for certain geometries under certain prescribed loads. From these solutions,we can sometimes formulate judicious assumptions as to deformation characteristics when more general loadings are applied. Using these approximations of behavior,along with portions of the basic and constitutive laws,we can at times establish some very useful and valuable formulations. Such is the case for the handling of beams,shafts,plates and shells. There is an extensive body of knowledge developed by this means forming a discipline which we call strength of materials. Knowing and working within the limitations of the formulations of strength of materials,we can make rapid computations of acceptable accuracy in many problems of engineering interest.

We shall develop some of the basic assumptions of strength of materials from the theory of elasticity. By approaching strength of materials from the theory of elasticity we shall have the means of viewing the simpler,handier formulations in proper perspective. That is,as we shall have means of knowing more realistically what are the limitations of the simpler formulations and the degree of error to be expected.

5.6 St. Venant's Principle

When solving problems in rigid-body mechanics, we found it profitable to employ the concept of a point force when we had a force distribution over a small area. At other times, we employed the rigid-body resultant force system of some distribution in the handling of a problem. Such replacements led to reasonably accurate and direct solutions.

Akin to this procedure, we have for the study of elastic behavior of bodies, the procedure stemming from St. Venant's principle. This principle states that the stresses and strains reasonably distant from an applied load are not significantly altered if this load is changed to another load which is equivalent from the viewpoint of rigid-body mechanics. We may call such a second load the statically equivalent load (Fig-

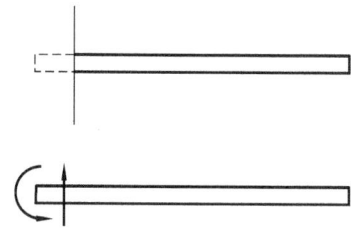

Figure 5-5 Cantilever beam and statically equivalent load

ure 5-5). Thus the reasonably distant effects from surface tractions over a part of the boundary can be thought of as dependent on the rigid-body resultant of the applied loads on this surface. By this principle, we can replace the complex supporting force system exerted by the wall on the cantilever beam by a single force and couple as shown in the diagram for the purpose of simplifying computations of stresses and strains in the domain to the right of the support.

Many of the problems to be undertaken in this book will permit us to use the St. Venant's principle to help simplify the computations without incurring serious error.

5.7 Uniqueness Condition

We shall now examine the important question as to the uniqueness of mathematical solutions of the first and second boundary-value problems. We assume for the moment that there are two solutions possible for either boundary-value problem and we give these solutions in the following way

Solution 1 $\qquad (\tau_{ij})_1 , (u_i)_1$

Solution 2 $\qquad (\tau_{ij})_2 , (u_i)_2$

$$(5\text{-}20)$$

Assuming the superposition principle holds, we can superpose possible deformations to form new possible deformations. Accordingly, we form τ_{ij} and u_i as follows

$$\tau_{ij} = (\tau_{ij})_1 - (\tau_{ij})_2 \qquad (5\text{-}21a)$$

$$u_i = (u_i)_1 - (u_i)_2 \qquad (5\text{-}21b)$$

Clearly the new solutions will correspond to a problem for which the body-force distribution is zero. Furthermore, if Solutions 1 and 2 correspond to a first boundary-value problem, the new solution will be one valid for zero surface tractions and, if Solutions 1 and 2 correspond to the second boundary-value problem, the new solution will be one valid for zero displacements of the boundary.

If, for our body, the vanishing of body forces and surface tractions or the vanishing of body forces and surface displacements means simultaneously the vanishing of stress and strain, we must conclude, for our new solution, that

$$\tau_{ij} = 0 \qquad (5\text{-}22a)$$

$$u_i = 0 \qquad (5\text{-}22b)$$

Accordingly

$$(\tau_{ij})_1 = (\tau_{ij})_2 \qquad (5\text{-}23a)$$

$$(u_i)_1 = (u_i)_2 \qquad (5\text{-}23b)$$

And so we see in this case that the two proposed solutions are indeed the same solution, thus proving the uniqueness condition.

Let us consider the superposition principle more carefully as it pertains to our present discussion. If we have large deformation, we have already made clear in Chapter 3 that the principle of superposition does not hold. However, it is to be pointed out that the superposition principle may fail for even small deformations. There is, for instance, the possibility that the induced deformation, however small, causes the external loads to develop a stress distribution appreciably different from that which would occur if there were no change in the initial geometry. For example, the stresses in the column shown in Figure 5-6 will be affected by the deformation δ even though this deflection might be small. For such cases the effect of an increment of load as to stress or deformation will definitely depend on the values of the loads already applied. In short we cannot compute the final deformation in terms of deformation due to increments of the load each acting on the original geometry as we in effect did in the proof. We shall see in future that for certain of these problems a continuum of solutions comes out of the

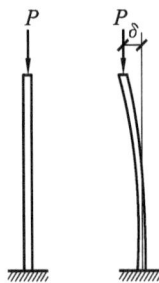

Figure 5-6　The stresses in the column
affected by the deformation δ

mathematics and not a unique solution.

Another condition in our proof was presented only tacitly. That is, we presumed that the admissible displacement fields $(u_i)_1$ and $(u_i)_2$ were single-valued functions which on being subtracted one from another led to single-valued functions u_i. We have pointed out in Chapter 3 that for simply connected bodies with one point completely fixed in space, the satisfaction of the compatibility equations was sufficient to insure single-valuedness of the displacement field functions u_i. For multiply connected regions, however, both the compatibility equations plus other requirements must be satisfied for guaranteeing single-valued displacement fields. We shall state here without proof that once the condition of single-valuedness of the displacement field has been insured, the uniqueness theorem goes through without further considerations of the connectivity of the body.

5.8   Elastic Problems of Prismatic Bar Hanging by Its Own Weight

We shall now analyze a simple problem using the fundamental equations and the St. Venant's principle. Thus, consider the straight prismatic homogeneous bar shown in Figure 5-7 hanging by its own weight. The body force distribution stems only from gravity and can then be given as

$$\boldsymbol{B} = -\gamma \boldsymbol{k} \qquad (5-24)$$

The surface tractions consist of uniform atmospheric pressure on all faces of the bar exposed to the atmosphere and a distribution at the support whose resultant must be $\gamma a L t$. If we do not get too close to the support we may employ this resultant by virtue of St. Venant's principle without requiring the exact knowledge of the nature of the distribution associated with this resultant. Thus, we see that we have here a boundary-value problem of the first kind if we wish to compute stresses, strains, and displacements inside the body in the domain away from the support. Consider as a possible stress distribution the following quantities

$$\sigma_z = \gamma z \quad \sigma_x = \sigma_y = \tau_{xy} = \tau_{xz} = \tau_{yz} = 0 \qquad (5-25)$$

Clearly the equations of equilibrium, Equations (5-3), are satisfied by such a distribution. Next, consider the boundary conditions of the problem. If we neglect the contribution of atmospheric pressure, we see that the surface traction on that part of the bar exposed to the atmosphere is zero at all points. In considering Equation (5-19), we

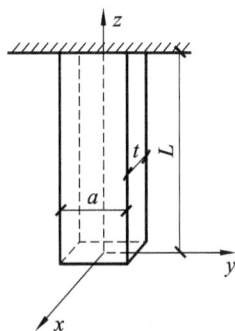

Figure 5-7 The straight prismatic homogeneous bar hanging by its own weight

see that we have zero stresses on the right side for all stresses except σ_z. Thus only Equation (5-19c) must be considered here. We have then

$$T_z = 0 = \sigma_z n \qquad (5\text{-}26)$$

Observing Figure 5-7, we see that n is zero for the surfaces of the bar parallel to the z axis. And from Equation (5-25), we see that $\sigma_z = 0$ at the bottom surface of the bar, that is, where $z = 0$. The boundary conditions are then satisfied on all surfaces exposed to the atmosphere. At the upper boundary we have from Equation (5-25) as the only nonzero stress, $\sigma_z = \gamma L$, and the resultant of this uniform stress at the upper face is

$$\int_A \sigma_z \, \mathrm{d}A = \int_A \gamma L \, \mathrm{d}A = \gamma a L t \qquad (5\text{-}27)$$

This is the proper resultant at the section and so we see that the proposed stress distribution also satisfies the upper boundary conditions for our purposes.

Next we ascertain the strain distributions by employing the stress-strain law. Thus assuming we are in the elastic range we get

$$\begin{cases} \varepsilon_{xx} = \dfrac{1}{E}(-\nu\gamma z) \\[2mm] \varepsilon_{yy} = \dfrac{1}{E}(-\nu\gamma z) \quad \gamma_{xy} = \gamma_{xz} = \gamma_{yz} = 0 \\[2mm] \varepsilon_{zz} = \dfrac{1}{E}\gamma z \end{cases} \qquad (5\text{-}28)$$

We can next check to see whether this strain distribution satisfies the compatibility Equations (5-9). Substituting the foregoing results into these equations, we see quite readily that the compatibility equations are satisfied, insuring us in this case that the strain stems from a single-valued, continuous displacement field.

We thus may be assured at this point that we have presented the unique solution to the problem at hand, at least away from the immediate vicinity of the support.

To complete the analysis, we now establish the displacement field for this problem. From Equations (5-7) and (5-28), we get

$$\frac{\partial u_x}{\partial x} = -\frac{\nu\gamma z}{E} \qquad (5\text{-}29a)$$

$$\frac{\partial u_y}{\partial y} = -\frac{\nu\gamma z}{E} \qquad (5\text{-}29b)$$

$$\frac{\partial u_z}{\partial z} = \frac{\gamma z}{E} \tag{5-29c}$$

$$\frac{\partial u_x}{\partial y} + \frac{\partial u_y}{\partial x} = 0 \tag{5-29d}$$

$$\frac{\partial u_y}{\partial z} + \frac{\partial u_z}{\partial y} = 0 \tag{5-29e}$$

$$\frac{\partial u_x}{\partial z} + \frac{\partial u_z}{\partial x} = 0 \tag{5-29f}$$

We shall begin by integrating Equation (5-29c). Thus

$$u_z = \frac{\gamma z^2}{2E} + f(x, y) \tag{5-30}$$

where $f(x, y)$ is an arbitrary function of the coordinates x and y. Substituting the foregoing result into Equations (5-29e) and (5-29f), we get

$$\frac{\partial u_y}{\partial z} = -\frac{\partial f}{\partial y} \tag{5-31a}$$

$$\frac{\partial u_x}{\partial z} = -\frac{\partial f}{\partial x} \tag{5-31b}$$

Integrate these equations, remembering that f is only a function of x and y. Thus

$$u_y = -\frac{\partial f}{\partial y} z + g(x, y) \tag{5-32a}$$

$$u_x = -\frac{\partial f}{\partial x} z + h(x, y) \tag{5-32b}$$

Where g and h are two more arbitrary functions of the coordinates x and y. Now substitute the preceding results into Equations (5-29a) and (5-29b). That is

$$-\frac{\partial^2 f}{\partial x^2} z + \frac{\partial h}{\partial x} = -\frac{\nu \gamma z}{E} \tag{5-33a}$$

$$-\frac{\partial^2 f}{\partial y^2} z + \frac{\partial g}{\partial y} = -\frac{\nu \gamma z}{E} \tag{5-33b}$$

Since f, g and h do not in any way depend on z, we may deduce the following relations from the preceding equations

$$\frac{\partial h}{\partial x} = 0 \tag{5-34a}$$

$$\frac{\partial g}{\partial y} = 0 \tag{5-34b}$$

$$\frac{\partial^2 f}{\partial x^2} = \frac{\nu \gamma}{E} \tag{5-34c}$$

$$\frac{\partial^2 f}{\partial y^2} = \frac{\nu \gamma}{E} \tag{5-34d}$$

We have yet to consider Equation (5-29d). Substituting from Equation (5-32), we get

$$-\frac{\partial^2 f}{\partial y \partial x}z+\frac{\partial h}{\partial y}-\frac{\partial^2 f}{\partial x \partial y}z+\frac{\partial g}{\partial x}=0 \tag{5-35}$$

This becomes, on combining terms

$$-2\frac{\partial^2 f}{\partial y \partial x}z+\frac{\partial h}{\partial y}+\frac{\partial g}{\partial x}=0 \tag{5-36}$$

Since $\partial h/\partial y$ and $\partial g/\partial x$ are functions of only x and y we can conclude from Equation (5-36) that

$$\frac{\partial h}{\partial y}+\frac{\partial g}{\partial x}=0 \tag{5-37a}$$

$$\frac{\partial^2 f}{\partial y \partial x}=0 \tag{5-37b}$$

Now from Equations (5-34a) and (5-34b), we see that g and h can be given as

$$\begin{cases}g=C_1\alpha(x)+C_2 \\ h=C_3\beta(y)+C_4\end{cases} \tag{5-38}$$

where C_1, C_2, C_3 and C_4 are arbitrary constants and α and β are functions of x and y, respectively.

Equation (5-37a) requires that

$$C_3\frac{\mathrm{d}\beta(y)}{\mathrm{d}y}+C_1\frac{\mathrm{d}\alpha(x)}{\mathrm{d}x}=0 \tag{5-39}$$

If $\beta(y)=y$ and $\alpha(x)=x$, and if $C_3=-C_1$, we see that the foregoing equation is satisfied. Thus we have, for g and h

$$g=C_1 x+C_2 \tag{5-40a}$$

$$h=-C_1 y+C_4 \tag{5-40b}$$

Equations (5-34c), (5-34d) and (5-37b) indicate that f has the following form

$$f=\frac{\nu\gamma}{2E}(x^2+y^2)+C_5 x+C_6 y+C_7 \tag{5-41}$$

You may verify by substitution. We can now give the displacement field in terms of six arbitrary constants. Substituting Equations (5-40) and (5-41) into Equations (5-30) and (5-32) we have

$$u_x=-\frac{\nu\gamma}{E}xz-C_5 z-C_1 y+C_4 \tag{5-42a}$$

$$u_y=-\frac{\nu\gamma}{E}yz-C_6 z+C_1 x+C_2 \tag{5-42b}$$

$$u_z=\frac{\gamma z^2}{2E}+\frac{\nu\gamma}{2E}(x^2+y^2)+C_5 x+C_6 y+C_7 \tag{5-42c}$$

We solve for the six constants of integration by insuring that there is no rigid-body translation or rotation of the bar itself. We know that for the point at $x=0, y=0, z=L$ we have $u_x=u_y=u_z=0$. This imposes the following conditions on our constants

$$\begin{cases} -C_5 L+C_4=0 \\ -C_6 L+C_2=0 \\ \dfrac{\gamma L^2}{2E}+C_7=0 \end{cases} \tag{5-43}$$

Also, there will be no rotation at this point so that we can also say, for position $(0,0,L)$

$$\begin{cases} \Phi_x=\dfrac{1}{2}\left(\dfrac{\partial u_z}{\partial y}-\dfrac{\partial u_y}{\partial z}\right)_{(0,0,L)}=0 \\[2mm] \Phi_y=\dfrac{1}{2}\left(\dfrac{\partial u_x}{\partial z}-\dfrac{\partial u_z}{\partial x}\right)_{(0,0,L)}=0 \\[2mm] \Phi_z=\dfrac{1}{2}\left(\dfrac{\partial u_y}{\partial x}-\dfrac{\partial u_x}{\partial y}\right)_{(0,0,L)}=0 \end{cases} \tag{5-44}$$

We then get the following additional conditions on the constants

$$\begin{cases} C_5=-C_6 \\ -C_5=C_5 \\ C_1=0 \end{cases} \tag{5-45}$$

The foregoing results indicate that $C_1=C_5=C_6=0$, and so going back to Equation (5-43), we see that

$$\begin{cases} C_4=0 \\ C_2=0 \\ C_7=-\dfrac{\gamma L^2}{2E} \end{cases} \tag{5-46}$$

Using these values for the constants then insures no rigid-body motion and the displacement field then becomes

$$u_x=-\frac{\nu\gamma}{E}xz \tag{5-47a}$$

$$u_y=-\frac{\nu\gamma}{E}yz \tag{5-47b}$$

$$u_z=\frac{\gamma z^2}{2E}+\frac{\nu\gamma}{2E}(x^2+y^2)-\frac{\gamma}{2E}L^2 \tag{5-47c}$$

Notice that points along the z axis have only vertical displacements given by the expression

$$u_z=\frac{\gamma}{2E}(z^2-L^2) \tag{5-48}$$

Other points have horizontal displacements due to contraction of the member. As an exercise, you may show that horizontal plane surfaces of the bar deform into paraboloidal surfaces.

5.9   Closure

In this chapter, we have formulated the basic laws of linear isothermal elasticity. We assumed that the properties of the material did not significantly vary with the temperature. Also we assumed that the development of stress in the body from external loads did not affect the temperature.

If we wish to include temperature-dependent properties and if we wish to include the contributions to temperature change due to loading we get involved in a very difficult undertaking. Newton's law is unchanged but now the first law of thermodynamics must be carefully considered. This law will include heat transfer, energy of deformation, etc. An equation of state for the material will also be needed. However, unlike fluids, little is known about equations of state for solids although physicists and engineers are actively conducting such investigations.

If we consider large deformation, we must redefine what is meant by normal and shear strain. We generally cannot use the Hooke's law employed in this chapter. Such considerations are well beyond the level of this book.

In the next chapter, we shall use the theory of elasticity to consider some useful plane-strain and plane-stress problems.

Problems

5-1   Given the following stress field

$$\sigma_x = 80x^3 + y \text{ Pa} \qquad \tau_{xy} = 1000 + 100y^2 \text{ Pa}$$
$$\sigma_y = 100x^3 + 1600 \text{ Pa} \quad \tau_{yz} = 0 \text{ Pa}$$
$$\sigma_z = 90y^2 + 100z^3 \text{ Pa} \quad \tau_{xz} = xz^3 + 100x^2 y \text{ Pa}$$

What is the body-force distribution that is required for equilibrium? What is the stress and body force at $(1,1,5)$?

5-2   In Problem 5-1 what is the strain at the position $(2,2,5)$ (Take $E = 2.1 \times 10^5$ MPa; $\nu = 0.3$)?

5-3   Does the distribution given by Problem 5-1 satisfy the compatibility equations?

5-4   Write the equations of equilibrium of the case of plane stress.

5-5 We are given the following stress distribution in polar coordinates for a case of plane stress

$$\sigma_r = \frac{(p_o - p_i)a^2 b^2}{b^2 - a^2} \frac{1}{r^2} + \frac{p_i a^2 - p_o b^2}{b^2 - a^2}$$

$$\sigma_\theta = -\frac{(p_o - p_i)a^2 b^2}{b^2 - a^2} \frac{1}{r^2} + \frac{p_i a^2 - p_o b^2}{b^2 - a^2}$$

$$\tau_{r\theta} = 0$$

Show that the equations of equilibrium are satisfied. Now show by considering the boundary conditions that this distribution corresponds to that in a long thick-walled cylinder subjected to a uniform pressure on the inner and outer surfaces as is shown in Figure 5-8.

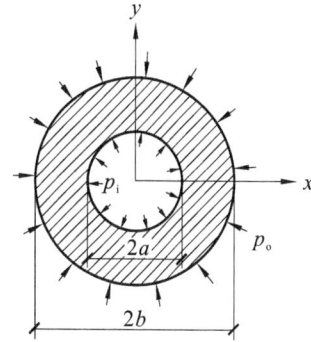

5-6 In Problem 5-5, take

$$p_o = 3.5 \text{ MPa} \quad a = 600 \text{ mm}$$
$$p_i = 0.35 \text{ MPa} \quad b = 1200 \text{ mm}$$

Figure 5-8 Figure of Problem 5-5

What is the maximum stress predicted by the theory of elasticity? Compare it with a stress computed by thin-shell theory, i. e. , where the tangential stress is assumed uniform for the whole cylinder.

5-7 Express Hooke's law for plane stress, first giving strains as functions of stresses. Next express Hooke's law for plane stress giving stresses in terms of strains.

5-8 Given the following stress distribution

$$\sigma_x = C_1$$
$$\sigma_y = C_2$$
$$\tau_{xy} = C_3$$
$$\sigma_z = \tau_{xz} = \tau_{yz} = 0$$

where the C_1, C_2, C_3 are constants. What is a boundary-value problem for which this is a solution?

5-9 Given the following stress distribution

$$\sigma_x = C_1 x + C_2 y$$
$$\sigma_y = C_3 x + C_4 y$$
$$\tau_{xy} = C_5 x + C_6 y$$
$$\sigma_z = \tau_{xz} = \tau_{yz} = 0$$

Under what restrictions does this stress distribution satisfy the basic equations of elasticity?

5-10　In Problem 5-9 take all constants but C_1 equal to zero, what is a boundary-value problem corresponding to such a state of stress?

5-11　Consider the case of the tip-loaded cantilever beam shown in Figure 5-9. Assume we have a case of plane stress where

$$\sigma_x = C_1 xy$$
$$\sigma_y = 0$$
$$\tau_{xy} = C_2 + C_3 y^2$$

Determine the constants C_1, C_2, and C_3 to satisfy boundary conditions and Newton's law. Explain the use of St. Venant's principle for this problem. Now determine the strains for this problem. Show that one of the compatibility equations is not satisfied by the computed strain distribution(Thus the foregoing solution is not an exact solution. However, we shall see in the next chapter that it is a good approximation).

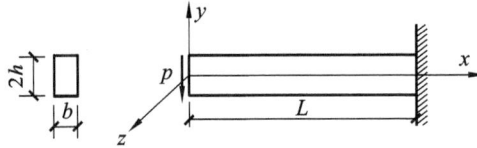

Figure 5-9　Figure of Problem 5-11

6 Problems in Plane Strain and Plane Stress from the Theory of Elasticity

6.1 Introduction

In Chapter 5, we presented the basic equations of the theory of elasticity and solved a three-dimensional linear-elastic problem. We wish now to consider the solution of additional classes of problems by use of the theory of elasticity. We shall attack the problem of plane strain first, and we shall see that we can proceed smoothly using the basic formulations. We shall then go to the plane-stress case. Here, rigorous analysis turns out to be so complicated that such treatment is beyond the scope of this book. However, we shall be able to set forth approximate procedures that are motivated by the theory.

6.2 Airy's Function

In Chapter 3, we defined plane strain as a state of strain which, for some axis, usually chosen to be the z axis, the strains γ_{xz}, γ_{yz} and ε_{zz} are zero in the domain of interest. Portions of prismatic bodies away from the end constraints may be considered to be in a state of plane strain under the following conditions.

(1) Taking the z direction to correspond to the axis of the prism, the surface tractions and body forces must be oriented normal to this axis.

(2) The surface tractions and body forces must be functions only of x and y and must have a zero resultant to avoid bending deformation.

Thus the dam shown in Chapter 3 may be considered under a state of plane strain in regions away from the rigid constraining walls. Also, the long thick pipe under either internal pressure or external pressure which is constrained at the ends (Figure 6-1) can be considered as a problem in plane strain away from the ends (We shall study this problem later). Now, we consider the basic equations for plane strain under the restrictions just stated.

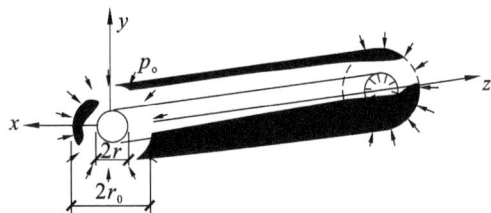

Figure 6-1 Thick-walled cylinder constrained at ends

Consider first Hooke's law for this case. We have

$$\varepsilon_{xx} = \frac{1}{E}[(\sigma_x - \nu(\sigma_y + \sigma_z)] \qquad (6\text{-}1\text{a})$$

$$\varepsilon_{yy} = \frac{1}{E}[(\sigma_y - \nu(\sigma_x + \sigma_z)] \qquad (6\text{-}1\text{b})$$

$$0 = \frac{1}{E}[(\sigma_z - \nu(\sigma_x + \sigma_y)] \qquad (6\text{-}1\text{c})$$

$$\gamma_{xy} = \frac{1}{G}\tau_{xy} \quad \tau_{xz} = \tau_{yz} = 0 \qquad (6\text{-}1\text{d})$$

Equation (6-1c) requires that

$$\sigma_z = \nu(\sigma_x + \sigma_y) \qquad (6\text{-}2)$$

Now Let us turn to Newton's law. We have

$$\frac{\partial \sigma_x}{\partial x} + \frac{\partial \tau_{xy}}{\partial y} + B_x = 0 \qquad (6\text{-}3\text{a})$$

$$\frac{\partial \tau_{yx}}{\partial x} + \frac{\partial \sigma_y}{\partial y} + B_y = 0 \qquad (6\text{-}3\text{b})$$

$$\frac{\partial \sigma_z}{\partial z} = 0 \qquad (6\text{-}3\text{c})$$

Note the body-force component B_z has been set equal to zero because of the earlier condition (1) on body forces. Equation (6-3c) then signifies that σ_z is not a function of z. It may be concluded furthermore, from the conditions imposed on the problem that σ_x, σ_y and τ_{xy} are also not functions of the coordinate z. We shall furthermore set forth at this time the condition that the body-force distribution be conservative, i. e.

$$\boldsymbol{B} = -\text{grad } \boldsymbol{V} \qquad (6\text{-}4)$$

where $\boldsymbol{V}$ is the so-called potential function. The equations for Newton's law then become for our case

$$\frac{\partial \sigma_x}{\partial x} + \frac{\partial \tau_{xy}}{\partial y} - \frac{\partial V}{\partial x} = 0 \qquad (6\text{-}5\text{a})$$

$$\frac{\partial \tau_{yx}}{\partial x} + \frac{\partial \sigma_y}{\partial y} - \frac{\partial V}{\partial y} = 0 \qquad (6\text{-}5\text{b})$$

$$\sigma_z = f(x, y) \qquad (6\text{-}5\text{c})$$

We have decided that all stresses τ_{ij} in our analysis are either constants or are functions of coordinates x and y. Hooke's law then requires the same condition for the strain terms ε_{ij}. If we keep this in mind when we consider the equations of compatibility Equation (5-9), we see that all but one equation of the set are identically satisfied. That equation is

$$\frac{\partial^2 \epsilon_{xx}}{\partial y^2} + \frac{\partial^2 \epsilon_{yy}}{\partial x^2} = \frac{\partial^2 \gamma_{xy}}{\partial x \partial y} \tag{6-6}$$

To get this equation in terms of stress, we substitute for the strains using Equation (6-1). Thus we get

$$\frac{1}{E}\frac{\partial^2 \sigma_x}{\partial y^2} - \frac{\nu}{E}\frac{\partial^2 \sigma_y}{\partial y^2} - \frac{\nu}{E}\frac{\partial^2 \sigma_z}{\partial y^2} + \frac{1}{E}\frac{\partial^2 \sigma_y}{\partial x^2} - \frac{\nu}{E}\frac{\partial^2 \sigma_x}{\partial x^2} - \frac{\nu}{E}\frac{\partial^2 \sigma_z}{\partial x^2} = \frac{1}{G}\frac{\partial^2 \tau_{xy}}{\partial x \partial y} \tag{6-7}$$

On the right side of the preceding equation, we replace G according to Chapter 4, we now rewrite

$$G = \frac{E}{2(1+\nu)} \tag{6-8}$$

Furthermore, we may also replace $\partial^2 \tau_{xy}/\partial x \partial y$ in Equation (6-7) in the following way. Take the partial derivative with respect to x of Equation (6-5a) and then take the partial derivative with respect to y of Equation (6-5b). Adding the two equations and solving for $\partial^2 \tau_{xy}/\partial x \partial y$, we get

$$\frac{\partial^2 \tau_{xy}}{\partial x \partial y} = \frac{1}{2}\left(\frac{\partial^2 V}{\partial x^2} + \frac{\partial^2 V}{\partial y^2} - \frac{\partial^2 \sigma_x}{\partial x^2} - \frac{\partial^2 \sigma_y}{\partial y^2}\right) \tag{6-9}$$

Now substituting Equations (6-8) and (6-9) into Equation (6-7), after cancelling $1/E$ and rearranging the terms we get

$$\mathbf{V}^2(\sigma_x + \sigma_y) = \frac{1}{1-\nu}\mathbf{V}^2 V \tag{6-10}$$

We now introduce a function Φ, called Airy's function, which we define in the following somewhat indirect manner

$$\sigma_x = V + \frac{\partial^2 \Phi}{\partial y^2} \tag{6-11a}$$

$$\sigma_y = V + \frac{\partial^2 \Phi}{\partial x^2} \tag{6-11b}$$

$$\tau_{xy} = -\frac{\partial^2 \Phi}{\partial x \partial y} \tag{6-11c}$$

Substituting these stresses into Equations (6-5a) and (6-5b), we see that these equations are identically satisfied. Thus, in putting the stresses in the foregoing form, we automatically satisfy the equations of equilibrium. Now substituting the stress formulations just given into Equation (6-10), we get

$$\mathbf{V}^2\left(\frac{\partial^2 \Phi}{\partial x^2} + \frac{\partial^2 \Phi}{\partial y^2}\right) = -\frac{1-2\nu}{1-\nu}\mathbf{V}^2 V \tag{6-12}$$

We may write Equation (6-12) in the form

$$\mathbf{V}^4 \Phi = -\frac{1-2\nu}{1-\nu}\mathbf{V}^2 V \tag{6-13}$$

where ∇^4, which we call a biharmonic operator, is defined in two dimensions as

$$\nabla^4 = \nabla^2(\nabla^2) = \frac{\partial^4}{\partial x^4} + 2\frac{\partial^4}{\partial x^2 \partial y^2} + \frac{\partial^4}{\partial y^4} \tag{6-14}$$

If there are no body forces, we have the following equation

$$\nabla^4 \Phi = 0 \tag{6-15}$$

We call this the biharmonic equation.

What have we accomplished by introducing the Airy's function? We have simplified our work, in that instead of having to solve for three unknowns σ_x, σ_y and τ_{xy}, we now need to solve only for Airy's function using Equation (6-15) if there are no body forces. Knowing Φ, we can then determine the stress distribution through Equation (6-11).

The boundary conditions for the plane-strain problem require that, on the lateral surface of the prism, we have for given surface traction components

$$T_x = \sigma_x l + \tau_{xy} m \tag{6-16a}$$

$$T_y = \tau_{yx} l + \sigma_y m \tag{6-16b}$$

$$T_z = 0 = \sigma_z n \tag{6-16c}$$

The last condition is identically satisfied since $n = 0$. Replacing σ_x, σ_y and τ_{xy}, using Equation (6-11), we have

$$\begin{cases} T_x = \left(V + \dfrac{\partial^2 \Phi}{\partial y^2}\right)l - \dfrac{\partial^2 \Phi}{\partial x \partial y}m \\ T_y = -\dfrac{\partial^2 \Phi}{\partial x \partial y}l + \left(V + \dfrac{\partial^2 \Phi}{\partial x^2}\right)m \end{cases} \tag{6-17}$$

For no body forces the boundary conditions become

$$\begin{cases} T_x = \dfrac{\partial^2 \Phi}{\partial y^2}l - \dfrac{\partial^2 \Phi}{\partial x \partial y}m \\ T_y = -\dfrac{\partial^2 \Phi}{\partial x \partial y}l + \dfrac{\partial^2 \Phi}{\partial x^2}m \end{cases} \tag{6-18}$$

6.3 Problems in Cylindrical Coordinates

There are problems of interest in plane strain that may be best investigated with the use of cylindrical coordinates rather than rectangular coordinates. Accordingly, we shall now set forth in cylindrical coordinates the basic equations presented in Section 6.2. In the interest of simplicity, we shall consider only the case of zero body force. We shall need the following transformation formulas

$$\begin{cases} x=r\cos\theta \quad r=(x^2+y^2)^{\frac{1}{2}} \\ y=r\sin\theta \quad \theta=\arctan\dfrac{y}{x} \end{cases} \tag{6-19}$$

As a first step we shall formulate σ_r in terms of the stresses σ_x, σ_y and τ_{xy}. Thus we have

$$\sigma_r=l^2\sigma_x+m^2\sigma_y+n^2\sigma_z+2(lm\tau_{yx}+mn\tau_{yz}+ln\tau_{zx}) \tag{6-20}$$

Noting that $n=0$, $l=\cos\theta$, and $m=\sin\theta$, we have

$$\sigma_r=\sigma_x\cos^2\theta+\sigma_y\sin^2\theta+2\tau_{xy}\sin\theta\cos\theta \tag{6-21}$$

Now replace the stresses on the right side of Equation (6-21), using Equation (6-11) for the case where there are no body forces

$$\sigma_r=\frac{\partial^2\Phi}{\partial y^2}\cos^2\theta+\frac{\partial^2\Phi}{\partial x^2}\sin^2\theta-2\frac{\partial^2\Phi}{\partial x\partial y}\sin\theta\cos\theta \tag{6-22}$$

To have σ_r expressed entirely in cylindrical coordinates in the foregoing equation, evaluate the partial derivatives of Φ in terms of cylindrical coordinates. Thus considering Φ not to depend on z and using Equation (6-19) we get

$$\frac{\partial\Phi}{\partial x}=\frac{\partial\Phi}{\partial r}\frac{\partial r}{\partial x}+\frac{\partial\Phi}{\partial\theta}\frac{\partial\theta}{\partial x}=\frac{\partial\Phi}{\partial r}\frac{x}{r}+\frac{\partial\Phi}{\partial\theta}\left(-\frac{y}{r^2}\right)=\frac{\partial\Phi}{\partial r}\cos\theta-\frac{\partial\Phi}{\partial\theta}\frac{\sin\theta}{r}$$

$$\begin{aligned} \frac{\partial^2\Phi}{\partial x^2}&=\left[\frac{\partial}{\partial r}\left(\frac{\partial\Phi}{\partial x}\right)\right]\frac{\partial r}{\partial x}+\left[\frac{\partial}{\partial\theta}\left(\frac{\partial\Phi}{\partial x}\right)\right]\frac{\partial\theta}{\partial x} \\ &=\left[\frac{\partial}{\partial r}\left(\frac{\partial\Phi}{\partial x}\right)\right]\cos\theta-\left[\frac{\partial}{\partial\theta}\left(\frac{\partial\Phi}{\partial x}\right)\right]\frac{\sin\theta}{r} \\ &=\frac{\partial^2\Phi}{\partial r^2}\cos^2\theta-2\frac{\partial^2\Phi}{\partial r\partial\theta}\frac{\sin\theta\cos\theta}{r}+2\frac{\partial\Phi}{\partial\theta}\frac{\sin\theta\cos\theta}{r^2}+\frac{\partial\Phi}{\partial r}\frac{\sin^2\theta}{r}+\frac{\partial^2\Phi}{\partial\theta^2}\frac{\sin^2\theta}{r^2} \end{aligned} \tag{6-23}$$

And by a similar procedure, we may also get

$$\frac{\partial^2\Phi}{\partial y^2}=\frac{\partial^2\Phi}{\partial r^2}\sin^2\theta+2\frac{\partial^2\Phi}{\partial r\partial\theta}\frac{\sin\theta\cos\theta}{r}-2\frac{\partial\Phi}{\partial\theta}\frac{\sin\theta\cos\theta}{r^2}+\frac{\partial\Phi}{\partial r}\frac{\cos^2\theta}{r}+\frac{\partial^2\Phi}{\partial\theta^2}\frac{\cos^2\theta}{r^2} \tag{6-24}$$

$$\frac{\partial^2\Phi}{\partial x\partial y}=\frac{\partial^2\Phi}{\partial r^2}\sin\theta\cos\theta+\frac{\partial^2\Phi}{\partial r\partial\theta}\frac{1-2\sin^2\theta}{r}-\frac{\partial^2\Phi}{\partial\theta^2}\frac{\sin\theta\cos\theta}{r^2}-\frac{\partial\Phi}{\partial r}\frac{\sin\theta\cos\theta}{r}-\frac{\partial\Phi}{\partial\theta}\frac{1-2\sin^2\theta}{r^2} \tag{6-25}$$

Substituting Equations (6-23), (6-24) and (6-25) into Equation (6-22), we may form the following result after considerable rearrangement and cancellation of terms

$$\sigma_r=\frac{1}{r}\frac{\partial\Phi}{\partial r}+\frac{1}{r^2}\frac{\partial^2\Phi}{\partial\theta^2} \tag{6-26}$$

By similar procedures, we may get corresponding relations for σ_θ and $\tau_{r\theta}$. We now present this set of relations

$$\sigma_r=\frac{1}{r}\frac{\partial\Phi}{\partial r}+\frac{1}{r^2}\frac{\partial^2\Phi}{\partial\theta^2} \tag{6-27a}$$

$$\sigma_\theta = \frac{\partial^2 \Phi}{\partial r^2} \tag{6-27b}$$

$$\tau_{r\theta} = \frac{1}{r^2}\frac{\partial \Phi}{\partial \theta} - \frac{1}{r}\frac{\partial^2 \Phi}{\partial r \partial \theta} \tag{6-27c}$$

We may arrive at the compatibility equation in cylindrical coordinates by employing the Laplacian operator in cylindrical coordinates in the following manner

$$\mathbf{V}^4 \Phi = \mathbf{V}^2(\mathbf{V}^2 \Phi) = \left(\frac{\partial^2}{\partial r^2} + \frac{1}{r}\frac{\partial}{\partial r} + \frac{1}{r^2}\frac{\partial^2}{\partial \theta^2}\right)\left(\frac{\partial^2 \Phi}{\partial r^2} + \frac{1}{r}\frac{\partial \Phi}{\partial r} + \frac{1}{r^2}\frac{\partial^2 \Phi}{\partial \theta^2}\right) = 0 \tag{6-28}$$

We now have available in cylindrical coordinates to the basic formulations for the computation of stress in plane-strain problems where the body forces are zero.

6.4　Axially-symmetric Stress Distributions

We shall limit our discussion here to the case of axial symmetry where the z axis will be considered the axis of symmetry. This means that all derivatives with respect to θ are zero. For no body forces, the basic equations for this case reduce to the following forms

$$\sigma_r = \frac{1}{r}\frac{\partial \Phi}{\partial r} \tag{6-29a}$$

$$\sigma_\theta = \frac{\partial^2 \Phi}{\partial r^2} \tag{6-29b}$$

$$\tau_{r\theta} = 0 \tag{6-29c}$$

$$\mathbf{V}^4 \Phi = \left(\frac{\partial^2}{\partial r^2} + \frac{1}{r}\frac{\partial}{\partial r}\right)\left(\frac{\partial^2 \Phi}{\partial r^2} + \frac{1}{r}\frac{\partial \Phi}{\partial r}\right) = \frac{\mathrm{d}^4 \Phi}{\mathrm{d}r^4} + \frac{2}{r}\frac{\mathrm{d}^3 \Phi}{\mathrm{d}r^3} - \frac{1}{r^2}\frac{\mathrm{d}^2 \Phi}{\mathrm{d}r^2} + \frac{1}{r^3}\frac{\mathrm{d}\Phi}{\mathrm{d}r} = 0 \tag{6-29d}$$

Let us now consider the biharmonic differential equation (6-29d) for Φ. You may recall from your course in differential equations that this is the well-known Euler-Cauchy equations. This equation may be transformed to one with constant coefficients by making the following transformation of the independent variable

$$r = \mathrm{e}^t \tag{6-30}$$

We can then say

$$\frac{\mathrm{d}\Phi}{\mathrm{d}r} = \frac{\mathrm{d}\Phi}{\mathrm{d}t}\frac{\mathrm{d}t}{\mathrm{d}r} = \mathrm{e}^{-t}\frac{\mathrm{d}\Phi}{\mathrm{d}t}$$

$$\frac{\mathrm{d}^2 \Phi}{\mathrm{d}r^2} = \mathrm{e}^{-t}\frac{\mathrm{d}}{\mathrm{d}t}\left(\mathrm{e}^{-t}\frac{\mathrm{d}\Phi}{\mathrm{d}t}\right) = \mathrm{e}^{-t}\left(\mathrm{e}^{-t}\frac{\mathrm{d}^2 \Phi}{\mathrm{d}t^2} - \mathrm{e}^{-t}\frac{\mathrm{d}\Phi}{\mathrm{d}t}\right) = \mathrm{e}^{-2t}\left(\frac{\mathrm{d}^2 \Phi}{\mathrm{d}t^2} - \frac{\mathrm{d}\Phi}{\mathrm{d}t}\right)$$

By continuing in this way, we get

$$\frac{d^3\Phi}{dr^3}=e^{-3t}\left(\frac{d^3\Phi}{dt^3}-3\frac{d^2\Phi}{dt^2}+2\frac{d\Phi}{dt}\right)$$

$$\frac{d^4\Phi}{dr^4}=e^{-4t}\left(\frac{d^4\Phi}{dt^4}-6\frac{d^3\Phi}{dt^3}+11\frac{d^2\Phi}{dt^2}-6\frac{d\Phi}{dt}\right)$$

Now substitute the foregoing results into Equation (6-29). We get

$$e^{-4t}\left(\frac{d^4\Phi}{dt^4}-6\frac{d^3\Phi}{dt^3}+11\frac{d^2\Phi}{dt^2}-6\frac{d\Phi}{dt}\right)+2e^{-4t}\left(\frac{d^3\Phi}{dt^3}-3\frac{d^2\Phi}{dt^2}+2\frac{d\Phi}{dt}\right)-$$

$$e^{-4t}\left(\frac{d^2\Phi}{dt^2}-\frac{d\Phi}{dt}\right)+e^{-4t}\left(\frac{d\Phi}{dt}\right)=0$$

Canceling out e^{-4t} and collecting terms, we now get the following differential equation with constant coefficients

$$\frac{d^4\Phi}{dt^4}-4\frac{d^3\Phi}{dt^3}+4\frac{d^2\Phi}{dt^2}=0 \tag{6-31}$$

The auxiliary equation for Equation (6-31) is

$$p^4-4p^3+4p^2=0 \tag{6-32}$$

We can factor the preceding equation as follows

$$p^2(p-2)^2=0 \tag{6-33}$$

The general solution to the differential equation (6-31) is then

$$\Phi=C_1+C_2t+C_3e^{2t}+C_4te^{2t} \tag{6-34}$$

Where we have four arbitrary constants of integration. Replacing t, using Equation (6-30), we then have the general solution for Φ in the case of axial symmetry about the z axis. Thus

$$\Phi=C_1+C_2\ln r+C_3r^2+C_4r^2\ln r \tag{6-35}$$

Using Equation (6-29) the corresponding stress distribution for the foregoing function is then

$$\begin{cases} \sigma_r=\dfrac{C_2}{r^2}+2C_3+C_4(1+2\ln r) \\[2mm] \sigma_\theta=-\dfrac{C_2}{r^2}+2C_3+C_4(3+2\ln r) \\[2mm] \tau_{r\theta}=0 \end{cases} \tag{6-36}$$

We shall now examine the particular case of a thick-walled cylinder (Figure 6-1) under internal and external pressures, p_i and p_0, respectively. The inner and outer radii are denoted as r_i and r_o, respectively. The boundary conditions [Equation (6-17)] reduce for this problem to the following conditions

$$-p_i=(\sigma_r)_{r_i} \tag{6-37a}$$

$$-p_o=(\sigma_r)_{r_o} \tag{6-37b}$$

Imposing these conditions on the stress distribution given by Equation (6-36), we have the following equations

$$-p_i = \frac{C_2}{r_i^2} + 2C_3 + C_4(1 + 2\ln r_i) \tag{6-38a}$$

$$-p_o = \frac{C_2}{r_o^2} + 2C_3 + C_4(1 + 2\ln r_o) \tag{6-38b}$$

We have two equations for the evaluation of three constants. Theory as we have developed it this far thus permits an infinity of solution. Consideration of displacement requires, for a unique solution, that the constant C_4 must be zero. We can then solve for the constants C_2 and C_3. Thus

$$C_2 = \frac{r_i^2 r_o^2 (p_o - p_i)}{r_o^2 - r_i^2} \tag{6-39a}$$

$$2C_3 = \frac{p_i r_i^2 - p_o r_o^2}{r_o^2 - r_i^2} \tag{6-39b}$$

We then have, for the stress distribution

$$\sigma_r = \frac{r_i^2 r_o^2 (p_o - p_i)}{r_o^2 - r_i^2} \frac{1}{r^2} + \frac{p_i r_i^2 - p_o r_o^2}{r_o^2 - r_i^2} \tag{6-40a}$$

$$\sigma_\theta = -\frac{r_i^2 r_o^2 (p_o - p_i)}{r_o^2 - r_i^2} \frac{1}{r^2} + \frac{p_i r_i^2 - p_o r_o^2}{r_o^2 - r_i^2} \tag{6-40b}$$

If we add the foregoing pair of equations, we see that the sum of σ_r and σ_θ is a constant, i. e.

$$\sigma_r + \sigma_\theta = 2\frac{p_i r_i^2 - p_o r_o^2}{r_o^2 - r_i^2} \tag{6-41}$$

But from Hooke's law we have

$$\varepsilon_{zz} = 0 = \frac{1}{E}[\sigma_z - \nu(\sigma_r + \sigma_\theta)] \tag{6-42}$$

Using Equation (6-41), we see that σ_z must then be a constant for this case with the following value

$$\sigma_z = \nu(\sigma_r + \sigma_\theta) = \frac{2\nu(p_i r_i^2 - p_o r_o^2)}{r_o^2 - r_i^2} \tag{6-43}$$

This completes the distribution of stress for the constrained thick-walled cylinder under internal and external pressures. In next section, we shall see that the solution is also valid for the unconstrained thick-walled cylinder with internal and external pressures. To do this, we next turn to the theory of plane stress.

6.5 Discussion of Basic Equations

We define plane stress, you may recall from Chapter 2, as a distribution where

$$\tau_{xz} = \tau_{yz} = \sigma_z = 0 \tag{6-44}$$

We pointed out that the simplest physical problems for which the foregoing stress formulations are good approximations are plates loaded in the plane of symmetry of the plate as shown in Figure 6-2. To handle such problems, we shall further assume that the nonzero stresses do not vary with the z coordinate normal to the plate. Thus the stresses will be taken as functions of x and y only. We now write Newton's law for this state of stress

$$\frac{\partial \sigma_x}{\partial x} + \frac{\partial \tau_{xy}}{\partial y} + B_x = 0 \tag{6-45a}$$

$$\frac{\partial \tau_{yx}}{\partial x} + \frac{\partial \sigma_y}{\partial y} + B_y = 0 \tag{6-45b}$$

$$B_z = 0 \tag{6-45c}$$

If we again require, as we did earlier, that the body-force field be a conservative field, Equations (6-45) become

$$\frac{\partial \sigma_x}{\partial x} + \frac{\partial \tau_{xy}}{\partial y} - \frac{\partial V}{\partial x} = 0 \tag{6-46a}$$

$$\frac{\partial \tau_{yx}}{\partial x} + \frac{\partial \sigma_y}{\partial y} - \frac{\partial V}{\partial y} = 0 \tag{6-46b}$$

$$\frac{\partial V}{\partial z} = 0 \tag{6-46c}$$

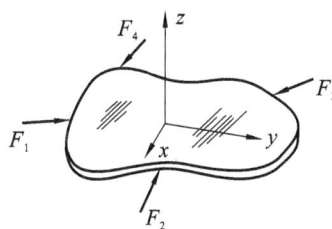

Figure 6-2 The plane loaded in the plane of symmetry of the plate

It is apparent from the Equation (6-46c) that V must be a function of only the coordinates x and y. Also, these equations are identically the same as the corresponding equations for plane strain problems and we can accordingly satisfy these equations by expressing the stresses in terms of the Airy's function. Thus, as before, we have

$$\begin{cases} \sigma_x = \dfrac{\partial^2 \Phi}{\partial y^2} + V \\[2mm] \sigma_y = \dfrac{\partial^2 \Phi}{\partial x^2} + V \\[2mm] \tau_{xy} = -\dfrac{\partial^2 \Phi}{\partial x \partial y} \end{cases} \tag{6-47}$$

We next turn to Hooke's law. We have, for the strains

$$
\begin{cases}
\varepsilon_{xx} = \dfrac{1}{E}(\sigma_x - \nu\sigma_y) \\[2mm]
\varepsilon_{yy} = \dfrac{1}{E}(\sigma_y - \nu\sigma_x) \\[2mm]
\varepsilon_{zz} = -\dfrac{\nu}{E}(\sigma_x + \sigma_y) \\[2mm]
\gamma_{xy} = \dfrac{1}{G}\tau_{xy} \quad \gamma_{xz} = 0 \quad \gamma_{yz} = 0
\end{cases}
\tag{6-48}
$$

We may now conclude that the strains also will not depend on the z coordinate. Now substitute from Equation (6-47) into Equation (6-48) to get the strains in terms of Φ and V. Thus

$$
\begin{cases}
\varepsilon_{xx} = \dfrac{1}{E}\left[\left(\dfrac{\partial^2\Phi}{\partial y^2} - \nu\dfrac{\partial^2\Phi}{\partial x^2}\right) + (1-\nu)V\right] \\[3mm]
\varepsilon_{yy} = \dfrac{1}{E}\left[\left(\dfrac{\partial^2\Phi}{\partial x^2} - \nu\dfrac{\partial^2\Phi}{\partial y^2}\right) + (1-\nu)V\right] \\[3mm]
\varepsilon_{zz} = -\dfrac{\nu}{E}\left[\left(\dfrac{\partial^2\Phi}{\partial x^2} + \dfrac{\partial^2\Phi}{\partial y^2}\right) + 2V\right] \\[3mm]
\gamma_{xy} = -\dfrac{1}{G}\dfrac{\partial^2\Phi}{\partial x\partial y} \\[3mm]
\gamma_{xz} = \gamma_{yz} = 0
\end{cases}
\tag{6-49}
$$

Let us now turn to the compatibility Equations (3-97). Examine the first of these. Substituting the preceding equations we get

$$
\frac{1}{E}\left[\frac{\partial^4\Phi}{\partial y^4} - \nu\frac{\partial^4\Phi}{\partial y^2\partial x^2} + (1-\nu)\frac{\partial^2 V}{\partial y^2} + \frac{\partial^4\Phi}{\partial x^4} - \nu\frac{\partial^4\Phi}{\partial x^2\partial y^2} + (1-\nu)\frac{\partial^2 V}{\partial x^2}\right] = -\frac{1}{G}\frac{\partial^4\Phi}{\partial y^2\partial x^2}
$$

Multiply through by E, replace E/G by $2(1+\nu)$ in accordance with Equation (6-8), and collect terms. We then get

$$
\frac{\partial^4\Phi}{\partial x^4} + 2\frac{\partial^4\Phi}{\partial y^2\partial x^2} + \frac{\partial^4\Phi}{\partial y^4} = -(1-\nu)\mathbf{\nabla}^2 V
\tag{6-50}
$$

However, not all the remaining compatibility equations are identically satisfied as was the case for plane strain. Thus Equations (b),(c) and (f) of the set of compatibility Equations (3-97), require that

$$
\frac{\partial^4\Phi}{\partial y^4} + \frac{\partial^4\Phi}{\partial y^2\partial x^2} + 2\frac{\partial^2 V}{\partial y^2} = 0
\tag{6-51a}
$$

$$
\frac{\partial^4\Phi}{\partial x^4} + \frac{\partial^4\Phi}{\partial x^2\partial y^2} + 2\frac{\partial^2 V}{\partial x^2} = 0
\tag{6-51b}
$$

$$\frac{\partial^4 \Phi}{\partial y \partial x^3} + \frac{\partial^4 \Phi}{\partial x \partial y^3} + 2\frac{\partial^2 V}{\partial x \partial y} = 0 \qquad (6\text{-}51c)$$

It may be concluded that the plane-stress problem is more difficult than the plane-strain problem. It is shown however, that if we disregard the compatibility requirements on Φ given by Equations (6-51) and consider only the compatibility requirements given by Equation (6-50), we get good approximations of stress distributions for thin plates. We can then conclude that the following equations

$$\sigma_x = \frac{\partial^2 \Phi}{\partial y^2} + V \qquad (6\text{-}52a)$$

$$\sigma_y = \frac{\partial^2 \Phi}{\partial x^2} + V \qquad (6\text{-}52b)$$

$$\tau_{xy} = -\frac{\partial^2 \Phi}{\partial x \partial y} \qquad (6\text{-}52c)$$

$$\nabla^4 \Phi = -(1-\nu)\nabla^2 V \qquad (6\text{-}52d)$$

Note that for no body forces, the equations of plane strain and plane stress become identical, with Φ in each case satisfying the biharmonic equation. Thus, for no body forces a solution to these equations leads to an exact solution of some problem in plane strain and an approximate solution of some problem in plane stress. We shall make use of this fact later. Finally, the results established in the aforementioned cylindrical coordinates in connection with plane strain are applicable for plane stress for the case of no body forces.

6.6   Plane-stress Solutions:Plate with a Hole

We have already presented a solution to the preceding set of equations expressed in cylindrical coordinates for the case of plane strain. The stress distribution, Equation (6-40), for the case studied corresponded to a thick-walled cylinder completely constrained in the direction of the axis of the cylinder. The plane strain requirement of having $\varepsilon_{zz} = 0$ was used to computer σ_z, which turned out to be a constant for the cylinder.

From our previous comments, the aforementioned solution given by Equation (6-40) is a good approximation for a particular plane-stress problem. There can be little question, after considering the boundary conditions, that the corresponding plane-stress problem is a thin disk with a concentric hole subjected to pressures p_o and p_i on the inside and outside peripheral surfaces, respectively. In short, we can say that our plane-stress problem is simply a thin slice of the thick-walled cylinder, completely uncon-strained in the direction normal to the slice and subject to pressures on the edges. For the thick-walled cylinder constrained at the ends, you will recall that $\varepsilon_{zz}=0$ and σ_z was a constant. In the case of a thin disk with a hole, we have $\sigma_z=0$ and ε_{zz} equal to a constant given according to Hooke's law as

$$\varepsilon_{zz}=\frac{1}{E}\left[-\nu(\sigma_r+\sigma_\theta)\right] \tag{6-53}$$

Substituting for σ_r and σ_θ from Equations (6-40), we get the following result

$$\varepsilon_{zz}=\frac{2\nu p_o r_o^2-p_i r_i^2}{E}\frac{}{r_o^2-r_i^2} \tag{6-54}$$

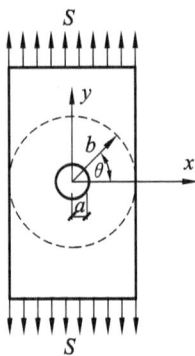

Figure 6-3   A thin plate under a tensile load

We now turn to a second related problem. Consider a thin plate under a tensile load, as shown in Figure 6-3. A small hole is at the center of this plate. We wish to compute the stress distribution for this geometry. For simplicity, we assume that the tensile load is applied as a uniform stress S at the end of the plate as is shown in the diagram.

If there were no hole in the plate, we would have a uniform stress field $\sigma_y=S, \sigma_x=\tau_{yx}=0$. The presence of the hole will cause a non-uniform stress distribution near the hole, but because of St. Venant's principle, the stress far from the hole should approach the fore-tasted uniform value. With this in mind, we consider the domain inside the hypothetical large circle of radius b shown dotted in Figure 6-3. We shall employ polar coordinates. To express σ_r and $\tau_{r\theta}$ at $r=b$ we may employ the transformation formulas for plane stress given by Equations (2-26). Thus considering the r direction to correspond to x and the θ direction to correspond to y we have

$$(\sigma_r)_{r=b}=\frac{S}{2}(1-\cos2\theta) \tag{6-55a}$$

$$(\tau_{r\theta})_{r=b}=\frac{S}{2}\sin2\theta \tag{6-55b}$$

The boundary conditions just presented can be profitably considered as being composed of two parts. First, there is uniform radial stress equal to $S/2$. For this part, we can think of the domain between the concentric circles as a slice of an unconstrained thick-walled cylinder having an inner pressure $p_i=0$ and outer pressure $p_o=-S/2$. We can accor-dingly use the solution for thick-walled cylinders as described earlier to take care of this boundary condition. The remaining stresses on the boundary give rise to the problem of a thin hollow disk having on the outside edge a variable radial stress $-(S/2)\cos2\theta$, and a variable shear stress, $(S/2)\sin2\theta$, whereas having on the inside edge zero radial and shear stresses. These two problems have been illustrated in Figure 6-4. The solution for the first problem can immediately be gotten from our previous results. Thus, from Equations (6-40) we have

$$\sigma'_r = -\frac{a^2b^2\frac{S}{2}}{b^2-a^2}\frac{1}{r^2} + \frac{\frac{S}{2}b^2}{b^2-a^2} \tag{6-56a}$$

$$\sigma'_\theta = \frac{a^2b^2\frac{S}{2}}{b^2-a^2}\frac{1}{r^2} + \frac{\frac{S}{2}b^2}{b^2-a^2} \tag{6-56b}$$

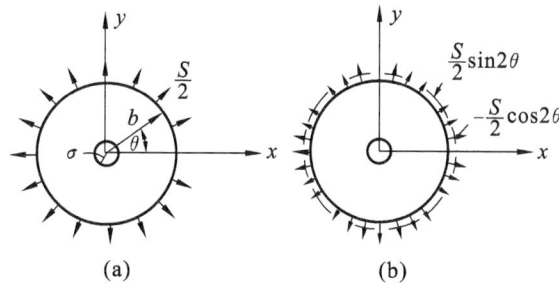

Figure 6-4   The enlarged of the domain inside the hypothetical large circle of radius b

Now let us turn to the second problem. First we shall assume the Airy's function for this problem to be of the following form

$$\Phi = [f(r)]\cos2\theta \tag{6-57}$$

where $f(r)$ is a function of r which is to be determined. The compatibility equation to be used for this case is Equation (6-28). Substituting the preceding form for Φ into this equation, we get the following equation

$$\left(\frac{\partial^2}{\partial r^2}+\frac{1}{r}\frac{\partial}{\partial r}+\frac{1}{r^2}\frac{\partial}{\partial\theta^2}\right)\cos2\theta\left(\frac{\partial^2 f}{\partial r^2}+\frac{1}{r}\frac{\partial f}{\partial r}-\frac{4}{r^2}f\right)=0 \tag{6-58}$$

Carrying out further differentiation in Equation (6-58) and collecting terms we get

$$\frac{d^4 f}{dr^4}+\frac{2}{r}\frac{d^3 f}{dr^3}-\frac{9}{r^2}\frac{d^2 f}{dr^2}+\frac{9}{r^3}\frac{df}{dr}=0 \tag{6-59}$$

We again have a Euler-Cauchy equation. The procedure to follow now is exactly the procedure taken in the beginning of Section 6.4. We therefore leave it as an exercise for you to demonstrate that the following is the general solution to the preceding equation

$$f(r)=C_1+C_2r^2+C_3r^4+\frac{C_4}{r^2} \tag{6-60}$$

The stress function is then

$$\Phi = \left(C_1 + C_2 r^2 + C_3 r^4 + \frac{C_4}{r^2} \right) \cos 2\theta \qquad (6\text{-}61)$$

Now substitute the foregoing value for Φ into Equation (6-27) to determine the corresponding stresses. We get, on collecting terms

$$\begin{cases} \sigma_r = -\left(2C_2 + \frac{4C_1}{r^2} + \frac{6C_4}{r^4} \right) \cos 2\theta \\[2mm] \sigma_\theta = \left(2C_2 + 12C_3 r^2 + \frac{6C_4}{r^4} \right) \cos 2\theta \\[2mm] \tau_{r\theta} = \left(2C_2 + 6C_3 r^2 - \frac{2C_1}{r^2} - \frac{6C_4}{r^4} \right) \sin 2\theta \end{cases} \qquad (6\text{-}62)$$

We now subject this stress distribution to the boundary conditions of this problem. Thus when $r = a$

$$\sigma_r = \tau_{r\theta} = 0$$

when $r = b$

$$\begin{cases} \sigma_r = -\frac{S}{2} \cos 2\theta \\[2mm] \tau_{r\theta} = \frac{S}{2} \sin 2\theta \end{cases} \qquad (6\text{-}63)$$

We then have

$$\frac{4}{a^2} C_1 + 2C_2 + 0 + \frac{6}{a^4} C_4 = 0 \qquad (6\text{-}64\text{a})$$

$$\frac{4}{b^2} C_1 + 2C_2 + 0 + \frac{6}{b^4} C_4 = \frac{S}{2} \qquad (6\text{-}64\text{b})$$

$$-\frac{2}{a^2} C_1 + 2C_2 + 6a^2 C_3 - \frac{6}{a^4} C_4 = 0 \qquad (6\text{-}64\text{c})$$

$$-\frac{2}{b^2} C_1 + 2C_2 + 6b^2 C_3 - \frac{6}{b^4} C_4 = \frac{S}{2} \qquad (6\text{-}64\text{d})$$

We may solve for the four constants of integration by purely algebraic means to get the following results

$$\left\{ \begin{aligned}
C_1 &= \cfrac{72S\left(\cfrac{a^2}{b^2}-\cfrac{b^2}{a^4}\right)}{-\cfrac{576}{b^4}+\cfrac{864}{a^2b^2}+144\,\cfrac{a^2}{b^6}+144\,\cfrac{b^2}{a^6}-\cfrac{576}{a^4}} \\[2em]
C_2 &= \cfrac{36S\left(\cfrac{4}{b^4}+\cfrac{3}{a^2b^2}+\cfrac{b^2}{a^6}\right)}{-\cfrac{576}{b^4}+\cfrac{864}{a^2b^2}+144\,\cfrac{a^2}{b^6}+144\,\cfrac{b^2}{a^6}-\cfrac{576}{a^4}} \\[2em]
C_3 &= \cfrac{24S\left(\cfrac{1}{a^6}-\cfrac{1}{b^2a^4}\right)}{-\cfrac{576}{b^4}+\cfrac{864}{a^2b^2}+144\,\cfrac{a^2}{b^6}+144\,\cfrac{b^2}{a^6}-\cfrac{576}{a^4}} \\[2em]
C_4 &= \cfrac{36S\left(\cfrac{b^2}{a^2}-\cfrac{a^2}{b^2}\right)}{-\cfrac{576}{b^4}+\cfrac{864}{a^2b^2}+144\,\cfrac{a^2}{b^6}+144\,\cfrac{b^2}{a^6}-\cfrac{576}{a^4}}
\end{aligned} \right. \tag{6-65}$$

Next by considering the ratio a/b to be zero, that is, by considering the radius b to become infinite, we then have for the constants of integration

$$C_1 = -\frac{a^2}{2}S \quad C_2 = \frac{S}{4} \quad C_3 = 0 \quad C_4 = \frac{a^4}{4}S \tag{6-66}$$

The solution to the second subsidiary problem is then

$$\left\{ \begin{aligned}
\sigma_r'' &= \frac{S}{2}\left(-1+4a^2\,\frac{1}{r^2}-3a^4\,\frac{1}{r^4}\right)\cos 2\theta \\[1em]
\sigma_\theta'' &= \frac{S}{2}\left(1+3a^4\,\frac{1}{r^4}\right)\cos 2\theta \\[1em]
\tau_{r\theta}'' &= \frac{S}{2}\left(1+2a^2\,\frac{1}{r^2}-3a^4\,\frac{1}{r^4}\right)\sin 2\theta
\end{aligned} \right. \tag{6-67}$$

We can now combine our subsidiary solutions to give the solution to the problem at hand. We must first adjust Equations (6-56) to reflect the fact that radius $b \rightarrow \infty$ in the formulations. Thus dividing through by b^2 and letting $a/b=0$ we get

$$\left\{ \begin{aligned}
\sigma_r' &= \frac{S}{2}\left(1-\frac{a^2}{r^2}\right) \\[1em]
\sigma_\theta' &= \frac{S}{2}\left(1+\frac{a^2}{r^2}\right) \\[1em]
\tau_{r\theta}' &= 0
\end{aligned} \right. \tag{6-68}$$

The total solution then becomes

$$\begin{cases} \sigma_r = \sigma_r' + \sigma_r'' = \frac{S}{2}\left[\left(1 - \frac{a^2}{r^2}\right) + \left(-1 + 4\frac{a^2}{r^2} - 3\frac{a^4}{r^4}\right)\cos2\theta\right] \\ \sigma_\theta = \sigma_\theta' + \sigma_\theta'' = \frac{S}{2}\left[\left(1 + \frac{a^2}{r^2}\right) + \left(1 + 3\frac{a^4}{r^4}\right)\cos2\theta\right] \\ \tau_{r\theta} = \tau_{r\theta}' + \tau_{r\theta}'' = \frac{S}{2}\left(1 + 2\frac{a^2}{r^2} - 3\frac{a^4}{r^4}\right)\sin2\theta \end{cases} \tag{6-69}$$

Let us now examine the foregoing stress distribution in regions far from the hole and at the hole itself. Note, as we get far from the hole, we can drop terms in the preceding equations having r in the denominator. We thus approach the following state of stress as r gets large

$$\begin{cases} \sigma_r = \frac{S}{2}(1 - \cos2\theta) \\ \sigma_\theta = \frac{S}{2}(1 + \cos2\theta) \\ \tau_{r\theta} = \frac{S}{2}\sin2\theta \end{cases} \tag{6-70}$$

This corresponds, you will recall, to the uniform stress field $\sigma_x = 0$, $\sigma_y = S$ and $\tau_{xy} = 0$. Next, let us consider the stress at the hole. Thus setting $r = a$ in Equations (6-69) we get

$$\begin{cases} (\sigma_r)_{r=a} = 0 \\ (\sigma_\theta)_{r=a} = S + 2S\cos2\theta \\ (\tau_{r\theta})_{r=a} = 0 \end{cases} \tag{6-71}$$

The radial and transverse stresses are clearly principal stresses over the entire periphery of the hole. The maximum normal stress then occurs when $\theta = 0$. We see that this gives a stress $\sigma_x = 3S$. Here is a vivid example of the stress-concentration danger in our discussion of fatigue. A small hole, probably quite harmless in appearance to the layman, causes a stress three times greater than the largest stress with no hole. Clearly great care must be taken for such situations, as we have pointed out in our earlier discussions.

In this problem, we have been able to compute the stress concentration factor K as equal to 3. Thus the stress $\sigma_{\max}$ is expressible as

$$\sigma_{\max} = K\sigma = 3S \tag{6-72}$$

where σ is the maximum stress without the hole.

You may find lists of concentration factors in handbooks for other common situations. Some have been computed from the theory, but most have been found by means of experimental stress analysis.

6.7   The Case of the Curved Beam

In the problems at the end of this chapter, you will have the opportunity to use the methods given in it to solve several straight, rectangular beam problems. We shall here consider the curved beam having upper and lower edges as concentric circular arcs and loaded at the ends by pure couples as has been shown in Figure 6-5. The beam is of rectangular cross section having a thickness t which is small compared to the other dimensions of the beam.

By considering free bodies of portions of the curved beam in the shape of radial segments, as is shown in Figure 6-6, we can conclude from equilibrium considerations that the resultants of the force distributions on the exposed internal surfaces must be pure couples with a magnitude M. Thus the resultants at internal sections are not a function of θ. And furthermore, it can be assumed that the stresses themselves on these exposed faces should not vary with θ. We thus have here a plane-stress problem with axial symmetry about O. Since we have zero body force, we may utilize the general solution to the biharmonic equation for the case of axial symmetry which was developed in Section 6.4. Hence, the stress distribution given by Equations (6-36) are to be subject to the boundary conditions of this problem; namely, at

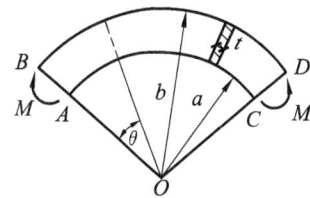

Figure 6-5   The curved beam for concentric circular arcs and loaded at the ends by pure couples

$$r=a \quad r=b \quad \sigma_r=0 \tag{6-73a}$$

$$r=a \quad r=b \quad \tau_{r\theta}=0 \tag{6-73b}$$

and the end surfaces

$$\int_A \sigma_\theta dA = 0 \tag{6-73c}$$

$$\int_A r\sigma_\theta dA = -M \tag{6-73d}$$

We then have, for conditions given by Equation (6-73a)

$$\frac{C_2}{a^2}+2C_3+C_4(1+2\ln a)=0 \tag{6-74a}$$

$$\frac{C_2}{b^2}+2C_3+C_4(1+2\ln b)=0 \tag{6-74b}$$

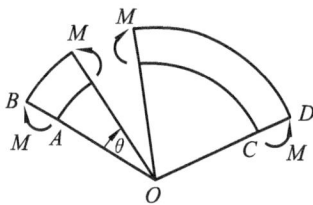

Figure 6-6   The free bodies of portions of the curved beam

Conditions given by Equation (6-73b) are identically satisfied since $\tau_{r\theta} = 0$ everywhere. Next we examine condition given by Equation (6-73c). We may state using Equation (6-27b)

$$\int_A \sigma_\theta dA = t \int_a^b \frac{d^2 \Phi}{dr^2} dr = 0$$

Canceling t, we have for the preceding equation

$$\int_a^b \frac{d^2 \Phi}{dr^2} dr = \frac{d\Phi}{dr}\Big|_a^b = b\left[\frac{C_2}{b^2} + 2C_3 + C_4(1 + 2\ln b)\right] - a\left[\frac{C_2}{a^2} + 2C_3 + C_4(1 + 2\ln a)\right] = 0 \tag{6-75}$$

But this equation adds nothing new since it will be satisfied if Equations (6-74) are satisfied. We now go to the last of our boundary condition given by Equation (6-73d). Using Equation (6-27b) again, we have

$$\int_A r\sigma_\theta dA = t \int_a^b r \frac{d^2 \Phi}{dr^2} dr = -M \tag{6-76}$$

Dividing through by t and integrating by parts, we get

$$\int_a^b r \frac{d^2 \Phi}{dr^2} dr = \int_a^b r d\left(\frac{d\Phi}{dr}\right) = r \frac{d\Phi}{dr}\Big|_a^b - \int_a^b \frac{d\Phi}{dr} dr = -\frac{M}{t} \tag{6-77}$$

Using Equations (6-74) and (6-75) we know that $d\Phi/dr$ is zero at the limits, so the foregoing formulation becomes

$$\int_a^b \frac{d\Phi}{dr} dr = \Phi\Big|_a^b = \frac{M}{t} \tag{6-78}$$

We then have using Equation (6-35)

$$C_2 \ln\left(\frac{b}{a}\right) + C_3(b^2 - a^2) + C_4(b^2 \ln b - a^2 \ln a) = \frac{M}{t} \tag{6-79}$$

We now solve for the constants C_2, C_3 and C_4 using Equations (6-74a), (6-74b) and (6-79). We get the following result

$$C_2 = -\frac{4M}{St} a^2 b^2 \ln \frac{b}{a} \tag{6-80a}$$

$$C_3 = \frac{M}{St}[b^2 - a^2 + 2(b^2 \ln b - a^2 \ln a)] \tag{6-80b}$$

$$C_4 = -\frac{2M}{St}(b^2 - a^2) \tag{6-80c}$$

where

$$S = (b^2 - a^2)^2 - 4a^2 b^2 \left(\ln \frac{b}{a}\right)^2 \tag{6-80d}$$

Substituting the foregoing values for the integration constants, we then get the following formulation for the state of stress at a point

$$\begin{cases} \sigma_r = -\frac{4M}{St}\Big(\frac{a^2b^2}{r^2}\ln\frac{b}{a}+b^2\ln\frac{r}{b}+a^2\ln\frac{a}{r}\Big) \\[2mm] \sigma_\theta = -\frac{4M}{St}\Big(-\frac{a^2b^2}{r^2}\ln\frac{b}{a}+b^2\ln\frac{r}{b}+a^2\ln\frac{a}{r}+b^2-a^2\Big) \\[2mm] \tau_{r\theta}=0 \end{cases} \qquad (6\text{-}81)$$

The foregoing stress distribution can be considered correct for the entire beam if the applied torques have the same distribution of stress corresponding to the solution. If the applied torques do not have this distribution (as will usually be the case), the results we have developed are valid in regions away from the ends of the beam in accordance with St. Venant's principle.

In Figure 6-7, we have plotted the stresses at a section. Notice that the maximum stress occurs at the lower fibers. We have also included the stresses that we can compute by approximate methods of strength of materials to be considered in later

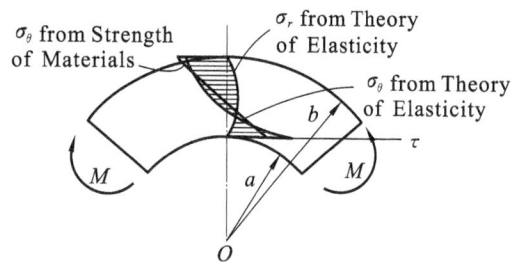

Figure 6-7   The stresses of the curved beam

chapters. Notice the more exact analysis indicates a larger stress than that given from the simpler formulations of strength of materials. Considerable error can be expected from strength of materials here if $(b-a)$ is not small compared to a. This is an example of how the more exact theory of elasticity serves to give the limits and degree of accuracy of the more simple, handy formulations of strength of materials.

6.8   Closure

In this chapter, we have been primarily concerned with the theory of elasticity. In Chapter 5, we examined a three-dimensional problem using the theory and in this chapter we have applied the theory to the plane-strain and plane-stress problems. Many other interesting problems can be examined using the theory of elasticity.

This brief introduction to the solution of problems via the theory of elasticity indicates that such procedures are not simple by any means. Because we often work with relatively simple body shapes in much of our structural work there are more available simple procedures, the totality of which we refer to as strength of materials.

By reviewing the details of strength of materials, we can note the relation between the theory of elasticity and strength of materials. The following pertinent remarks may help us keep a proper perspective:

(1) In strength of materials we make certain simple assumptions concerning the deformation of classes of bodies,and using some of the laws of the theory of elasticity,we arrive at useful working formulas serves as a check on these assumptions. The theory of elasticity gives us insight on the range of validity and the general accuracy of the approximate formulation of strength of materials.

(2) The formulation of strength of materials has a limited range of applications. When problems fall outside this limited range,we must resort to the full theory. Quite often the use of numerical methods and computers is then required. Thus the concentration factor K used in the oversimplified strength of materials computations are computed sometimes from the more general theory as was illustrated in this chapter for the tension plate with a small hole.

(3) There is a large range of experimental methods which are used for the evaluation of stress and strain in complicated geometries. These include photoelastic methods, strain-gauge methods,brittle-lacquer methods,and a variety of useful analog techniques. To be able to employ these sophisticated methods effectively for other than trivial problems requires a knowledge of the theory of elasticity.

Problems

6-1   Consider a function Φ given as the product of two functions f and g. Thus

$$\Phi = fg$$

Show that

$$\mathbf{V}^2\Phi = (\mathbf{V}^2 f)g + 2\left(\frac{\partial f}{\partial x}\frac{\partial g}{\partial x} + \frac{\partial f}{\partial y}\frac{\partial g}{\partial y}\right) + (\mathbf{V}^2 g)f$$

Suppose g is a harmonic function and $f = x$. Show that $\mathbf{V}^4\Phi = 0$; that is,the product of x times a harmonic function gives a biharmonic function.

6-2   In Problem 6-1 take the function $f = r^2 = x^2 + y^2 + z^2$.

(1)Show that

$$\mathbf{V}^2\Phi = 6g + 4\left(x\frac{\partial g}{\partial x} + y\frac{\partial g}{\partial y} + z\frac{\partial g}{\partial z}\right)$$

(2)Show that $\mathbf{V}^4\Phi = 0$.

That is,the product $(r^2)(g)$ with g a harmonic function is a biharmonic function.

6-3   Consider a thick-walled cylinder with an inner radius of 300 mm and an outer radius of 450 mm. A pressure of 140 Pa is maintained inside the cylinder. If the cylinder is completely constrained along its axis, what is the maximum normal stress? What is the maximum shear stress?

6-4 Consider a cylinder with an outside diameter of 1500 mm. Call the inner diameter d_i. If a pressure of 350 Pa is maintained in the cylinder, what is the inside diameter d_i at which thin-wall cylinder theory gives a maximum stress within 10 percent of the correct maximum stress?

6-5 Show that Equation (6-60) forms a general solution to the differential equation given by Equation (6-59).

6-6 Show that the constants C_1, C_2, C_3 and C_4 in Equations (6-65) simplify to those given by Equations (6-66) when b is considered infinite.

6-7 Show that the stress concentration for a small hole in a plate loaded as shown in Figure 6-8 is 4.

6-8 Consider a curved beam, such as is shown in Figure 6-9, for which

$$a = 1200 \text{ mm}$$
$$b = 1500 \text{ mm}$$
$$M = 1.40 \text{ kN} \cdot \text{m}$$
$$t = 90 \text{ mm}$$

Determine the maximum stress arising from the action of the couples. Compare this result with the maximum stress from the flexure formula to be studied in material strength. This formula is

$$\sigma_\theta = \frac{My}{I}$$

where y is the distance from the centroidal axis at any section and I is the second moment of inertia of the cross section about this axis.

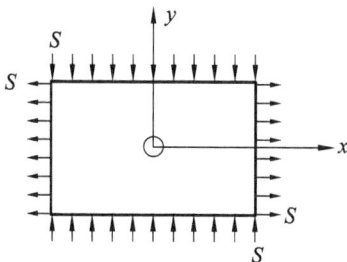

Figure 6-8   Figure of Problem 6-7

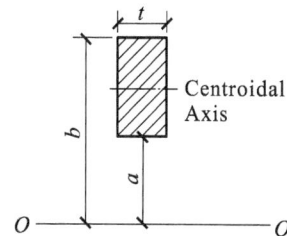

Figure 6-9   Figure of Problem 6-8

6-9 Given the polynomial

$$\Phi = C_1 x^2 + C_2 xy + C_3 y^2$$

If used as an Airy's function, what boundary problem may it be considered a solution of in plane stress and in plane strain?

6-10   Using the following polynomial as an Airy's function

$$\Phi = C_1 x^3 + C_2 x^2 y + C_3 x y^2 + C_4 y^3$$

evaluate the constants so that you have a solution to the beam under pure bending as shown in Figure 6-10.

6-11   Using the polynomial given in Problem 6-10, find other boundary-value problems for which you have a solution.

6-12   Given the following polynomial

$$\Phi = C_1 x^4 + C_2 x y + C_3 x y^3 + C_4 y^4$$

Make any adjustments in the coefficient so that the function is a biharmonic function. Now adjust the constants so that you have a solution for the cantilever-beam problem shown in Figure 6-11. Where uniform shear-stress tractions are applied at the upper and lower edges of the beam and a load P is applied at the tip.

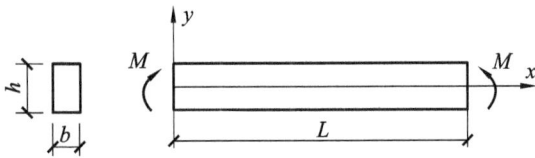

Figure 6-10   Figure of Problem 6-10   Figure 6-11   Figure of Problem 6-12

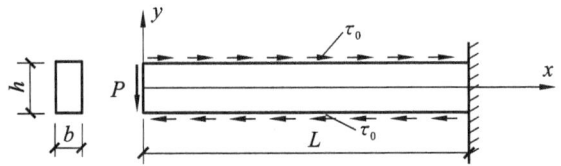

7 The Theory of Plasticity

7.1 Introduction

Two basic factors govern the developments on elasticity. One implied complete reversibility of the loading process; thus, when the forces that caused the body to be strained were removed, complete recovery to the initially unstrained state occurred immediately. The other factor implied that deformation or strain of the body under load was dependent only upon the end stresses and not upon the stress history or strain path. Elastic behavior may be viewed as a point function since any induced strain can be determined from the initial and final stresses and particular proportionality constants. It's the same that two factors are not evident when plastic or permanent deformations are induced.

To produce plastic deformation or flow, a certain level of stress, henceforth called the yield stress, must be exceeded. With many solids (e. g. ductile metals) such deformation or shape change can continue to a large degree if the yield stress is greatly exceeded; in addition, as the final deformation is produced, a particular strain element could undergo different histories prior to the body reaching its end state. Thus, not only is complete reversibility not found upon load removal, as in elasticity, but the final strain is found to depend upon the history of loading rather than the beginning and final stresses alone. Such a finding means that plastic behavior is a path function and requires the use of incremental strains summed up over the strain path whenever the total induced strain is to be determined.

There are certainly at least three fairly distinct approaches that have been taken in the study of plasticity; these are:

(1) The mathematical approach which employs idealized models of material behavior and is concerned primarily with the stress and strain distributions that satisfy prescribed boundary conditions. This might properly be called the macroscopic theory of plasticity and is most analogous to the longer standing topic known as the theory of elasticity.

（2）The approach utilized in metal physics. Here, the manner of deformation in single crystals of real solids forms the basis of study with one objective being to relate and extend the basic behavior of single crystals to polycrystalline aggregates that make up the solids usually employed by engineers. This might properly be called the microscopic theory of plasticity.

（3）The technological approach, which attempts to unite experimentally observed behavior of real solids on a macroscopic scale with mathematical expressions by invoking certain phenomenological rules. This enables useful predictions to be made in the general area of design; it might properly be called macroscopic engineering plasticity. It is this approach that will receive major attention in this chapter.

7.2 Comparison of Elasticity and Plasticity

For convenience, many of the above comments are summarized in tabular form in some texts. In this way a direct comparison, indicating the major differences in these two types of behavior, can be seen.

Since the onset of yielding and the behavior that might follow is of primary concern, various models will be used in order to illustrate the physical processes involved. With any of the models presented below, several assumptions are invoked:

（1）The solid is isotropic and homogeneous.

（2）The onset of yielding in tension and compression is identical. This implies that there is no Bauschinger effect.

（3）Volume changes are negligible. Thus, the dilatation is zero and Poisson's ratio is one-half. Although this ratio is an elastic constant, no confusion should result by extending this meaning to plastic deformation.

（4）The magnitude of the mean normal stress or hydrostatic component of the stress state does not influence yielding.

（5）Effects of strain rate are negligible.

（6）Temperature effects are not considered.

Note that assumptions (3) and (4) have been reasonably substantiated by experiments with many ductile metals. This is not the case with many solid polymers.

7.3   Models for Plastic Deformation

7.3.1   A Rigid Perfectly Plastic Solid Model

Rigid perfectly plastic behavior has been widely used in many analytical studies. It implies that no deformation occurs until a certain level of stress is reached (i. e. , E is infinite),then deformation proceeds indefinitely as long as the necessary flow stress is applied. Note that nothing is implied in regard to potential fracture of the solid. A satisfactory model is shown in Figure 7-1.

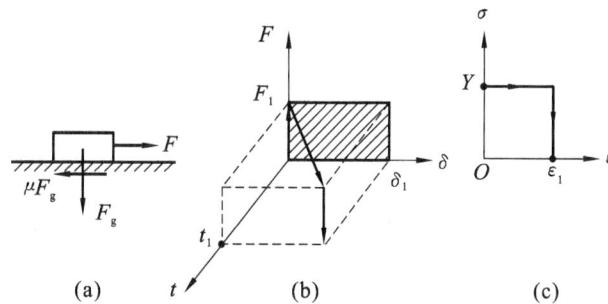

Figure 7-1   A satisfactory model

(a) Description of a rigid perfectly plastic solid showing the model;

(b) the force-displacement-time plot;(c) the stress-strain plot

Note the following:

(1) As the applied load F is increased,no displacement occurs until some critical force F_1 is reached. Once this happens,deformation proceeds continuously with time. The force F_1 is related directly to the yield or flow stress Y.

(2) Upon removal of load F_1,there is no recovery of plastic work (shown by the shaded area in the F-δ plane). Rather,a permanent deformation given by δ_1 remains.

(3) The solid does not become stronger during deformation. This implies there is no strain-hardening effect.

7.3.2   A Rigid Linear Strain-hardening Plastic Solid Model

A rigid linear strain-hardening plastic solid model is somewhat more realistic than the previous model since it incorporates the influence of strain-hardening observed in many solids, especially ductile metals. Again, a certain critical stress level must be reached before plastic deformation commences,but continued deformation demands an increasing applied stress. This is shown in Figure 7-2. The following effects are noted:

(1) As F is applied, displacement begins only when a critical force F_0 is reached and produces the initial flow stress Y_0.

(2) Displacement continues only under an increase in applied stress Y where $Y = Y_0 + f(\varepsilon)$ and $f(\varepsilon)$ is related to the slope of the line. Its similarity to the modulus E should be noted. In this model, strain hardening occurs and implies that plastic deformation causes an increase in the stress required for further deformation.

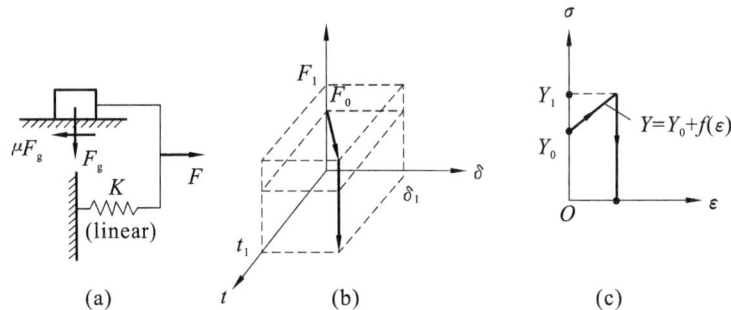

Figure 7-2 Description of a rigid linear strain-hardening solid model with plots as described in Figure 7-1

7.3.3 A Rigid Nonlinear Strain-hardening Plastic Solid Model

A rigid strain-hardening plastic solid model where such hardening follows a power law form of behavior provides an even better description for many solids. Figure 7-3 portrays this model.

Here it is noted that: the behavior is identical with the previous model except that strain hardening occurs at a nonlinear rate, the exponent n being greater than zero but less than unity.

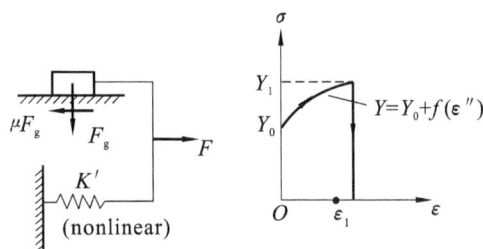

Figure 7-3 Description of a rigid strain-hardening solid following a power law hardening behavior

Finally, elastic effects can be included in any of the three models by adding on a straight-line section in the initial stages of deformation where the slope would indicate an elastic modulus of an appropriate value that is less than infinity. Because many situations of concern involve plastic strains that are orders of magnitude greater than the elastic strains, it is convenient to ignore the latter. In so doing, three facts should be realized:

(1) Volume changes can only be determined by including elastic effects where ν is

less than one-half. By ignoring such effects, the concept of volume constancy may be introduced.

(2) Recovery upon loading generally does occur and is tied in with elastic recovery. Thus, if such a result is of immediate concern, the above models would not describe such behavior. Note also that in situations involving the effects of elastic recovery, continually increasing elastic strains accompany an ever-increasing plastic flow.

(3) If elastic and plastic strains are of the same order of magnitude, the above models would not be useful unless the elastic portion were included as mentioned above.

7.4 The Yield Locus and Surface

With the assumptions of isotropy, no Bauschinger effect, incompressibility during plastic flow and yielding being uninfluenced by hydrostatic effects, several inherent conditions must prevail in any criterion that is to be used to predict the onset of yielding.

A plot of two-dimensional stress space is introduced to indicate some of the results of the above assumptions. Here it is envisioned that the individual stresses may be treated as components of the total stress and are handled as vectors for this purpose. This has nothing to do with transformations to new axes and must be so understood. This discussion is also restricted to the use of principal stresses where in all case, one of these stresses is zero. A plot in σ_1-σ_2 space will be used. Suppose a tensile stress is applied in one direction and $0 < \sigma_1 < Y$ describes elastic behavior only. With equivalence in tension and compression, the elastic range is extended to $-Y < \sigma_1 < Y$ and because of isotropy, $-Y < \sigma_2 < Y$. Thus, there exist four points in σ_1-σ_2 stress space that indicate the onset of yielding. To develop an acceptable theory of yielding more complex stress states must be included. This requires a generalization of what is meant by the elastic range and yield point and involves the use of certain stress limits. Figure 7-4 shows how this is started.

The four points shown at $\pm Y$ fall on a yield locus in this two-dimensional stress space. Suppose now that the material is stressed to point A as shown and that stress is maintained while a stress σ_2 is added. At some point, such as B, elastic behavior ends and we refer to B as a yield point in stress space. Note that the simultaneous loading in the one and two directions could have proceeded along the line OB such that yielding again occurred at B. Thus, to reach B, numerous loading paths might be followed and until that yield point is reached, all behavior is elastic. Using a number of loading paths, the locus described by the resulting yield points divides elastic behavior (inside the locus) and the onset of yielding which is the locus itself. Considering the models that included strain-hardening, these implied that such an effect tended to increase the subsequent yield strength or new flow stress. In this book any such tendencies will be assumed to enlarge the initial yield locus in a uniform manner; this is called isotropic hardening.

It is appropriate here to introduce the concept of three-dimensional stress space. In Figure 7-5, the combinations of stress a, b and c, acting in the 1, 2 and 3 directions are assumed to just cause yielding. This total stress state is defined by σ which originates at the origin and its tip in space provides a yield point. If enough experiments were conducted, all such points would lie on a yield surface. Any stress state, described by a single vector such as σ, that lies within the surface causes elastic effects only. As the tip of such a vector approaches the surface, yielding is incipient. Note that a yield locus is described by passing a plane through the surface with one of the three principal stresses being a constant (e. g. , $\sigma_3 = 0$ in the earlier development).

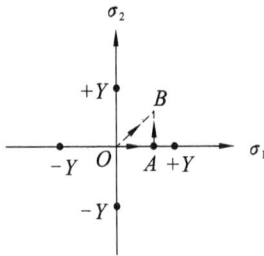

Figure 7-4　Yield points in two-dimensional stress space　　　**Figure 7-5　Stress resultant in three-dimensional stress space**

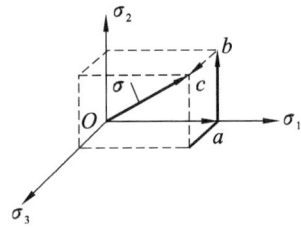

Considering that the magnitude of the mean normal stress σ_m, does not influence yielding, the concept of a yield surface can be explained more fully. Reference to Figure 7-6 will clarify the meaning of σ_m where an applied stress state is indicated on the left element. As shown, σ_m is equal to one third the algebraic sum of the three normal stresses; compressive stresses would count negative in this summation. If the mean stress is subtracted from each applied stress, the deviatoric stresses result. Literally, these deviate from the mean and if yielding is uninfluenced by σ_m then these deviatoric stresses must in some way induce yielding. It will be shown that they are nothing more than functions of shear stresses.

Thus, if a combination of stresses $(\sigma_1, \sigma_2, \sigma_3)$ just causes yielding, then $(\sigma_1 + \sigma_0, \sigma_2 + \sigma_0, \sigma_3 + \sigma_0)$ must also cause yielding and all combinations of $(\sigma_1, \sigma_2, \sigma_3)$ for various σ_0 therefore generate a line on the yield surface that is parallel to the line $\sigma_1 = \sigma_2 = \sigma_3$. This line is defined by equal direction cosines with respect to the 1, 2, 3 system. Now since isotropy and no Bauschinger effect are assumed, the rotation of such a line around the space diagonal ($\sigma_1 = \sigma_2 = \sigma_3$) must generate a prism which is the yield surface. To fully specify this prism, both its cross-sectional shape and size must be defined.

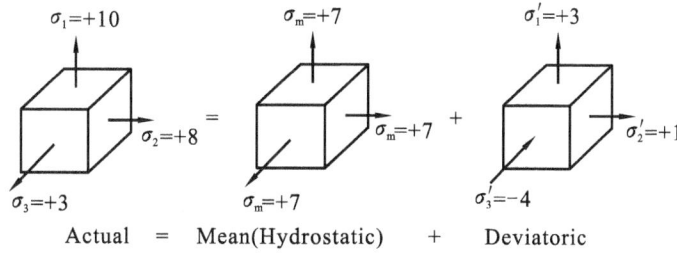

Figure 7-6 Stress state separated into mean and deviatoric components

All planes perpendicular to the space diagonal are defined by the equation $\sigma_1 + \sigma_2 + \sigma_3 =$ constant, this being $3\sigma_m$ for any one group of normal stresses. If the constant is set equal to zero, that plane passes through the origin at right angles to the axis of the prism, it is often called the π plane and its intersection with the yield surface is referred to as the C curve. This indicates that any point on the yield surface can be reduced to its equivalent point on the C curve simply by applying a proper value of σ_0 which simply moves the original point up or down the yield surface. Finally, consider a stress state as follows:

$$(\sigma_1, \sigma_2, \sigma_3) = (6, -2, 1) \quad \text{and} \quad \sigma_m = \frac{5}{3}$$

If this stress state is "reduced" by σ_m, then

$$(\sigma_1', \sigma_2', \sigma_3') = \left(\frac{13}{3}, -\frac{11}{3}, -\frac{2}{3}\right) \quad \text{and} \quad \sum \sigma_i' = 0$$

Thus, the three-dimensional stress vector is composed of the deviatoric stresses that lie in the π plane and the mean or hydrostatic component that is perpendicular to the π plane. Now that the physical meaning of a yield locus and surface has been developed, we consider some possible surfaces that have been proposed.

7.5 Yield Criteria

As discussed in Chapter 2, for any three-dimensional stress state there exists a cubic equation whose three roots are the principal stresses. A useful form of this equation is

$$\sigma_p^3 - I_1\sigma_p^2 - I_2\sigma_p - I_3 = 0 \tag{7-1}$$

where the invariants, I_1, I_2 and I_3, may be expressed as functions of principal stresses as follows

$$\begin{cases} I_1 = \sigma_1 + \sigma_2 + \sigma_3 \\ I_2 = -(\sigma_1\sigma_2 + \sigma_2\sigma_3 + \sigma_3\sigma_1) \\ I_3 = \sigma_1\sigma_2\sigma_3 \end{cases} \tag{7-2}$$

It may be noted immediately that $I_1 = 3\sigma_m$; thus the first invariant is a function of the hydrostatic or mean stress component and should not influence yielding. Therefore, any acceptable yield criterion should not include any reference to I_1 for those solids whose yield behavior has been found to be independent of σ_m.

Suppose that a yield criterion is proposed as follows: When $\sigma_1 - \sigma_2 - \sigma_3 = \text{constant} = +10$, yielding will occur. If this were an acceptable criterion then $\sigma_1 = +5, \sigma_2 = -2, \sigma_3 = -3$, provides a stress state that would lead to yielding. Now superimpose a stress, $\sigma_0 = +10$. This means the new stress state is $(15, +8, +7)$, which according to the proposed criterion, would not cause yielding since $\sigma_1 - \sigma_2 - \sigma_3$ is zero and not $+10$. Yet only the mean stress was varied. Such a criterion would not agree with experimental observations so it could not be considered to possess the generality required. The two most widely used criteria both satisfy independence of I_1 and have found best agreement when experiments have utilized ductile metals.

7.6 Tresca Yield Criterion

This criterion proposes that yielding will occur when some function of the maximum shear stress reaches a critical value. Whenever possible, the convention $\sigma_1 > \sigma_2 > \sigma_3$ will be used but there are cases when this relative comparison is not known a priori. In addition, this convention cannot be maintained rigorously when plots in two-dimensional or three-dimensional stress space are considered.

It is useful to recall that when the three Mohr's circles are plotted it is the radius of the largest circle that gives the maximum shear stress. Accounting for algebraic signs, this criterion is written as

$$|\sigma_{\max} - \sigma_{\min}| = \text{constant} = |\sigma_1 - \sigma_3| \qquad \text{if } \sigma_1 > \sigma_2 > \sigma_3 \tag{7-3}$$

If such a criterion finds reasonably universal acceptance regardless of the applied stress state, then the constant should be readily determined from simple standard tests.

(1) For uniaxial tension, yielding occurs when σ_1 reaches the uniaxial yield stress, Y. Thus

$$\sigma_1 = Y, \quad \sigma_2 = \sigma_3 = 0 \quad \text{and} \quad \tau_{\max} = \frac{Y}{2}$$

Using Equation (7-3), this means

$$|\sigma_1 - 0| = \text{constant} = Y$$

(2) For pure shear, $\sigma_1 = -\sigma_3 = \tau_{max}$, $\sigma_2 = 0$. For convenience, let the maximum allowable shear stress be designated as k, the shear yield stress. Using Equation (7-3), we get

$$|\sigma_1 - (-\sigma_1)| = \text{constant} = 2\sigma_1 = 2k$$

Thus, the Tresca criterion may be expressed as

$$|\sigma_{max} - \sigma_{min}| = Y = 2k = |\sigma_1 - \sigma_3| \quad \text{if } \sigma_1 > \sigma_2 > \sigma_3 \tag{7-4}$$

If a solid obeyed this criterion exactly, then the tensile and shear yield stresses would relate in a two-to-one ratio. This does not mean this ratio must be observed; rather, it is predicated by this criterion.

Example 7-1　A material whose tensile yield strength Y is 50 MPa is subjected to a uniaxial compressive stress of 30 MPa(Figure 7-7). Determine the magnitude of the tensile stress applied at right angles to the initial compressive stress, that would cause yielding according to the Tresca yield criterion. Plot the Mohr's circle for this situation.

Figure 7-7　Figure of Example 7-1 (Unit: MPa)

Solution　$|\sigma_1 - \sigma_3| = Y$　where　$\sigma_1 > \sigma_2 > \sigma_3$

Here σ_2 is indicated as zero, σ_3 is negative, and σ_1, the unknown stress, is positive.

$$|\sigma_1 - (-30)| = 50, \quad \text{so} \quad \sigma_1 = 20 \text{ MPa}$$

Note that the diameter of the circle is 50 MPa.

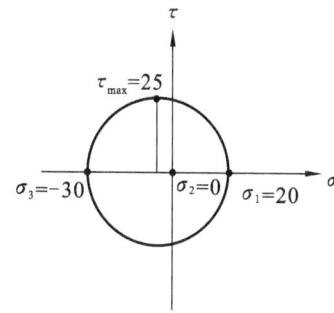

7.7　Von Mises Yield Criterion

Perhaps because the Tresca criterion produces a yield surface having corners on the prism that results in stress space and because the intermediate stress σ_2 is ignored, a criterion that involves a smooth function was proposed by von Mises. Although the mathematical statement reduces to $6J_2 = \text{constant}$, this gives little concrete understanding. J_2 is called the second tensor invariant of deviatoric stress. Where $J_2 = -(\sigma_x'\sigma_y' + \sigma_y'\sigma_z' + \sigma_x'\sigma_z' - \tau_{xy}^2 - \tau_{yz}^2 - \tau_{xz}^2)$ or $J_2 = -[(\sigma_x - \sigma_m)(\sigma_y - \sigma_m) + (\sigma_y - \sigma_m)(\sigma_z - \sigma_m) + (\sigma_x - \sigma_m)(\sigma_z - \sigma_m) - (\tau_{xy}^2 + \tau_{yz}^2 + \tau_{xz}^2)]$, Thus $6J_2 = (\sigma_x - \sigma_y)^2 + (\sigma_y - \sigma_z)^2 + (\sigma_x - \sigma_z)^2 + 6(\tau_{xy}^2 + \tau_{yz}^2 + \tau_{xz}^2)$, or $6J_2 = (\sigma_1 - \sigma_2)^2 + (\sigma_2 - \sigma_3)^2 + (\sigma_3 - \sigma_1)^2$. Several physical interpretations have been suggested; these are called the distortion energy and octahedral shear stress theories. It is to be emphasized that they are interpretations which followed the proposed mathematical postulate, but there is no point in indulging in historical sequencing as far as the use of this criterion is concerned. In its most widely used form, the von Mises yield criterion, in terms of principal stresses, predicts that yielding occurs when

$$(\sigma_1-\sigma_2)^2+(\sigma_2-\sigma_3)^2+(\sigma_3-\sigma_1)^2=\text{constant} \tag{7-5}$$

In a more general form

$$(\sigma_x-\sigma_y)^2+(\sigma_y-\sigma_z)^2+(\sigma_z-\sigma_x)^2+6(\tau_{xy}^2+\tau_{yz}^2+\tau_{zx}^2)=\text{constant} \tag{7-6}$$

The proof of the equivalence of these equations constitutes an exercise for the interested reader.

With Equation (7-5) there is no need to know at the outset how the principal stresses relate algebraically since all are equally weighted. Note also that each stress difference is a shear stress connected with one of the three Mohr's circles, shear stresses tie in with the onset of yielding so as with the Tresca yield criterion. To determine the constant, the same procedure used in Section 7. 6 is used.

(1) For uniaxial tension, yielding occurs when $\sigma_1=Y, \sigma_2=\sigma_3=0$. Using Equation (7-5)

$$2\sigma_1^2=\text{constant}=2Y^2$$

(2) For pure shear, yielding occurs when $\sigma_1=-\sigma_3=k, \sigma_2=0$ and using Equation (7-5)

$$\sigma_1^2+\sigma_1^2+4\sigma_1^2=\text{constant}=6\sigma_1^2=6k^2$$

Thus, the von Mises yield criterion may be written as:

$$(\sigma_1-\sigma_2)^2+(\sigma_2-\sigma_3)^2+(\sigma_3-\sigma_1)^2=2Y^2=6k^2 \tag{7-7}$$

According to this criterion, the tensile and shear yield stresses are related as

$$Y=\sqrt{3}k$$

which is the first inkling that the two criteria may lead to different predictions.

It is convenient to consider each criterion as a function of an effective stress denoted as $\bar{\sigma}$ where $\bar{\sigma}$ is a function of the applied stresses. Whenever its magnitude reaches the yield strength in uniaxial tension, then that applied stress state should cause yielding to occur (i. e. , it has reached an effective level). Thus

von Mises

$$\bar{\sigma}=\frac{1}{\sqrt{2}}[(\sigma_1-\sigma_2)^2+(\sigma_2-\sigma_3)^2+(\sigma_3-\sigma_1)^2]^{\frac{1}{2}} \tag{7-8}$$

Tresca

$$\bar{\sigma}=|\sigma_{\max}-\sigma_{\min}| \tag{7-9}$$

When $\bar{\sigma}$ reaches a value of Y because of the effects of σ_1, σ_2 and σ_3, either criterion predicts yielding. However, according to the von Mises yield criterion when $\bar{\sigma}$ reaches a value of $\sqrt{3}k$ yielding is predicted whereas the Tresca yield criterion requires $\bar{\sigma}$ to reach $2k$ before yielding is expected.

There are many ways these two rules can be expressed but it is of importance to realize they are not natural laws as such and their initial postulation was due to mathematical rather than physical reasoning. It is a bit surprising that these two postulates have provided such close agreement with physical observations.

Example 7-2　Repeat Example 7-1 using the von Mises yield criterion for predictive purposes(Figure 7-8).

Solution　Using Equation (7-7)

$$(\sigma_1-0)^2+[0-(-30)]^2+(-30-\sigma_1)^2=2\times50^2$$

$$\sigma_1^2+900+900+60\sigma_1+\sigma_1^2=5000$$

$$\sigma_1^2+30\sigma_1-1600=0$$

$$\sigma_1=\frac{-30\pm(900+6400)^{\frac{1}{2}}}{2}=\frac{-30\pm85.44}{2}$$

so

$$\sigma_1=-57.72 \text{ MPa or } 27.72 \text{ MPa}.$$

In view of the question the tensile value of 27.72 MPa is the correct answer. The negative value of -57.72 MPa is the compressive stress that is required to cause yielding if such a loading were of concern. Note that $\tau_{\max}$ equals $Y/\sqrt{3}$ here whereas $\tau_{\max}$ equals $Y/2$ in Example 7-1.

That neither criterion is influenced by σ_m can be shown by reference to Figure 7-9 where the deviatoric stresses (as

Figure 7-8　Figure of Example 7-2

(Unit: MPa)

functions of shear stresses) are seen to be crucial. Suppose that upon the original stress state, a stress component of $+9$ is superimposed, then $\sigma_1=48$ MPa, $\sigma_2=17$ MPa, $\sigma_3=13$ MPa and the new $\sigma_m=26$ MPa. The center of the largest circle is at $\sigma=30.5$ MPa, and $\tau_{\max}=17.5$ MPa. Note that $\sigma_1'=22$ MPa, $\sigma_2'=-9$ MPa, and $\sigma_3'=-13$ MPa, as above. Thus, the change in σ_m displaces the circles but neither changes their size nor influences the magnitudes of the deviatoric stresses. Note that

$$\sigma_1'=\sigma_1-\sigma_m=\frac{(\sigma_1-\sigma_2)+(\sigma_1-\sigma_3)}{3}$$

and similarly for σ_2' and σ_3', so the deviatoric stresses are functions of shear stresses.

Many practical problems can be approximated by or reduced to a biaxial (plane)

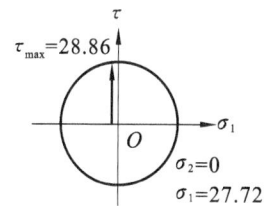

stress situation. Consider that $\sigma_2 = 0$. If a plot is made in $\sigma_1 - \sigma_3$ stress space, Figure 7-10 results for the Tresca yield criterion and Figure 7-11 shows von Mises yield locus. If the two plots are combined, Figure 7-12 results.

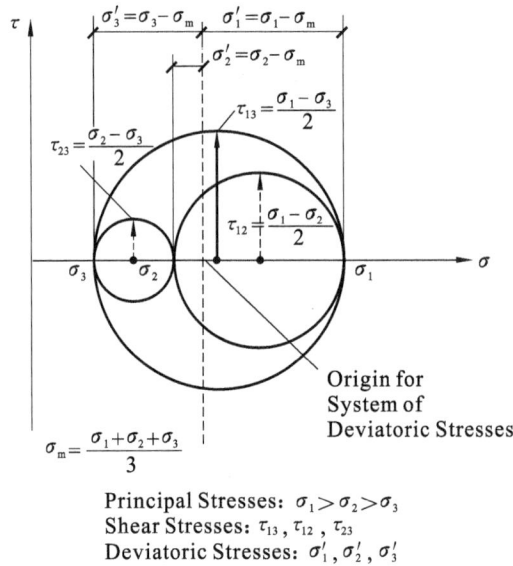

Suppose $\sigma_1 = 39$, $\sigma_2 = 8$, and $\sigma_3 = 4$(all in MPa), then $\sigma_m = 17$, center of largest circle at $\sigma = 21.5$, and $\tau_{max} = 17.5$. Note that $\sigma_1' = 22$, $\sigma_2' = -9$, and $\sigma_3' = -13$

Figure 7-9 Mohr's circle for three-dimensional stress state showing

principal stresses, mean normal stress, and deviatoric components(Unit: MPa)

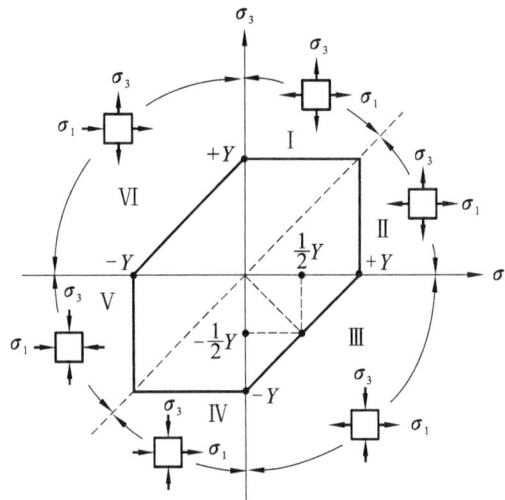

Zone	Condition	Boundary	Any stress state for σ_1, σ_3
I	$\sigma_3 > \sigma_1 > 0$	$\sigma_3 = +Y$	that lies within the heavy lines (boundary)above,will not cause yielding.
II	$\sigma_1 > \sigma_3 > 0$	$\sigma_1 = +Y$	
III	$\sigma_1 = -\sigma_3 = k = Y/2$	45° line as shown	
IV	$0 > \sigma_1 > \sigma_3$	$\sigma_3 = -Y$	
V	$0 > \sigma_3 > \sigma_1$	$\sigma_1 = -Y$	
VI	see III above		

Figure 7-10 Tresca yield locus

The various constant stress ratio loading paths in Figure 7-12 show that the maximum predicted difference between these criteria occurs when $\alpha = -1$ (pure shear) or $\alpha = 1/2$ or 2. It is often overlooked that the two criteria coincide at particular points, where the ellipse circumscribes the hexagon, only because each was defined to predict yielding when the uniaxial tensile or compressive yield stress was used to define the constants in Equations (7-8) and (7-9).

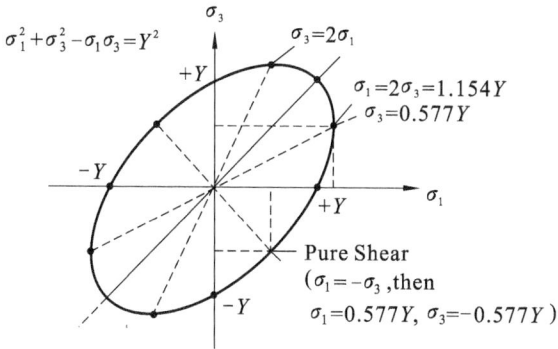

Figure 7-11   Von Mises yield locus

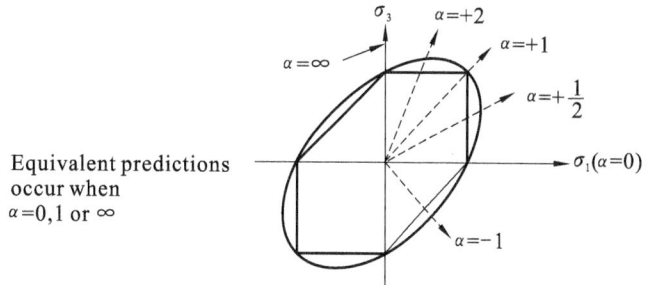

Figure 7-12   Yield loci for the Tresca and von Mises criteria showing certain loading paths

7.8   Distortion Energy

One interpretation of the Mises yield criterion is that yielding occurs when the elastic energy causing distortion reaches a critical value. This strain energy is found in a general way by subtracting the dilatational strain energy from the total elastic strain energy. The developments showed the total strain energy per unit volume to be

$$W_v = \frac{1}{2}(\sigma_x \varepsilon_{xx} + \sigma_y \varepsilon_{yy} + \sigma_z \varepsilon_{zz} + \tau_{xy}\gamma_{xy} + \tau_{yz}\gamma_{yz} + \tau_{zx}\gamma_{zx}) \tag{7-10}$$

or for the case of principal stresses

$$W_v = \frac{1}{2}(\sigma_1 \varepsilon_{11} + \sigma_2 \varepsilon_{22} + \sigma_3 \varepsilon_{33}) \tag{7-11}$$

To express Equation (7-11) as a function of stresses, the generalized form of Hooke's law given by Equations (4-23) is used to give

$$W_v = \frac{1}{2E}(\sigma_1^2 + \sigma_2^2 + \sigma_3^2) - \frac{\nu}{E}(\sigma_1\sigma_2 + \sigma_2\sigma_3 + \sigma_3\sigma_1) \tag{7-12}$$

Since only normal stresses cause a volume change, the dilatation strain is

$$\Delta = \varepsilon_{11} + \varepsilon_{22} + \varepsilon_{33} = \frac{1-2\nu}{E}(\sigma_1 + \sigma_2 + \sigma_3) = \frac{3}{E}(1-2\nu)\sigma_m \tag{7-13}$$

Now the normal strains associated with σ_m must be equivalent and since $\Delta = 3\varepsilon_m$

$$\varepsilon_m = \frac{1-2\nu}{E}\sigma_m \qquad (7\text{-}14)$$

Observing that the work due to dilatation, $W_d = 3\sigma_m\varepsilon_m/2$, then $W_d = 3(1-2\nu)\sigma_m^2/(2E)$ and finally

$$W_d = \frac{1-2\nu}{6E}(\sigma_1+\sigma_2+\sigma_3)^2 \qquad (7\text{-}15)$$

By subtracting Equation (7-15) from Equation (7-12) to give the shear strain energy, W_s, the result after much manipulation is

$$W_s = \frac{1}{12G}[(\sigma_1-\sigma_2)^2+(\sigma_2-\sigma_3)^2+(\sigma_3-\sigma_1)^2] \qquad (7\text{-}16)$$

Now the shear strain energy induced during uniaxial tension where $\sigma_2 = \sigma_3 = 0$ is

$$W_s = \frac{\sigma_1^2}{6G} \qquad (7\text{-}17)$$

The critical value W_{sc}, that must be developed to cause yielding will result when $\sigma_1 = Y$. Setting Equation (7-16) equal to this critical value leads to

$$\frac{1}{12G}[(\sigma_1-\sigma_2)^2+(\sigma_2-\sigma_3)^2+(\sigma_3-\sigma_1)^2] = \frac{Y^2}{6G} \qquad (7\text{-}18)$$

which is identical to Equation (7-7). This explains why the von Mises yield criterion is often called the distortion energy theory whose physical meaning is that yielding will occur when the elastic energy causing distortion reaches a critical value.

7.9　Octahedral Shear Stress

A second physical interpretation of the von Mises yield criterion has also been proposed. For simplicity consider a coordinate system defined by principal stress directions and a line from the origin having direction cosines where $l = m = n$. The planes normal to this line and equivalent lines in other regions of space are called octahedral planes where the intersections of these eight equivalent planes form an octahedron. For this physical situation, it was shown earlier by Chapter 2 that

$$\sigma_n = \sigma_1 l^2 + \sigma_2 m^2 + \sigma_3 n^2 \qquad (7\text{-}19)$$

Since $l = m = n = \cos 54°44' = 1/\sqrt{3}$, $\sigma_n = 1/3(\sigma_1+\sigma_2+\sigma_3)$, thus the stress normal to the octahedral planes is σ_m, and since σ_m has no influence upon yielding, it has been proposed that the shear stresses acting on this plane (τ_0) must reach a critical value for yielding to occur. Although not fully derived here, this stress can be shown to be

$$\tau_0 = \frac{1}{3}[(\sigma_1-\sigma_2)^2+(\sigma_2-\sigma_3)^2+(\sigma_3-\sigma_1)^2]^{\frac{1}{2}} \tag{7-20}$$

and a comparison with Equation (7-8) shows that

$$\tau_0 = \frac{\sqrt{2}}{3}\sigma \tag{7-21}$$

With the use of a proper multiplying factor, this is just another version of the von Mises yield criterion. Throughout the remainder of this book, Equation (7-8) will be used whenever this particular criterion is to be employed.

7.10 Flow Rules or Plastic Stress-strain Relationships

Just as the generalized Hooke's law in the elastic region can be expressed by

$$\varepsilon_{11}^e = \frac{1}{E}[\sigma_1-\nu(\sigma_2+\sigma_3)]$$

analogous relations for the plastic region are needed.

These flow rules can be developed in a simple way as follows, recalling that path or history dependency requires the use of incremental strains during plastic deformation.

Consider plastic flow under uniaxial tension as indicated in Figure 7-13.

$$\sigma_m = \frac{\sigma_1+\sigma_2+\sigma_3}{3} = \frac{1}{3}\sigma_1$$

Now the deviatoric stress in the 1 direction is $\sigma_1' = \sigma_1 - \sigma_m$ and at the particular instant represented in Figure 7-13.

$$\sigma_1' = \frac{2}{3}\sigma_1 \quad \text{and} \quad \sigma_2' = \sigma_3' = 0 - \frac{1}{3}\sigma_1 = -\frac{1}{3}\sigma_1$$

so

$$\sigma_1' = -2\sigma_2' = -2\sigma_3'$$

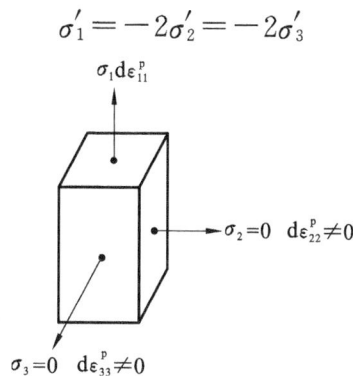

Figure 7-13 Stresses and incremental plastic strains for uniaxial tension

For volume constancy, the sum of the plastic strain increments must be zero; therefore

$$d\varepsilon_{11}^p + d\varepsilon_{22}^p + d\varepsilon_{33}^p = 0$$

And because of symmetry in this instance $d\varepsilon_{22}^p = d\varepsilon_{33}^p$, therefore, $d\varepsilon_{11}^p = -2d\varepsilon_{22}^p = -2d\varepsilon_{33}^p$. This leads to $d\varepsilon_{11}^p / d\varepsilon_{22}^p = -2 = \sigma_1' / \sigma_2'$, and so forth, which can be written as

$$\frac{d\varepsilon_{11}^p}{\sigma_1'} = \frac{d\varepsilon_{22}^p}{\sigma_2'} = \frac{d\varepsilon_{33}^p}{\sigma_3'} = \text{constant} = d\lambda \quad \text{(in general)} \tag{7-22}$$

i. e. , the constant isn't always -2 but these ratios are always in some constant proportion.

The implication is that the ratio of the current incremental plastic strain increments to the current deviatoric stresses is a constant; nothing is implied regarding the magnitudes of either. Several comments are worth noting:

(1) The above development uses a simple method that produces the flow rules. In effect, it may be viewed as a necessary but not sufficient condition since no real proof is offered to indicate that other stress states would lead to the same finding.

(2) Equation (7-22) expresses the Prandtl-Reuss flow rules where the elastic strain increments have been omitted.

(3) Equation (7-22) is identical to the Levy-Mises equations (proposed earlier than Prandtl) where the total strain increment is assumed to be equivalent to the plastic strain increment. Thus, the Levy-Mises equations may be viewed as a special form of the more general expression.

For greater convenience, the flow rules may be expressed in various forms other than Equation (7-22). These are

$$\frac{d\varepsilon_{11}^p - d\varepsilon_{22}^p}{\sigma_1 - \sigma_2} = d\lambda \tag{7-23}$$

$$d\varepsilon_{11}^p = \frac{2}{3} d\lambda \left[\sigma_1 - \frac{1}{2}(\sigma_2 + \sigma_3) \right] \tag{7-24}$$

$$d\varepsilon_{11}^p = \frac{d\overline{\varepsilon}^p}{\overline{\sigma}} \left[\sigma_1 - \frac{1}{2}(\sigma_2 + \sigma_3) \right] \tag{7-25}$$

Note the great similarity between Equation (7-25) and the generalized Hooke's law, where $1/E$ is replaced by $d\overline{\varepsilon}^p/\overline{\sigma}$ and ν is replaced by $1/2$ as a consequence of incompressibility. The coefficient $d\overline{\varepsilon}^p/\overline{\sigma}$ is not a constant in the sense that E is; rather it is a variable proportionality factor. The incremental effective strain, $d\overline{\varepsilon}^p$, requires definition and the form to be used from this point is

$$d\overline{\varepsilon}^p = \frac{\sqrt{2}}{3} \left[(d\varepsilon_{11}^p - d\varepsilon_{22}^p)^2 + (d\varepsilon_{22}^p - d\varepsilon_{33}^p)^2 + (d\varepsilon_{33}^p - d\varepsilon_{11}^p)^2 \right]^{\frac{1}{2}} \tag{7-26}$$

Note its great similarity with the effective stress function given by Equation (7-8). The coefficient $\sqrt{2}/3$ is chosen so that the value of $d\bar{\varepsilon}^P$ is equal to $d\varepsilon_{11}^P$ under uniaxial tension in the same way that $\bar{\sigma}$ was made equivalent to σ_1 by the use of coefficient $1/\sqrt{2}$ in Equation (7-8). It is crucial to realize that from Equations (7-8) and (7-26), both $\bar{\sigma}$ and $d\bar{\varepsilon}^P$ are always positive, so their ratio in Equation (7-25) must also be positive.

Flow rules for any yield criterion may be derived by using the concept of a plastic potential. This method proposes that the incremental strains resulting from a stress σ_{ij} are found by using

$$d\varepsilon_{ij}^P = \frac{\partial f}{\partial \sigma_{ij}} d\lambda' \tag{7-27}$$

Where f is taken as the yield function. If the von Mises yield criterion is used,

$$f(\sigma_{ij}) = (\sigma_1 - \sigma_2)^2 + (\sigma_2 - \sigma_3)^2 + (\sigma_3 - \sigma_1)^2 = \text{constant}$$

then

$$\frac{\partial f}{\partial \sigma_1} = 2(\sigma_1 - \sigma_2) - 2(\sigma_3 - \sigma_1) = 4\sigma_1 - 2(\sigma_2 + \sigma_3)$$

$$3\sigma_m = \sigma_1 + \sigma_2 + \sigma_3$$

so

$$\sigma_2 + \sigma_3 = 3\sigma_m - \sigma_1$$

Now,

$$\frac{\partial f}{\partial \sigma_1} = 6(\sigma_1 - \sigma_m) = 6\sigma_1'$$

Finally,

$$d\varepsilon_{11}^P = 6\sigma_1' d\lambda' \quad \text{or} \quad \frac{d\varepsilon_{11}^P}{\sigma_1'} = d\lambda$$

as in Equation (7-22). Flow rules associated with the Tresca yield criterion have found little use.

The plastic flow rule corresponding to any particular choice of the plastic potential may be readily obtained.

$$d\varepsilon_{ii}^P = \sigma_i' d\lambda, \quad d\gamma_{ij}^P = \tau_{ij} d\lambda$$

or

$$\frac{d\varepsilon_{xx}^P}{\sigma_x'} = \frac{d\varepsilon_{yy}^P}{\sigma_y'} = \frac{d\varepsilon_{zz}^P}{\sigma_z'} = \frac{d\gamma_{xy}^P}{\tau_{xy}} = \frac{d\gamma_{yz}^P}{\tau_{yz}} = \frac{d\gamma_{zx}^P}{\tau_{zx}} = d\lambda \tag{7-28}$$

The stress-strain relation in this form was suggested independently by Levy and von Mises, who used the plastic strain increments instead of the total strain increments. The modified Equation (7-28) which allow for the elastic strain increments, were proposed by Prandtl for plane strain and by Reuss for an arbitrary state of strain. Since the plastic shear strain increments, according to Equation (7-28), vanish with the corresponding shear stresses, the principal axes of the stress and the plastic strain increment coincide. Equation (7-28) also indicates that the Mohr's circle for the stress can be used for the plastic strain increment, provided that the origin is moved in the appropriate direction of the σ axis by an amount equal to the hydrostatic stress.

The complete Prandtl-Reuss equation for an elastic/plastic material may be expressed as

$$d\varepsilon = d\varepsilon^e + d\varepsilon^p$$

$$d\varepsilon_{ii} = \frac{1}{E}[(1+\nu)d\sigma_i - \nu\delta_{ii}d\sigma_k] + \left(\sigma_i - \frac{1}{3}\sigma_{ii}\delta_k\right)d\lambda \qquad (7\text{-}29)$$

$$d\varepsilon_{ij} = \frac{1}{E}[(1+\nu)d\tau_{ij}] + \frac{\tau_{ij}}{2}d\lambda$$

which $d\lambda = 3d\bar{\varepsilon}^p/(2\bar{\sigma})$ from Equations (7-23) and (7-24). For a non-hardening material, $d\lambda$ may be treated as a basic unknown of the problem. Equation (7-29) consists of three equations of each of the two types

$$d\varepsilon_{xx} = \frac{1}{E}[d\sigma_x - \nu(d\sigma_y + d\sigma_z)] + \frac{2}{3}d\lambda\left[\sigma_x - \frac{1}{2}(\sigma_y + \sigma_z)\right]$$

$$d\varepsilon_{yy} = \frac{1}{E}[d\sigma_y - \nu(d\sigma_x + d\sigma_z)] + \frac{2}{3}d\lambda\left[\sigma_y - \frac{1}{2}(\sigma_x + \sigma_z)\right]$$

$$d\varepsilon_{zz} = \frac{1}{E}[d\sigma_z - \nu(d\sigma_x + d\sigma_y)] + \frac{2}{3}d\lambda\left[\sigma_z - \frac{1}{2}(\sigma_x + \sigma_y)\right]$$

$$d\varepsilon_{xy} = \frac{d\tau_{xy}}{2G} + \frac{\tau_{xy}}{2}d\lambda$$

$$d\varepsilon_{yz} = \frac{d\tau_{yz}}{2G} + \frac{\tau_{yz}}{2}d\lambda$$

$$d\varepsilon_{xz} = \frac{d\tau_{xz}}{2G} + \frac{\tau_{xz}}{2}d\lambda$$

In a number of practical problems, the loading paths are such that the elastic strains are small in comparison with the plastic strains. The Prandtl-Reuss equations may then be replaced by the more tractable Levy-Mises equations, which correspond to Equation (7-28) with the superscripts omitted. This is equivalent to assuming the material to both rigid and plastic.

Note that plastic strains are related to total stresses whereas elastic strains are associated with incremental stress changes. Figure 7-14 demonstrates this point. In Figure 7-14(a) an equal incremental change in stress causes the same incremental change in strain, since a linear relation exists. However, in Figure 7-14(b) an equal incremental stress change does not produce the same incremental change in strain; instead, these are related to a changing slope, so a particular increment depends upon the total stress at a given instant.

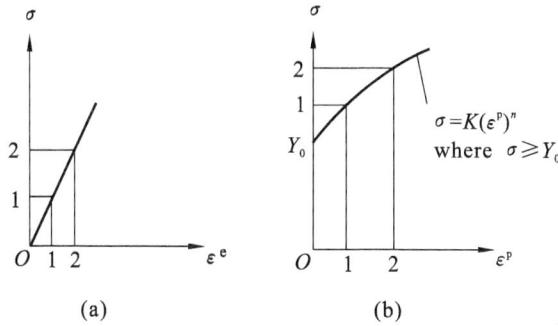

Figure 7-14   Stress-strain curves for elastic and plastic deformation

Example 7-3   Consider a thin-walled tube having closed ends that is internally pressurized to 7 MPa; it is assumed that this causes plastic deformation to occur and end effects are ignored. With this loading state, the following stress relationships are acceptable.

$$\sigma_\theta = \sigma_1 = \frac{pr}{t}, \quad \sigma_l = \sigma_2 = \frac{pr}{2t}, \quad \sigma_r = \sigma_3 = 0 \tag{7-30}$$

where p is internal pressure and r is the radius of the thin-walled tube.

Consider the r/t ratio to be 20 and note that $\sigma_2 = \sigma_1/2 = \sigma_m$. The total strain in the hoop direction is to be found.

Solution   With the use of Equation (7-25)

$$d\varepsilon_{11}^P = -d\varepsilon_{33}^P, \quad d\varepsilon_{22}^P = 0 \tag{7-31}$$

In incremental form, the total strain in the "1" direction is

$$d\varepsilon_{11} = d\varepsilon_{11}^P + d\varepsilon_{11}^e \quad \text{(i. e. ,plastic plus elastic)}$$

Using Equations (7-25) and Hooke's law

$$d\varepsilon_{11} = \frac{d\bar{\varepsilon}^P}{\bar{\sigma}}\left[\sigma_1 - \frac{1}{2}(\sigma_2 + \sigma_3)\right] + \frac{1}{E}\left[d\sigma_1 - \nu(d\sigma_2 + d\sigma_3)\right]$$

Which reduces to

$$d\varepsilon_{11} = \frac{d\bar{\varepsilon}^P}{\bar{\sigma}}\left(\frac{3}{4}\sigma_1\right) + \frac{d\sigma_1}{E}\left(1 - \frac{\nu}{2}\right)$$

Using the relations in Equation (7-30). Substituting the stress relations of Equation (7-30) into Equation (7-8) and the strain relations of Equations (7-31) into Equation (7-26) gives

$$\bar{\sigma} = \frac{\sqrt{3}}{2}\sigma_1 \quad \text{and} \quad d\bar{\varepsilon}^P = \frac{2}{\sqrt{3}}d\varepsilon_{11}^P \tag{7-32}$$

For the values in the problem statement, $\sigma_1 = 140$ MPa, so $\bar{\sigma} = 70\sqrt{3}$ MPa. What is now essential is an effective stress-effective strain relationship. Assume for now that the form $\bar{\sigma} = K(\overline{\varepsilon^{\mathrm{p}}})^n$ is appropriate for the plastic portion of deformation, where $K = 175$ MPa and $n = 0.25$. Then the relation

$$70\sqrt{3} = 175\,(\overline{\varepsilon^{\mathrm{p}}})^{0.25}$$

results

$$\overline{\varepsilon^{\mathrm{p}}} = (0.693)^4 = 0.23$$

which, is from Equation (7-26), $\int_0^{\overline{\varepsilon^{\mathrm{p}}}} \mathrm{d}\overline{\varepsilon^{\mathrm{p}}}$. Therefore,

$$\int_0^{\varepsilon_{11}^{\mathrm{p}}} \mathrm{d}\varepsilon_{11}^{\mathrm{p}} = \int_0^{\overline{\varepsilon^{\mathrm{p}}}} \frac{\sqrt{3}}{2} \mathrm{d}\overline{\varepsilon^{\mathrm{p}}}$$

so

$$\varepsilon_{11}^{\mathrm{p}} = 0.199$$

To compute the elastic portion of strain, note that $\mathrm{d}\sigma_1$ equals $20\mathrm{d}p$ where $\mathrm{d}p$ equals 7 MPa. For aluminum (Whose values of K and n are reasonably well represented by the numbers used above) take $E = 7 \times 10^4$ MPa and $\nu = \frac{1}{3}$. Since $\mathrm{d}\varepsilon_{11}^{\mathrm{p}} = \mathrm{d}\sigma_1(1-\nu/2)/E, \varepsilon_{11}^{\mathrm{p}} \approx 0.002$, so $\varepsilon_{11} = 0.199 + 0.002 = 0.201$.

Several points are pertinent.

(1) Where large plastic strains are encountered, ignoring elastic strains introduces little error and often greatly simplifies an analysis.

(2) Many real problems must invoke the use of approximations since the resulting deformation does not follow a simple loading path as used in this example.

(3) Some type of $\bar{\sigma}$-$\bar{\varepsilon}$ relationship must be available if numerical answers are required.

(4) The lower limits on the above integrals were taken as zero. Physically this implies that the pressure was zero at the outset and no elastic or plastic strains had been induced in the material prior to the application of pressure.

7.11　Deformation Theory of Plasticity (Total Strain Theory)

So far, the stress-strain relationships given in this chapter have been in incremental form, so it is called incremental theories of plasticity. The basis of these incremental theories is that the plastic deformation depends on the loading path. Therefore, in order to obtain the final deformation state, we must integrate the incremental strain-stress relationship along the loading path.

However, there are some plastic theories that provide relationships between the total components of the stress and strain. These theories are called deformation theory of plasticity or total strain theory. Hencky first put forward the total theory of plasticity in 1924, and Ilyushin further developed and improved the theory in 1943 and 1947 respectively. Its contents are briefly described as follows.

The total strain theory indicates that the total plastic strain component is directly proportional to the total deviatoric stress component:

$$\frac{\varepsilon_{xx}^p}{\sigma_x'} = \frac{\varepsilon_{yy}^p}{\sigma_y'} = \frac{\varepsilon_{zz}^p}{\sigma_z'} = \frac{\gamma_{xy}^p}{\tau_{xy}} = \frac{\gamma_{yz}^p}{\tau_{yz}} = \frac{\gamma_{xz}^p}{\tau_{xz}} = \lambda \tag{7-33}$$

$$\begin{cases} \varepsilon_{xx}^p = \lambda\sigma_x' \\ \varepsilon_{yy}^p = \lambda\sigma_y' \\ \varepsilon_{zz}^p = \lambda\sigma_z' \end{cases} \quad \begin{cases} \gamma_{xy}^p = \lambda\tau_{xy} \\ \gamma_{yz}^p = \lambda\tau_{yz} \\ \gamma_{xz}^p = \lambda\tau_{xz} \end{cases} \tag{7-34}$$

In addition to using the plastic strain component ε_{ij}^p instead of the plastic strain increment $d\varepsilon_{ij}^p$, this equation is similar to the Prandtl-Reuss equation, which means that the plastic strain component is coaxial with the stress deviation component. Note that $\sigma_x', \sigma_y', \sigma_z'$ given in this equation is the deviatoric stress component and $\tau_{xy}, \tau_{yz}, \tau_{xz}$ is the stress component. If the equivalent stress and equivalent plastic strain are

$$\bar{\sigma} = \frac{1}{\sqrt{2}}[(\sigma_x - \sigma_y)^2 + (\sigma_y - \sigma_z)^2 + (\sigma_z - \sigma_x)^2 + 6(\tau_{xy}^2 + \tau_{yz}^2 + \tau_{xz}^2)]^{\frac{1}{2}}$$

$$= \frac{1}{\sqrt{2}}[(\sigma_1 - \sigma_2)^2 + (\sigma_2 - \sigma_3)^2 + (\sigma_3 - \sigma_1)^2]^{\frac{1}{2}} \tag{7-35}$$

$$\bar{\varepsilon}^p = \frac{\sqrt{2}}{3}[(\varepsilon_{xx}^p - \varepsilon_{yy}^p)^2 + (\varepsilon_{yy}^p - \varepsilon_{zz}^p)^2 + (\varepsilon_{zz}^p - \varepsilon_{xx}^p)^2 + 6(\gamma_{xy}^2 + \gamma_{yz}^2 + \gamma_{xz}^2)]^{\frac{1}{2}}$$

$$= \frac{\sqrt{2}}{3}[(\varepsilon_{11}^p - \varepsilon_{22}^p)^2 + (\varepsilon_{22}^p - \varepsilon_{33}^p)^2 + (\varepsilon_{33}^p - \varepsilon_{11}^p)^2]^{\frac{1}{2}} \tag{7-36}$$

The proportional function λ can be solved according to $\bar{\sigma}$ and $\bar{\varepsilon}^p$:

$$\lambda = \frac{3\bar{\varepsilon}^p}{2\bar{\sigma}} \tag{7-37}$$

Equation (7-34) becomes

$$\begin{cases} \varepsilon_{xx}^p = \frac{3\bar{\varepsilon}^p}{2\bar{\sigma}}\sigma_x' \\ \varepsilon_{yy}^p = \frac{3\bar{\varepsilon}^p}{2\bar{\sigma}}\sigma_y' \\ \varepsilon_{zz}^p = \frac{3\bar{\varepsilon}^p}{2\bar{\sigma}}\sigma_z' \end{cases} \quad \begin{cases} \gamma_{xy}^p = \frac{3\bar{\varepsilon}^p}{2\bar{\sigma}}\tau_{xy} \\ \gamma_{yz}^p = \frac{3\bar{\varepsilon}^p}{2\bar{\sigma}}\tau_{yz} \\ \gamma_{xz}^p = \frac{3\bar{\varepsilon}^p}{2\bar{\sigma}}\tau_{xz} \end{cases} \tag{7-38}$$

As a part of this theory, it is assumed that there is a generalized function between equivalent stress and equivalent plastic strain

$$\bar{\sigma} = \bar{\sigma}(\bar{\varepsilon^{p}}) \tag{7-39}$$

This function does not depend on the loading path, so it can be obtained by uniaxial tensile or compression tests.

The elastic strain is controlled by Hooke's law.

$$\frac{\varepsilon_{ii}^{e'}}{\sigma_i'} = \frac{1}{2G} \tag{7-40}$$

where $\varepsilon_{ii}^{e'} = \varepsilon_{ii}^{e} - \varepsilon_{m}^{e}$ and ε_{m}^{e} is the mean elastic strain.

Thus, the total strain can be considered as the sum of elastic strain and plastic strain:

$$\varepsilon_{ii} = \frac{\sigma_i'}{2G} + \frac{\sigma_{m}}{3K} + \frac{3}{2}\frac{\bar{\varepsilon^{p}}}{\bar{\sigma}}\sigma_i' \tag{7-41}$$

$$\gamma_{ij} = \frac{\tau_{ij}}{G} + \frac{3}{2}\frac{\bar{\varepsilon^{p}}}{\bar{\sigma}}\tau_{ij} \tag{7-42}$$

where $K = \dfrac{E}{3(1-2)\nu}$ is the elastic bulk modulus. If the elastic strain is small compared to plastic strain, the elastic strain can be ignored. So the basic equation of the total strain theory is

$$\varepsilon_{ii} = \frac{3}{2}\frac{\bar{\varepsilon^{p}}}{\bar{\sigma}}\sigma_i' \tag{7-43}$$

$$\gamma_{ij} = \frac{3}{2}\frac{\bar{\varepsilon^{p}}}{\bar{\sigma}}\tau_{ij} \tag{7-44}$$

and

$$\bar{\sigma} = \bar{\sigma}(\bar{\varepsilon^{p}}) \tag{7-45}$$

Because plastic deformation generally depends on the loading path, the application of the total strain theory is limited. As described above, plastic deformation has two characteristics: nonlinearity of loading and irreversibility of plastic flow. Irreversibility is a necessary response to load path dependence, so there is a need for incremental constitutive equations. However, for proportional loading paths, the total strain theory, which satisfies the nonlinear relationship between stress and strain, provides a satisfactory description of the plastic deformation. In fact, the incremental theory of plasticity is reduced to the total strain theory when all the stress components load in the same ratio. Increasing the loading stress at the same rate

$$\sigma_{ij} = k\sigma_{ij}^{0} \tag{7-46}$$

Where k is the monotonically increasing scale factor, σ_{ij}^{0} is an arbitrary stress state.

So we can get

$$\overline{\sigma} = k\,\overline{\sigma^0} \tag{7-47}$$

The Prandtl-Reuss equation of incremental theory can be written as follows

$$d\varepsilon_{ii}^{p} = \frac{3d\,\overline{\varepsilon^{p}}}{2\overline{\sigma}}\sigma_{i}' = \frac{3d\,\overline{\varepsilon^{p}}}{2\,\overline{\sigma^0}}\sigma_{i}^{0'} \tag{7-48a}$$

$$d\gamma_{ij}^{p} = \frac{3d\,\overline{\varepsilon^{p}}}{2\,\overline{\sigma^0}}\tau_{ij} = \frac{3d\,\overline{\varepsilon^{p}}}{2\,\overline{\sigma^0}}\tau_{ij}^{0} \tag{7-48b}$$

The above formula can easily be intergrated to become

$$\varepsilon_{ii}^{p} = \frac{3\,\overline{\varepsilon^{p}}}{2\,\overline{\sigma^0}}\sigma_{i}^{0'} = \frac{3\,\overline{\varepsilon^{p}}}{2\,\overline{\sigma}}\sigma_{i}' \tag{7-49a}$$

$$\gamma_{ij}^{p} = \frac{3\,\overline{\varepsilon^{p}}}{2\,\overline{\sigma^0}}\tau_{ij}^{0} = \frac{3\,\overline{\varepsilon^{p}}}{2\,\overline{\sigma^0}}\tau_{ij} \tag{7-49b}$$

This is the basic model of Hencky-Ilyushin total strain theory, which shows that the plastic strain is a function of the final stress state and does not depend on the loading path. The application of the total strain theory is not limited to proportional loading, but in fact, there are still some problems in the application scope of the total strain theory. It should be emphasized that due to the irreversibility of plastic deformation, the total strain theory must be used cautiously. When there is a significant difference between the real loading path and the proportional loading path, it will lead to serious errors.

7.12　Simultaneous Solution Method with both Equilibrium Differential Equation and Yield Criterion

When solving plastic problems such as plane axisymmetric, the stress distribution of the body when it has been plastically deformed can be obtained by solving both the equilibrium differential equation and the yield criterion simultaneously. The integration constant is determined according to the boundary conditions on the free surface and the contact surface.

Stress calculation for a plastic cylinder under internal pressure is as follows.

The sizes of the cylinder are shown in Figure 7-15. The inner wall of the cylinder is subjected to uniform pressure p. As the cylinder is long, such as pressure vessels and pipes, it can be simplified as a plane strain problem. Specifically, it is an axisymmetric plane problem. Therefore, τ_{rz} and $\tau_{\theta r}$ equal to zero. σ_{θ} and σ_{r} are the principal stresses, and they vary with r. As a result, Equation (5-12) which is expressed in the partial differential form becomes an ordinary differential equation. The equilibrium differential equation under neglecting body force in the cylindrical coordinate can be written as

$$\frac{d\sigma_r}{dr} + \frac{\sigma_r - \sigma_\theta}{r} = 0 \tag{7-50}$$

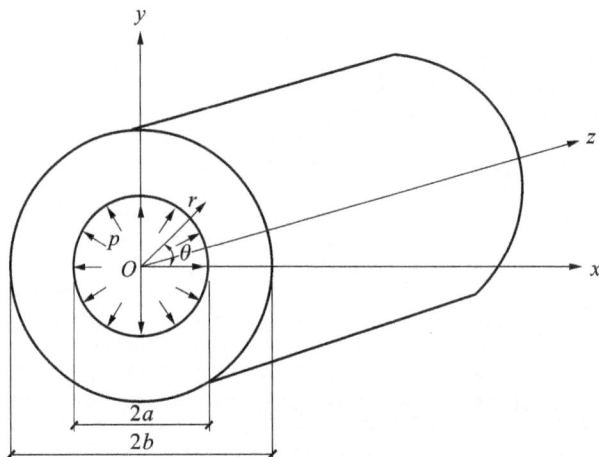

Figure 7-15   A plastic cylinder under internal pressure

According to the von Mises yield criterion

$$\bar{\sigma} = \frac{1}{\sqrt{2}}\sqrt{(\sigma_\theta - \sigma_r)^2 + (\sigma_\theta - \sigma_z)^2 + (\sigma_r - \sigma_z)^2} = Y \tag{7-51}$$

For the plane strain problem $\sigma_z = \frac{1}{2}(\sigma_r + \sigma_\theta)$, substituting it into Equation (7-51)

$$\sigma_\theta - \sigma_r = \frac{2}{\sqrt{3}}Y \tag{7-52}$$

Substituting Equation (7-52) into Equation (7-50)

$$d\sigma_r = \frac{2}{\sqrt{3}}Y\frac{dr}{r} \tag{7-53}$$

After integral, $\sigma_r = \frac{2}{\sqrt{3}}Y\ln Cr$. The integration constant C is calculated by means of

the boundary conditions. When $r = b$ and $\sigma_r = 0$, $C = \frac{1}{b}$. Combining with Equation

(7-52), the stress of the plastic cylinder is obtained

$$\begin{cases} \sigma_r = \frac{2}{\sqrt{3}}Y\ln\frac{r}{b} \\ \sigma_\theta = \frac{2}{\sqrt{3}}Y\left(1 + \ln\frac{r}{b}\right) \end{cases} \tag{7-54}$$

According to Equation (7-54), it is found that σ_θ is always the tensile stress and σ_r is the compressive stress. When $r = a$, the maximum stress occurs and is

$$p = \frac{2}{\sqrt{3}}Y\ln\frac{b}{a} \tag{7-55}$$

7.13 Principal Stress Method of Calculating the Deformation Force for Cylinder Upsetting

In fact, the principal stress method is the simultaneous solution with both equilibrium equation and yield criterion. In engineering applications, some basic assumptions need to be used to simplify calculation.

The principal stress method involves the following steps:

(1) Simplify the research object as axisymmetric problem or plane problem, and determine the corresponding coordinate system.

(2) At a certain time, according to the deformation trend of the deformed body, select the typical element containing the contact surfaces, which have the normal stresses and the shear stresses from friction. Assuming that there are only uniformly distributed normal stresses (principal stresses) on the other non-contact surfaces. Thus, the number of equilibrium equations is reduced to one, and the partial differential equation becomes ordinary differential equation.

(3) When determining the plastic conditions of the element, ignoring the influence of friction tangential stress, assuming that the normal stress on the element surface is the principle stress, obtaining a simplified yield criterion.

(4) By solving the equilibrium equation and yield criterion simultaneously, obtaining the stress distribution on the contact surface.

The schematic diagram of cylinder upsetting between parallel templates is shown in Figure 7-16. Because the upsetting object is axisymmetric in geometry and the load is axisymmetric, it belongs to axisymmetric problem.

Considering the radial flow of metal, the element shown in Figure 7-16 is selected. For the blank material of a fan shape with the instantaneous height h and the thickness dr, the stresses are σ_z, σ_r,

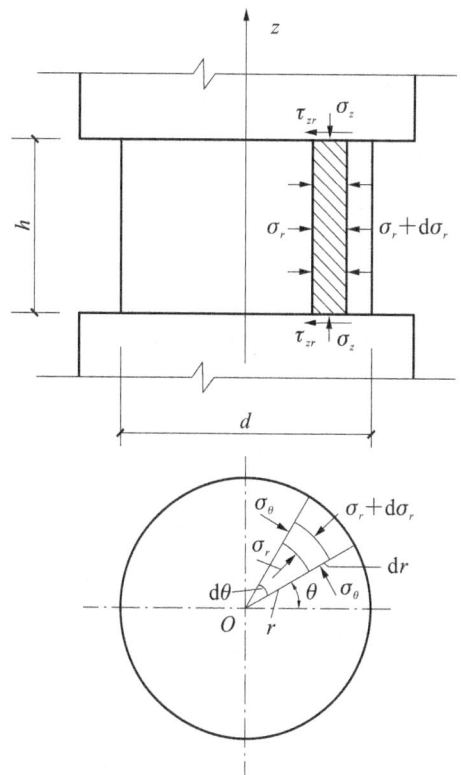

Figure 7-16　Element stress analysis of the
upsetting cylinder

σ_θ and τ_{zr} respectively.

The equilibrium equation along the radial direction is

$$(\sigma_r + \mathrm{d}\sigma_r)(r + \mathrm{d}r)\mathrm{d}\theta h - \sigma_r \mathrm{d}\theta h - 2\sigma_\theta \sin\frac{\mathrm{d}\theta}{2}\mathrm{d}rh + 2\tau_{zr}r\mathrm{d}\theta\mathrm{d}r = 0 \qquad (7\text{-}56)$$

Rearranging and ignoring high-order items, Equation (7-56) can be written as

$$\frac{\mathrm{d}\sigma_r}{\mathrm{d}r} + \frac{2\tau_{zr}}{h} + \frac{\sigma_r - \sigma_\theta}{r} = 0 \qquad (7\text{-}57)$$

For solid cylinder upsetting, the radial strain is $\varepsilon_r = \mathrm{d}r/r$ and the tangential strain is

$$\varepsilon_\theta = \frac{2\pi(r + \mathrm{d}r) - 2\pi r}{2\pi r} = \frac{\mathrm{d}r}{r} \qquad (7\text{-}58)$$

Obviously, $\varepsilon_r = \varepsilon_\theta$. According to the stress-strain relationship, we have

$$\sigma_r = \sigma_\theta \qquad (7\text{-}59)$$

Substituting Equation (7-59) into Equation (7-57), then

$$\frac{\mathrm{d}\sigma_r}{\mathrm{d}r} + \frac{2\tau_{zr}}{h} = 0 \qquad (7\text{-}60)$$

The friction condition is assumed to be $\tau_{zr} = \begin{cases} \mu Y \\ 0 \\ \mu\sigma_z \end{cases}$.

if $\tau_{zr} = \dfrac{Y}{2}$, then Equation (7-60) becomes

$$\frac{\mathrm{d}\sigma_r}{\mathrm{d}r} = -\frac{Y}{h} \qquad (7\text{-}61)$$

Because $\varepsilon_r = \varepsilon_\theta > 0, \varepsilon_z < 0$, considering the sign of stress, it is obvious that $(-\sigma_r) = (-\sigma_\theta) > (-\sigma_z)$. Ignoring friction τ_{zr}, Tresca yield criterion $\sigma_{\max} - \sigma_{\min} = Y$ becomes

$$(-\sigma_r) - (-\sigma_z) = Y \qquad (7\text{-}62)$$

Rearranging

$$\sigma_z - \sigma_r = Y \qquad (7\text{-}63)$$

Then

$$\frac{\mathrm{d}\sigma_z}{\mathrm{d}r} = \frac{\mathrm{d}\sigma_r}{\mathrm{d}r} \qquad (7\text{-}64)$$

Based on both Equation (7-61) and Equation (7-64), we have

$$\mathrm{d}\sigma_z = -\frac{Y}{h}\mathrm{d}r \qquad (7\text{-}65)$$

Integrating Equation (7-65), Equation (7-66) is deduced

$$\sigma_z = -\frac{Y}{h}r + C \qquad (7\text{-}66)$$

If plastic deformation exists in the whole cylinder upsetting, according to the boundary conditions, when $r=d/2$, $\sigma_r=0$, Equation (7-63) becomes

$$\sigma_z = Y \tag{7-67}$$

Then

$$C = Y + \frac{Y}{h}\frac{d}{2} \tag{7-68}$$

Substituting C into Equation (7-66), the stress distribution on the contact surface of the cylinder upsetting is shown as Equation (7-69)

$$\sigma_z = Y\left[1 + \frac{1}{h}\left(\frac{d}{2} - r\right)\right] \tag{7-69}$$

If the boundary friction condition $\tau_{zr} = \mu\sigma_z$ is taken, the stress distribution on the contact surface for the cylinder upsetting is shown as Equation (7-70)

$$\sigma_z = Y\exp\left[\frac{2\mu(0.5d - r)}{h}\right] \tag{7-70}$$

If the boundary friction condition $\tau_{zr} = \mu Y$ is taken, the stress distribution on the contact surface of the cylinder upsetting is shown as Equation (7-71)

$$\sigma_z = Y\left[1 + \frac{2\mu}{h}(0.5d - r)\right] \tag{7-71}$$

7.14   Normality and Yield Surfaces

A physical interpretation of the flow rules is that the axes of principal stresses and strains coincide. To clarify this point, it is useful to consider these components as vectors which may be plotted in three-dimensional space (as a yield surface plot) or in two-dimensional space (as a yield locus plot). It is again emphasized that this approach has nothing to do with tensor transformations; rather, the direction and magnitude of each component of stress or strain define a vector in regard to the plots that are developed.

Several approaches to the subject of normality have been presented which lead to the conclusion that the total strain vector must be normal to the yield surface. The most rigorous proof is attributed to Drucker and is based upon the concept that plastic work is positive. As a consequence, any acceptable yield surface must be convex around its origin, or a straight line passing through the surface cannot cut it at more than two points.

Two simple physical examples will assist here. First consider Figure 7-17. Assume σ_1 is a principal stress and the material is isotropic. Two possible shape changes are shown above. On the basis of intuition alone, the change shown in Figure 7-17(a) is far more likely to occur than that described in Figure 7-17(b). Therefore, the stress and strain directions would coincide.

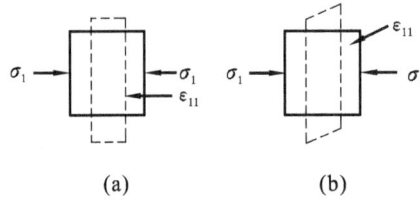

Figure 7-17 Possible conditions illustrating directions of principal stress and strain

Now consider a block resting on a surface as shown in Figure 7-18 where a friction prevails at the interface as indicated in part (a). The top view (b) indicates F acting at some angle θ with respect to the block. As F must overcome the friction force, the maximum work will result when θ reduces to zero in regard to motion of the block. Thus, it would be expected that the block would move in the x direction and the resulting work would be maximized when the force and motion have the same direction. This is a simple example of the principle of maximum work which results in a maximum dissipation of energy, thereby lowering the total energy of the system(In a thermodynamic sense, the external system of stresses does work which is considered positive in the usual sense). A corollary might state that the material being deformed offers maximum resistance to plastic deformation. For this to occur, the strain vector must be normal to the yield surface.

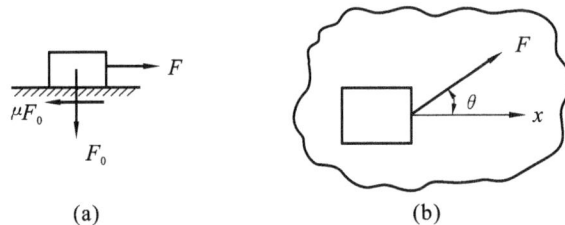

Figure 7-18 An illustration of the principle of maximum work

It is now prudent to indicate, in reasonable detail, a physical development of a yield surface in the three dimensional stress space. The coordinate system relates to principal directions and is shown in Figure 7-19. The state of stress at P is caused by the principal stresses where $\sigma_1 = OP_1, \sigma_2 = OP_2, \sigma_3 = OP_3$ and OP is the total stress in this stress space.

In this coordinate system, consider a line OH having the same direction cosines (i. e. , $l = m = n = 1/\sqrt{3}$) such that $\alpha = \beta = \gamma = 54°44'$.

Project OP on OH to N such that angle $ONP = 90°$, then

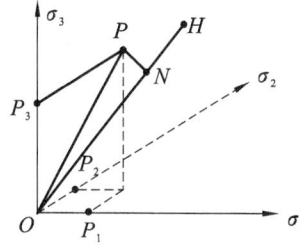

Figure 7-19   Three-dimensional stress space showing total stress OP, ON

related to hydrostatic effects and NP related to the deviatoric stresses

$$ON = \frac{1}{\sqrt{3}}(\sigma_1 + \sigma_2 + \sigma_3)$$

$$(NP)^2 = (OP)^2 - (ON)^2 = (\sigma_1^2 + \sigma_2^2 + \sigma_3^2) - \left(\frac{\sigma_1 + \sigma_2 + \sigma_3}{\sqrt{3}}\right)^2$$

$$= \sigma_1^2 + \sigma_2^2 + \sigma_3^2 - \frac{1}{3}[\sigma_1^2 + \sigma_2^2 + \sigma_3^2 + 2(\sigma_1\sigma_2 + \sigma_2\sigma_3 + \sigma_3\sigma_1)]$$

$$= \frac{2}{3}(\sigma_1^2 + \sigma_2^2 + \sigma_3^2) - \frac{2}{3}(\sigma_1\sigma_2 + \sigma_2\sigma_3 + \sigma_3\sigma_1)$$

$$= \frac{1}{3}[(\sigma_1 - \sigma_2)^2 + (\sigma_2 - \sigma_3)^2 + (\sigma_3 - \sigma_1)^2]$$

With the von Mises yield criterion, yielding occurs when $\sum (\sigma_1 - \sigma_2)^2 = 2Y^2$, so

$$3(NP)^2 = 2Y^2 \quad \text{or} \quad NP = \frac{\sqrt{2}Y}{\sqrt{3}}$$

Thus, the yield locus can be expressed as a circle of radius $NP = \sqrt{2}Y/\sqrt{3}$ and in principal stress space this is a circular cylinder of radius $\sqrt{2}Y/\sqrt{3}$ whose axis is a line passing through the origin and equally inclined to the three coordinate axes.

Figure 7-20 shows the three-dimensional plot of both the von Mises "cylinder" and the Tresca "hexagon". Note that the plane where $\sigma_2 = 0$ cuts the surfaces results in the more familiar yield loci of these two criteria. By superimposing the principal strains upon the same coordinate system, the vector sum of the incremental strains is shown as $d\varepsilon_v^p$; it is this quantity that is normal to the yield surface. It is always composed of components $d\varepsilon_{11}^p$, $d\varepsilon_{22}^p$ and $d\varepsilon_{33}^p$ which have the same axes as σ_1, σ_2 and σ_3. Thus, the direction of the total strain vector, $d\varepsilon_v^p$, depends only upon the shape of the yield surface. So long as P falls within the surface, yielding will not occur, but as P approaches the yield surface, flow becomes imminent. Note that ON represents the hydrostatic component and moving along OH (i. e., increasing or decreasing σ_m) has no influence on yielding. NP represents the deviatoric stress which governs yielding.

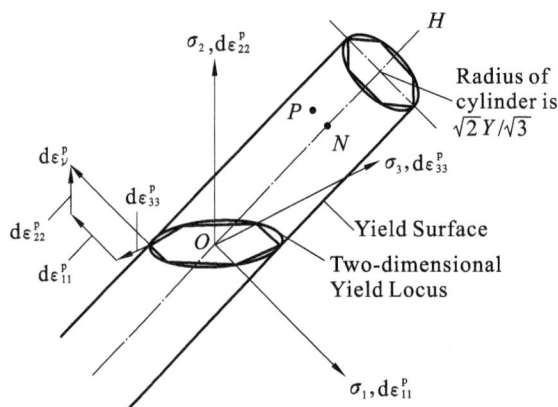

Figure 7-20 Tresca and von Mises yield surfaces

If the stress is projected onto a plane through O and normal to OH (i. e. , look down from the top of the cylinder in a direction parallel to OH), Figure 7-21 results. As mentioned earlier, this is called the π plane and has the equation

$$\sigma_1 + \sigma_2 + \sigma_3 = 0 \quad (\text{i. e. }, \sigma_m = 0)$$

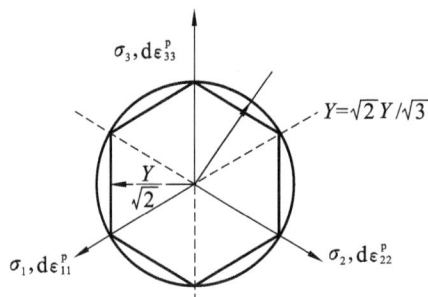

Figure 7-21 π plane projected by Tresca and Von Mises yield surfaces

and the original coordinate axes are 120° apart. Any hydrostatic component is perpendicular to the π plane so it cannot induce plastic work since there is no tendency for the yield surface to expand. However, the deviatoric components, being perpendicular to the surface, would cause plastic work if they caused the yield surface to expand. Referring back to Figure 7-20, maximum work occurs when the total strain vector ($d\varepsilon_v^p$ in Figure 7-20) is normal to the yield surface and in line with the deviatoric component of the total stress.

Now consider the meaning of the above comments in terms of the yield locus shown in Figure 7-22. Note that although $\sigma_2 = 0$, $d\varepsilon_{22}^p$ is not zero in most cases. What is shown as $d\varepsilon_v^p$ as shown in Figure 7-22. Note too that $\boldsymbol{\sigma}$ represents the vector sum of σ_1 and σ_3 and could again be broken down into another pair of components representing σ_m and the deviatoric stress.

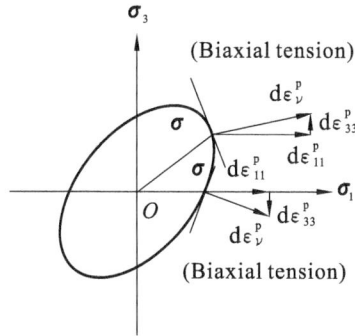

Figure 7-22 Illustration of the principle of normality as related to a yield locus

One of the valuable aspects of normality is its use in the construction of yield surfaces and loci based upon experimental findings. Backofen demonstrates this for a number of situations.

7.15 Plastic Work

Consider a bar of length l_0 subjected to a tensile force F acting on an area $(w_0 t_0)$ with a resulting plastic extension dl. The work done is Fdl and on a unit volume is

$$dW_p = \frac{Fdl}{w_0 t_0 l_0} = \frac{F}{w_0 t_0} \cdot \frac{dl}{l_0} = \sigma d\varepsilon^p \tag{7-72}$$

If a shear force caused deformation, a similar argument would show that $\tau d\gamma$ expresses the work per unit volume done by that force. These individual contributions could be summed up in a manner that produced Equation (7-10) noting that the coefficient of one-half does not appear in Equation (7-72). In terms of principal stresses

$$dW_p = \sigma_1 d\varepsilon_{11}^p + \sigma_2 d\varepsilon_{22}^p + \sigma_3 d\varepsilon_{33}^p \tag{7-73}$$

In defining the incremental effective stress and strain function by Equations (7-8) and (7-26), no mention was made about plastic work, but one might now expect that such definitions could be used in this regard. This is indeed the case and the resulting expression is

$$dW_p = \bar{\sigma} d\overline{\varepsilon^p} \tag{7-74}$$

Two examples will demonstrate this equivalence. Consider uniaxial tension where $\sigma_1 \neq 0, \sigma_2 = \sigma_3 = 0$ and $d\varepsilon_{22}^p = d\varepsilon_{33}^p = -d\varepsilon_{11}^p / 2$ (since $d\varepsilon_{11}^p + d\varepsilon_{22}^p + d\varepsilon_{33}^p = 0$). Using Equation (7-73) gives $dW_v = \sigma_1 d\varepsilon_{11}$ since the other terms vanish. The incremental effective stress and strain functions show that

$$\bar{\sigma} = \sigma_1 \quad \text{and} \quad d\overline{\varepsilon^p} = d\varepsilon_{11}^p$$

So equivalence results.

Next consider pure shear where $\sigma_1 = -\sigma_3, \sigma_2 = 0$ and $d\varepsilon_{11}^p = -d\varepsilon_{33}^p, d\varepsilon_{22}^p = 0$. From Equation (7-73)

$$dW_p = \sigma_1 d\varepsilon_{11}^p + (-\sigma_1)(-d\varepsilon_{11}^p) = 2\sigma_1 d\varepsilon_{11}^p$$

The relations in terms of effective values show that

$$\bar{\sigma} = \sqrt{3}\sigma_1 \quad \text{while} \quad d\bar{\varepsilon}^p = \frac{2}{\sqrt{3}} d\varepsilon_{11}^p$$

Thus

$$dW_p = (\sqrt{3}\sigma_1)\left(\frac{2}{\sqrt{3}} d\varepsilon_{11}^p\right) = 2\sigma_1 d\varepsilon_{11}^p$$

as before.

Thus, besides finding use in regard to yielding predictions and the flow rules, the concepts of incremental effective stress and strain provide a convenient way to calculate the work due to plastic deformation.

7.16 Comparison of Mohr's Circles for Stress and Plastic Strain Increments

Since the hydrostatic component of the total stress causes no plastic flow, this implies that the value of σ_m on a circle plot of stresses should coincide with the origin of the incremental strain circle (i. e., $d\varepsilon^p = 0$). Under uniaxial tension, $\sigma_m = \sigma_1/3, \sigma_2 = \sigma_3 = 0$, while $d\varepsilon_{22}^p = d\varepsilon_{33}^p = -d\varepsilon_{11}^p/2$. Figure 7-23(a) shows the physical meaning that relates the stress circle to the scaled results in terms of strains; note the negative two-to-one correspondence of increment strains. Figure 7-23(b) shows a more general case where $\sum d\varepsilon_{ii}^p$

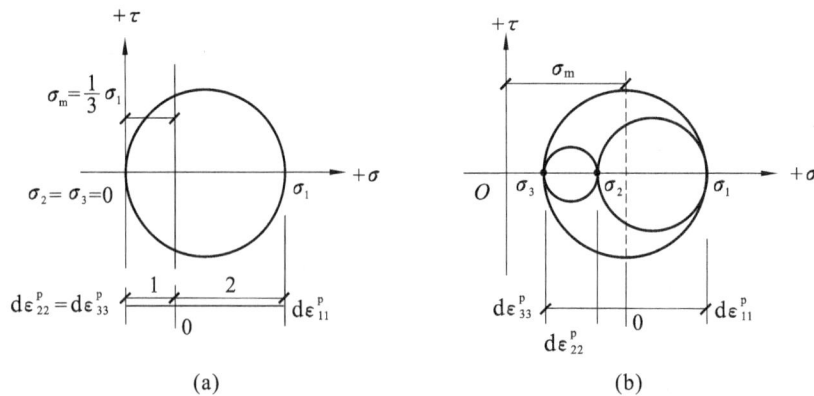

Figure 7-23   Relationship of Mohr's circle of stress and incremental plastic strains for uniaxial tension (a) and a general triaxial situation (b)

must equal zero. It is also instructive to consider plane strain

$$\sigma_2 = \frac{1}{2}(\sigma_1 + \sigma_3) \quad \text{since} \quad \nu = \frac{1}{2} \text{ for plastic flow}$$

For this case, $\sigma_m = \sigma_2$ so the zero of the incremental strain circle coincides with σ_2 or $d\varepsilon_{22}^p = 0$. This result can be readily checked using Equation (7-25).

It should be realized that changing σ_m by superimposing an additional uniform normal stress would displace the stress circle but would have no effect on the size of either set of circles. Thus, a knowledge of strain ratios does not uniquely define the current stresses.

7.17 A Geometrical Considerations

The deviatoric stress vector **OQ** may be regarded as the sum of the projections on the deviatoric plane of the component vectors of **OL** along the axes of reference (Figure 7-24). Since each stress axis is inclined to the deviatoric plane at an angle $\arcsin(1/\sqrt{3})$ in the original stress space, each projected length along the axes is $\sqrt{2/3}$ times the actual length. Hence the lengths of the component vectors in the deviatoric plane are $OL = \sqrt{2/3}\sigma_1, LM = \sqrt{2/3}\sigma_2, MQ = \sqrt{2/3}\sigma_3$.

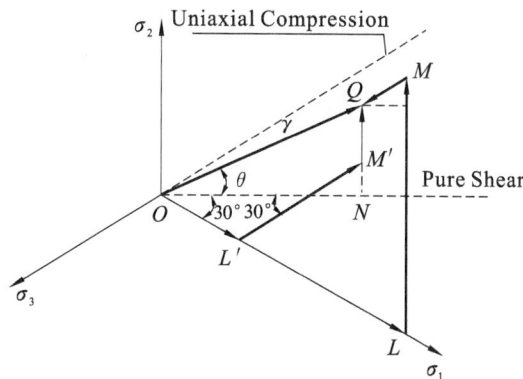

Figure 7-24 Deviatoric stress vector and its components along the projected axes

From geometry, the rectangular components of **OQ** with respect to the horizontal and vertical through O are

$$\begin{cases} ON = \dfrac{\sigma_1 - \sigma_3}{\sqrt{2}} = r\cos\theta \\[2mm] QN = \dfrac{2\sigma_2 - \sigma_3 - \sigma_1}{\sqrt{6}} = r\sin\theta \end{cases} \tag{7-75}$$

where (r, θ) are the polar coordinates of Q. For the experimental determination of the yield criterion, it is convenient to introduce the Lode stress parameter μ defined as

$$\mu = \frac{2\sigma_2 - \sigma_3 - \sigma_1}{\sigma_3 - \sigma_1} = -\sqrt{3}\tan\theta \quad (\sigma_1 > \sigma_2 > \sigma_3) \tag{7-76}$$

To obtain the yield locus, it is only necessary to apply stresses for which θ covers the range from 0 to $\pm\pi/6$, μ varying from 0 to ∓ 1. When $\mu = 0$, $\sigma_2 = (\sigma_3 + \sigma_1)/2$ and we have a state of pure shear denoted by $\frac{1}{2}(\sigma_3 - \sigma_1, \sigma_1 - \sigma_3, 0)$ together with a hydrostatic stress $\frac{1}{2}(\sigma_3 + \sigma_1)$. When $\mu = -1$, $\sigma_1 = \sigma_2$ and the state of stress is equivalent to a uniaxial compression $(\sigma_3 - \sigma_1, 0, 0)$ and a hydrostatic stress σ_1. Let the yield stresses in pure shear and uniaxial tension (or compression) be denoted by k and Y respectively. Then it follows from Equations (7-75) that $r = \sqrt{2}k$ at $\theta = 0$ and $r = \sqrt{2/3}Y$ at $\theta = \pi/6$. For most metals k lies between $Y/2$ and $Y/\sqrt{3}$. Equations (7-75) may be written alternatively in the form

$$\sigma_1' - \sigma_3' = \sqrt{2}r\cos\theta; \quad \sigma_1' + \sigma_3' = -\sigma_2' = -\sqrt{2/3}r\sin\theta$$

from which the deviatoric principal stresses can be expressed as

$$\begin{cases} \sigma_1' = \sqrt{\dfrac{2}{3}}r\cos\left(\dfrac{\pi}{6} + \theta\right) \\[2mm] \sigma_2' = \sqrt{\dfrac{2}{3}}r\sin\theta \\[2mm] \sigma_3' = -\sqrt{\dfrac{2}{3}}r\cos\left(\dfrac{\pi}{6} - \theta\right) \end{cases} \tag{7-77}$$

When the yield locus is given, r is a known function of θ, and Equations (7-77) define the yield criterion parametrically through θ.

When the vector $\boldsymbol{OQ}$ is given, the deviatoric stresses can be obtained graphically by noting the fact that $\sigma_2' = \sqrt{2/3}\,QN$. Hence if a point M' is located on $\boldsymbol{QN}$ such that $QM'/NM' = 2$, then $QM' = \sqrt{2/3}\sigma_2'$. Let $L'M'$ be drawn parallel to $O\sigma_3$. Since $\boldsymbol{L'M'} = \boldsymbol{OL'} + \boldsymbol{QM'}$ by geometry, it follows that $OL' = \sqrt{2/3}\sigma_1'$ and $L'M' = -\sqrt{2/3}\sigma_3'$. The points L' and M' therefore define the deviatoric stresses. The hydrostatic stress is of magnitude $\sqrt{2/3}LL'$, but it is not defined by the vector $\boldsymbol{OQ}$. The ratios of the deviatoric stresses are

$$\sigma_1' : \sigma_2' : \sigma_3' = (\sqrt{3} - \tan\theta) : 2\tan\theta : -(\sqrt{3} + \tan\theta) \tag{7-78}$$

in view of Equations (7-77). Multiplying the three equations in Equations (7-77) and remembering that $\sigma_1'\sigma_2'\sigma_3' = J_3$ and $r = \sqrt{2J_2}$, we obtain

$$J_3^2 = \frac{4}{27}J_2^3 \sin^2 3\theta \tag{7-79}$$

If the polar equation of the yield locus is given, J_2 is a known function of θ.

7.18 Isotropic Hardening

We have seen that an element of material yields when the magnitude of the deviatoric stress vector is increased to a value such that the stress point reaches the yield locus. Unless the locus is a circle (as for the Von Mises yield criterion), the magnitude of the stress vector causing yielding depends on its final direction in the deviatoric plane. If the material is non-hardening, the plastic stress state can change in such a way that the stress point always lies on a constant yield locus. For a strain-hardening material, the size and shape of the yield locus depend on the complete history of plastic deformation since the previous annealing. It is assumed that the material is isotropic at the annealed state and that the anisotropy and the Bauschinger effect developed during the cold work may be neglected. The preceding discussion of the yield criterion is then appropriate for any given state of hardening of the material.

A convenient mathematical formulation for strain-hardening is obtained by assuming further that the yield surface uniformly expands without change in shape, as the state of stress changes along a certain path P_0P in the stress space(Figure 7-25), the amount of hardening being given by the final plastic state. Since the yield locus merely increases in size, any given state of hardening may be defined by the current yield stress in uniaxial tension. It is, therefore, necessary to relate the current yield stress to the amount of plastic deformation following a given initial state of yielding. To this end, we replace Y in the yield criterion by $\bar{\sigma}$, which is known as the equivalent stress, effective stress, or generalized stress. Referring to the von Mises yield criterion, we write

$$\bar{\sigma} = \sqrt{\frac{3}{2}}\,(\sigma_x'^2 + \sigma_y'^2 + \sigma_z'^2 + 2\tau_{xy}^2 + 2\tau_{yz}^2 + 2\tau_{zx}^2)^{\frac{1}{2}}$$

$$= \sqrt{\frac{1}{2}}\,[(\sigma_x-\sigma_y)^2 + (\sigma_y-\sigma_z)^2 + (\sigma_z-\sigma_x)^2 + 6\tau_{xy}^2 + 6\tau_{yz}^2 + 6\tau_{zx}^2]^{\frac{1}{2}} \qquad (7\text{-}80)$$

Consider first the hypothesis in which the amount of hardening is taken as a function of the total plastic work per unit volume. This assumption is obviously consistent with the fact that no hardening is produced by purely elastic strains. If the plastic part of the strain increment tensor is denoted by $\mathrm{d}\varepsilon_{ij}^{\mathrm{p}} = \dot{\varepsilon}_{ij}^{\mathrm{p}}\,\mathrm{d}t$, where $\dot{\varepsilon}_{ij}^{\mathrm{p}}$ is the rate of plastic deformation and $\mathrm{d}t$ is the time element, the increment of plastic work per unit volume is

$$\mathrm{d}W_{\mathrm{p}} = \tau_{ij}\,\mathrm{d}\varepsilon_{ij}^{\mathrm{p}} = (\tau_{ij}' + \sigma_{\mathrm{m}}\delta_{ij})\,\mathrm{d}\varepsilon_{ij}^{\mathrm{p}} = \tau_{ij}'\,\mathrm{d}\varepsilon_{ij}^{\mathrm{p}}$$

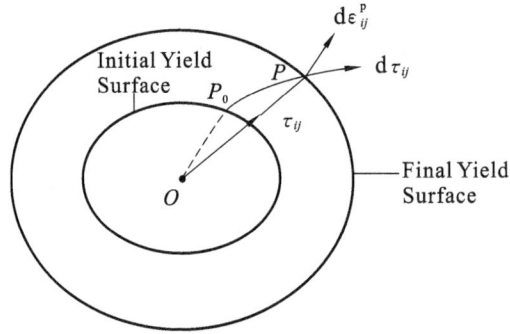

Figure 7-25 Geometrical representation of the isotropic hardening rule

where $\tau_{ij}=\sigma_i$ if $i=j$ and $\tau'_{ij}=\tau_{ij}$ if else; the last step follows from the condition $d\varepsilon^p_{ij}=0$, implying that there is no plastic volume change. Plastic incompressibility of metals is in close agreement with experimental observations, and is also consistent with the fact a uniform hydrostatic stress produces no plastic strain. The strain-hardening hypothesis may be stated mathematically as

$$\bar{\sigma} = \Phi\left(\int d\omega_p\right) = \Phi\left(\int \tau_{ij}\,d\varepsilon_{ij}\right) \tag{7-81}$$

where the integral is taken over the actual strain path starting from some initial state. The function Φ can be determined from the true stress-strain curve in uniaxial tension or compression. If the true stress σ is plotted against the plastic part of the strain, then W_p is equal to the area under the curve up to the ordinate σ. Since $\bar{\sigma}=\sigma$ in this case, the area directly gives the argument of Φ in Equation (7-81).

In an alternative hypothesis, more frequently in use, $\bar{\sigma}$ is regarded a function of a certain measure of the total plastic strain. Considering the second invariant of the plastic strain increment tensor, an equivalent (or generalize) plastic strain increment is defined as

$$d\bar{\varepsilon}^p = \sqrt{\frac{2}{3}}(d\varepsilon^p_{ij}\,d\varepsilon^p_{ij})^{\frac{1}{2}}$$
$$= \sqrt{\frac{2}{3}}\left[(d\varepsilon^p_{xx})^2 + (d\varepsilon^p_{yy})^2 + (d\varepsilon^p_{zz})^2 + 2(d\gamma^p_{xy})^2 + 2(d\gamma^p_{yz})^2 + 2(d\gamma^p_{zx})^2\right]^{\frac{1}{2}} \tag{7-82}$$

where only the positive root is implied. The numerical factor in the above expression is so chosen that in uniaxial tension, $d\bar{\varepsilon}^p$ equals the longitudinal plastic strain increment. This follows from the fact that the magnitude of the lateral compressive plastic strain in the tensile test of an isotropic bar is half the longitudinal tensile plastic strain. The strain-hardening hypothesis may now be expressed as

$$\bar{\sigma} = F\left(\int d\,\bar{\varepsilon}^{\mathrm{p}}\right) = F\left[\sqrt{\frac{2}{3}\,d\varepsilon_{ij}^{\mathrm{p}}\,d\varepsilon_{ij}^{\mathrm{p}}}\right] \tag{7-83}$$

where the integral is taken along the strain path as before. The integrated strain, known as the total equivalent plastic strain, provides a suitable measure of the plastic deformation. The function F is given by the relationship between the true stress and the plastic strain in uniaxial tension or compression. Equation (7-83) implies that the amount of hardening is determined by every infinitesimal plastic distortion leading to the final shape of an element, and not merely by the difference between the initial and final shapes of the element.

7.19 Closure

In this chapter, we first introduced the difference between elastic and plastic behavior and three fairly distinct approaches in the study of plasticity. Models for plastic deformation were explained. With the assumptions of isotropy, no Bauschinger effect, incompressibility during plastic flow and yielding being uninfluenced by hydrostatic effects, there have existed two criterions that are to be used to predict the onset of yielding such as Tresca and Von Mises yield criterion. We interpreted that Von Mises yield criterion is the elastic energy causing distortion, and octahedral shear stress. The Prandtl-Reuss and the Levy-Mises equations are expressed, and the Levy-Mises equations may be viewed as a special form of the more general expression. There was also another plastic theory for the relationship between the total components of the stress and strain which is called deformation theory of plasticity or total strain theory. Several approaches to the subject of normality have been presented which lead to the conclusion that the total strain vector must be normal to the yield surface. The concepts of effective stress and strain provide a convenient way to calculate the work due to plastic deformation. We also gave geometrical considerations to the deviatoric stress vector and discussed the yield criterion under the isotropic hardening.

Problems

7-1 Show that the "reduced stress invariant" J_2, equals

$$\frac{1}{6}\left[(\sigma_1 - \sigma_2)^2 + (\sigma_2 - \sigma_3)^2 + (\sigma_3 - \sigma_1)^2\right]$$

7-2 The tensile yield strength of a metal is given as Y MPa. If a specimen of this material were subjected to two normal compressive stresses of $-\frac{1}{4}Y$ and $-\frac{1}{2}Y$ acting along x and y directions, what tensile stress applied in the z direction would cause yielding according to:

(1) the Von Mises yield criterion?

(2) the Tresca yield criterion?

7-3　A cube of metal, 25 mm, on a side, is subjected to a compressive load of 250 kN at which point the metal yields. Assume that friction at the loading interface is negligible (i. e., no shear stresses). If an identical cube were first constrained by compressive loads of 90 kN and 135 kN on the other pairs of faces (again, friction is zero), what compressive load in the third coordinate direction would be necessary to cause yielding? Use the Von Mises yield criterion.

7-4　A long, thin-walled tube, capped on the ends is made of a metal whose yield strength in uniaxial tension is 280 MPa. The tube is 1500 mm long, has a wall thickness of 0. 38 mm and a diameter 50 mm. Under service conditions the tube experiences an axial tensile load of 4. 5 kN, a torque of 0. 115 kN • m, and is to be pressurized internally. At what internal pressure is yielding predicted by:

(1) the Tresca yield criterion?

(2) the Von Mises yield criterion?

7-5　A pressure vessel, in the form of a cylinder with hemispherical ends, has a radius of 600 mm and is to be made from a metal whose $k=560$ MPa. The maximum internal pressure intended during use is 35 MPa. If no section of the vessel is to yield, what minimum wall thickness should be specified according to:

(1) the Tresca yield criterion?

(2) the Von Mises yield criterion?

7-6　Consider the cases of (a) uniaxial tension, (b) pure shear, and (c) the triaxial case where $\sigma_1 > \sigma_2 > \sigma_3$ respectively. For each case compare the ratio of the effective stress, $\bar{\sigma}$ and the maximum shear stress, τ_{max}.

7-7　A thin-walled cylinder (diam=80 mm, wall thickness=3. 5 mm) just yields when a uniform axial stress of 200 MPa is applied. If an identical cylinder is loaded in bending to a maximum axial normal stress of 140 MPa, calculate the internal pressure required to cause yielding, using the Von Mises yield criterion.

7-8　A metal whose yield in pure shear is 105 MPa is formed into a thin-walled tube capped on the ends where: diam=150 mm, wall thickness=1. 9 mm, length=500 mm. The elastic modulus is 1. 4×10⁵ MPa and $\nu=0. 25$. If the tube is pressurized to a level of

3. 5 MPa above that at which initial yielding occurred according to the Von Mises yield criterion, determine the final length of the tube.

7-9 Consider plane strain plastic deformation ($d\varepsilon_{22}^p = 0$). Using the Von Mises yield criterion, show that $\sigma_1 - \sigma_3 = 2\bar{\sigma}/\sqrt{3}$.

7-10 A metal flows at a constant yield stress, i. e. , $\bar{\sigma} = Y =$ constant. If this metal is deformed by uniaxial tension to a strain ε that does not induce necking, demonstrate that the plastic work per unit volume is $\bar{\sigma}\overline{\varepsilon^p}$.

7-11 A material whose yield strength $= 350$ MPa is made in the form of a cube and subjected to a tensile stress, σ_1, along one axis and a stress $\sigma_3 = -\sigma_1/2$ along a second set of axes.

(1) Determine the ratio of the principal strain increment $d\varepsilon_{11}^p/d\varepsilon_{22}^p$.

(2) Using the Von Mises yield criterion, determine the magnitude of τ_{max} at the onset of yielding.

(3) Repeat (2) where $\sigma_3 = +\dfrac{1}{2}\sigma_1$.

(4) Repeat (2) and (3) using the Tresca yield criterion.

7-12 A stress state is described in (MPa) by $\sigma_1 = 30, \sigma_2 = 15, \sigma_3 = 0$.

(1) Determine the strain ratio $d\varepsilon_{11}^p/d\varepsilon_{33}^p$.

(2) If a hydrostatic stress of 20 MPa is superimposed by fluid pressure upon the initial stress state, how does the ratio in (1) change? Explain it.

7-13 A thin-walled cylinder (diam$=80$ mm, wall thickness$=3. 5$ mm) just yields when a uniform axial stress of 200 MPa is applied. If an identical cylinder is loaded in bending to a maximum axial normal stress of 140 MPa, calculate the internal pressure required to cause yielding, using the Tresca yield criterion (See Problem 7-7).

7-14 A thin-walled cylinder just yields when the torsional shear stress reaches 306 MPa. If an identical cylinder is loaded to a torsional shear stress of 270 MPa, calculate the applied axial compressive stress necessary to just cause yielding. Assume the Tresca yield criterion holds.

7-15 A rigid-plastic cube is subjected to σ_1 on one pair of opposite faces, to $\sigma_2 = 0. 2\sigma_1$ on a second pair, and to $\sigma_3 = -0. 4\sigma_1$ on the third pair of faces. The stresses are gradually increased, maintaining the above ratios. Using the Von Mises criterion for yielding (for $Y = 300$ MPa), calculate the magnitudes of the principal stresses and the ratios of the three principal strain increments at the moment of yielding.

7-16 The Von Mises yield criterion is to be used for yield predictions of a given metal.

Under plane strain deformation where σ_3 is zero, make a sketch of the Von Mises ellipse for the first quadrant in σ_1-σ_2 stress space and

(1) indicate how the strain increments $d\varepsilon_{11}^p$ and $d\varepsilon_{22}^p$ would look on the plot where normality prevails.

(2) determine the plastic work increment in terms of the principal stresses and strain increments.

(3) explain your answer in (2) with reference to the plot in (1).

8　The Examples of Elastoplastic Analysis

8.1　Introduction

In a deformable body subjected to external loads of gradually increasing magnitude, plastic flow begins at a stage when the yield criterion is first satisfied in the most critically stresses element. Further increase in loads causes spreading of the plastic zone which is separated from the elastic material by an elastic/plastic boundary. The position of this boundary is an unknown of the problem, and is generally so complicated in shape that the solution of the boundary-value problem often involves numerical methods. The solution must be carried out in a succession of small increments of strain even when the deformation is restricted to an elastic order of magnitude. It is necessary to ensure at each stage that the calculated stresses and displacements in the elastic and plastic regions satisfy the conditions of continuity across the elastic/plastic boundary. In this chapter, we shall be concerned mainly with problems of bending and torsion in the elastic/plastic range, assuming the deformation to be sufficiently small.

8.2　Plane Strain Compression of a Block

As a simple application of the Prandtl-Reuss theory, consider the frictionless compression of a rectangular block of metal between a pair of rigid overlapping platens (Figure 8-1). The edges of the block are parallel to the rectangular axes, with the x axis taken in the direction of compression.

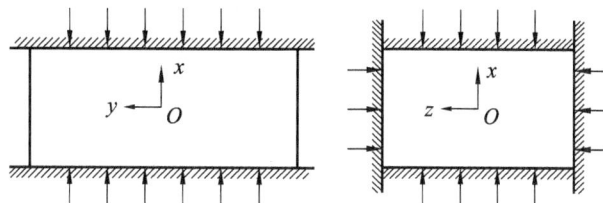

Figure 8-1　Plastic compression of a block between smooth rigid platens under conditions of plane strain

A condition of plane strain is achieved by suppressing lateral expansion in the z direction with the help of rigid dies. It is therefore a case of homogeneous compression in which $\sigma_y = 0$ throughout the deformation, and $\sigma_z = \nu\sigma_x$ while the block is still elastic. If Tresca's yield criterion is adopted, yielding begins when $\sigma_x = -Y$ in each element of the block. The relevant stress-strain equations in the plastic range are

$$
\begin{cases}
\mathrm{d}\varepsilon_{xx} = \dfrac{1}{E}(\mathrm{d}\sigma_x - \nu\mathrm{d}\sigma_z) + \dfrac{1}{3}(2\sigma_x - \sigma_z)\mathrm{d}\lambda \\
\mathrm{d}\varepsilon_{zz} = \dfrac{1}{E}(\mathrm{d}\sigma_z - \nu\mathrm{d}\sigma_x) + \dfrac{1}{3}(2\sigma_z - \sigma_x)\mathrm{d}\lambda = 0
\end{cases}
\tag{8-1}
$$

If the material is non-hardening, $\sigma_x = -Y$ throughout the plastic compression. The elimination of $\mathrm{d}\lambda$ from Equation (8-1) then gives

$$
E\mathrm{d}\varepsilon_{xx} = \left(\frac{1}{2} - \nu\right)\mathrm{d}\sigma_x + \frac{3\mathrm{d}\sigma_z}{2(2\sigma_z + Y)}
$$

At the initial yielding, $\sigma_z = -\nu Y$ and $\varepsilon_{xx} = -(1-\nu^2)Y/E$. Under these initial conditions, the above equation integrates to

$$
\frac{E}{Y}\varepsilon_{xx} = \left(\frac{1}{2} - \nu\right)\left(\frac{\sigma_z}{Y} + \nu\right) - \frac{3}{4}\ln\frac{1 - 2\nu}{1 + \dfrac{2\sigma_z}{Y}} - (1 - \nu^2)
\tag{8-2}
$$

Equation (8-2) gives the variation of σ_z with the amount of compression. As the deformation proceeds, the first term becomes increasingly unimportant, while σ_z rapidly approaches the limiting value $-(1/2)Y$. Taking $\nu = 0.3$, for instance, σ_z is found to have the value $-0.49Y$ when ε_{xx} is only 3.5 times that at the initial yielding.

When the material yields according to the von Mises yield criterion $\sigma_x^2 - \sigma_x\sigma_z + \sigma_z^2 = Y^2$, the initial yielding of the block corresponds to $\sigma_x = \sigma_x^0$ and $\sigma_z = \nu\sigma_x^0$, where

$$
\sigma_x^0 = -\frac{Y}{\sqrt{1 - \nu + \nu^2}}
$$

During the subsequent compression, the yield criterion can be identically satisfied by writing the stresses in terms of a parameter θ as

$$
\begin{cases}
\sigma_x = -\dfrac{2Y}{\sqrt{3}}\cos\theta \\
\sigma_z = -\dfrac{2Y}{\sqrt{3}}\sin\left(\dfrac{\pi}{6} - \theta\right)
\end{cases}
\tag{8-3}
$$

The condition $\sigma_z = \nu\sigma_x$ at the initial yielding furnishes the initial value of θ as

$$
\theta_0 = \arctan\frac{1 - 2\nu}{\sqrt{3}}
\tag{8-4}
$$

When $\nu = 0.3$, we get $\sigma_x^0 \approx -1.127Y$ and $\theta_0 \approx 13°$. Substitution from Equation (8-3) into the stress-strain relation Equation (8-1) gives

$$d\varepsilon_{xx} = \frac{2Y}{\sqrt{3}E}\left[\sin\theta - \nu\cos\left(\frac{\pi}{6} - \theta\right)\right]d\theta - \frac{2Y}{\sqrt{3}}\cos\left(\frac{\pi}{6} - \theta\right)d\lambda$$

$$0 = \frac{2Y}{\sqrt{3}E}\left[\cos\left(\frac{\pi}{6} - \theta\right) - \nu\sin\theta\right]d\theta + \frac{2Y}{\sqrt{3}}\sin\theta d\lambda$$

Since $d\lambda$ must be positive, the second equation indicates that θ decreases as the compression proceeds. Eliminating $d\lambda$ from the above equations, we get

$$Ed\varepsilon_{xx} = \frac{2Y}{\sqrt{3}}\left[(1 - 2\nu)\cos\left(\frac{\pi}{6} - \theta\right) + \frac{3}{4}\frac{1}{\sin\theta}\right]d\theta$$

Using the initial condition $\varepsilon_{xx} = -(1 - \nu^2)\sigma_x^0/E$ when $\theta = \theta_0$, the above equation is readily integrated to obtain

$$-\frac{E}{Y}\varepsilon_{xx} = \frac{2}{\sqrt{3}}(1 - 2\nu)\sin\left(\frac{\pi}{6} - \theta\right) + \frac{\sqrt{3}}{2}\ln\left(\cot\frac{\theta}{2}\tan\frac{\theta_0}{2}\right) + \sqrt{1 - \nu + \nu^2} \qquad (8\text{-}5)$$

As the deformation continues, the first term on the right-hand side soon becomes negligible. The angle θ rapidly approaches the limiting value zero, the corresponding values of σ_x and σ_z being $-2Y/\sqrt{3}$ and $-Y/\sqrt{3}$ respectively. It is found that σ_z is within 1 percent of its limiting value when ε_x is only four times that at the initial yielding, for $\nu = 0.3$. Owing to the rapid initial change in stress, the elastic and plastic strain increments are comparable up to a total strain which is three to four times that at the elastic limit. A graphical comparison of the solutions based on the Tresca and Mises yield criteria is made in Figure 8-2.

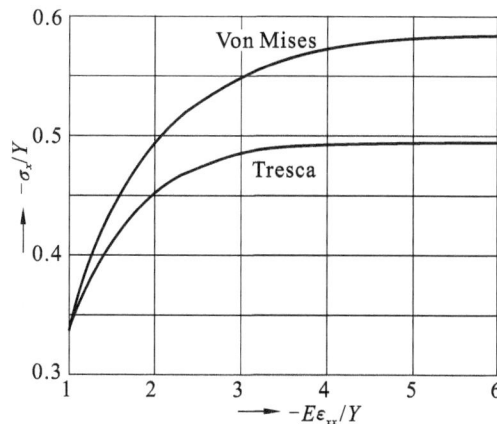

Figure 8-2   The plane strain compression of a block in the plastic range

8.3   Plane Strain Bending of a Beam

A related problem is the bending of a uniform rectangular beam by terminal couples under conditions of plane strain (Figure 8-3). The radius of curvature of the bent beam is assumed large compared to its depth $2h$, so that transverse stresses may be neglected. The neutral fibre coincides with Ox, and is bent into a circular of radius R. All fibres above this line are extended and those below this line are compressed during the bending. So long as the beam remains elastic, the longitudinal stress σ_x is distributed linearly across the depth of the beam according to the relationship

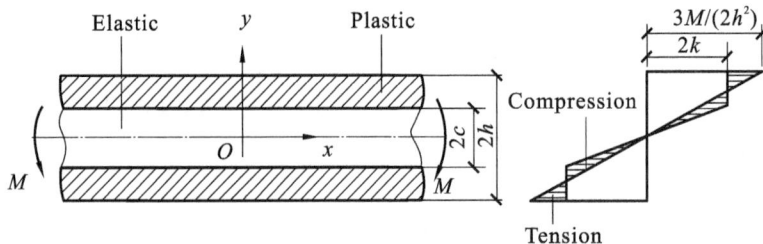

Figure 8-3   Geometry and stress distribution (approximate) in the plane strain bending of a beam(The residual stress is given by the shaded triangles)

$$\sigma_x = \frac{Ey}{(1-\nu^2)R} = \frac{My}{I_z}$$

where M is the bending couple per unit width of the beam, and $I_z = (2/3)h^3$ the moment of inertia of the cross section per unit width about the z axis. The factor $(1-\nu^2)$ arises from the condition of plane strain $(\varepsilon_{zz}=0)$ during the bending. Plastic yielding begins at the boundaries $y=\pm h$ when the longitudinal stress attains the value $\pm Y/\sqrt{1-\nu+\nu^2}$. The bending moment M_e at the initial yielding, and the corresponding radius of curvature R_e of the neutral surface, are

$$\begin{cases} M_e = \dfrac{2h^2Y}{3\sqrt{1-\nu+\nu^2}} \\ R_e = \dfrac{Eh\sqrt{1-\nu+\nu^2}}{Y(1-\nu^2)} \end{cases} \tag{8-6}$$

where subscript e represents the elastic limit. For $\nu=0.3$, the numerical values of M_e and R_e are $0.751Yh^2$ and $0.934Eh/Y$ respectively.

If the bending moment is increased further, plastic zones spread inward from the outer surfaces, the depth of the elastic part at any stage being denoted by $2c$. The stresses in the elastic region are

$$\sigma_x = \frac{Yy}{c\sqrt{1-\nu+\nu^2}}; \quad \sigma_z = \nu\sigma_x \quad (-c \leqslant y \leqslant c) \tag{8-7}$$

The longitudinal strain at a generic point of the cross section is y/R throughout the bending. The application of Hooke's law to the elastic part of the beam gives

$$R = \frac{Ec\sqrt{1-\nu+\nu^2}}{Y(1-\nu^2)} = \left(\frac{c}{h}\right) R_e$$

during the elastic/plastic bending. In the lower plastic region, the stresses are given by Equation (8-3), where θ depends on y according to Equation (8-5) with $\varepsilon_{xx} = y/R$. The stresses and strains in the upper plastic region are identical in magnitude but opposite in sign. The applied couple per unit width is

$$M = 2\int_0^h \sigma_x y \, \mathrm{d}y = \frac{2Yc^2}{3\sqrt{1-\nu+\nu^2}} + \frac{2Y}{\sqrt{3}}\int_c^h y\cos\theta \, \mathrm{d}y \tag{8-8}$$

Using Equation (8-5), with $-\varepsilon_{xx}$ replaced by y/R, the above integral can be evaluated numerically to obtain $M/(h^2 Y)$ for any assumed value of c/h.

For practical purposes, it is sufficiently accurate to replace the Von Mises yield criterion by the modified Tresca yield criterion $\sigma_x = \pm 2Y/\sqrt{3}$. Then the magnitude of the longitudinal stress increases from zero at the neutral surface to $2Y/\sqrt{3}$ at the elastic/plastic boundary. The integration in this case is straightforward, and the result is

$$M \approx \frac{2Y}{\sqrt{3}}\left(h^2 - \frac{1}{3}c^2\right) = \frac{1}{2}M_e\left[3 - \left(\frac{R}{R_e}\right)^2\right] \tag{8-9}$$

where

$$M_e \approx \frac{4Yh^2}{3\sqrt{3}}; \quad R_e \approx \frac{\sqrt{3}Eh}{2Y(1-\nu^2)}$$

The maximum error in this approximation is about 2 percent, occurring at the initial yielding. The bending moment M rapidly approaches the asymptotic value $(2/\sqrt{3})h^2 Y$ or $(3/2)M_e$, which is the fully plastic or collapse moment per unit width of the beam. The limiting plastic state involves a stress discontinuity of amount $4Y/\sqrt{3}$ across the neutral surface.

If the beam is unloaded from the partly plastic state, there is a certain distribution of residual stress left in the beam. The residual stress can be calculated on the assumption that the change in stress during the unloading is purely elastic. It is therefore necessary to superpose an elastic stress distribution due to an opposite moment equal in magnitude to that which is released. Subtracting My/I_z from the existing stress in the elastic/plastic beam, where M is given by Equation (8-9), we obtain the residual stress on complete unloading as

$$\begin{cases} \dfrac{\sigma_x}{Y} = \dfrac{2}{\sqrt{3}}\left[\dfrac{y}{c} - \dfrac{y}{2h}\left(3 - \dfrac{c^2}{h^2}\right)\right] & (|y| \leqslant c) \\[3mm] \dfrac{\sigma_x}{Y} = \dfrac{2}{\sqrt{3}}\left[1 - \dfrac{y}{2h}\left(3 - \dfrac{c^2}{h^2}\right)\right] & (|y| \geqslant c) \end{cases} \tag{8-10}$$

The distribution is shown diagrammatically by the shaded triangles in Figure 8-3. The residual stress changes sign in the region $c < |y| < h$, vanishing at a distance $2h/(3-c^2/h^2)$ from the neutral surface. The stress attains its greatest magnitude at the outer surface for $c/h \geqslant \sqrt{2}-1$, and at the plastic boundary for $c/h \leqslant \sqrt{2}-1$. As the beam is rendered increasingly plastic, the residual stress at $y = \pm h$ approaches the limiting value $\mp Y/\sqrt{3}$.

The curvature of the unloaded beam is obtained by subtracting from the elastic/plastic curvature $h/(cR_e)$ the amount of elastic spring-back equal to $(1-\nu^2)M/(EI_z)$. Substituting for R_e, M and I_z, the residual curvature may be expressed as

$$\frac{1}{R} = \frac{2}{\sqrt{3}}(1-\nu^2)\frac{Y}{Ec}\left(1 - \frac{3c}{2h} + \frac{c^3}{2h^3}\right) \tag{8-11}$$

The factor outside the bracket is the curvature of the beam at the moment of unloading. The expression within the bracket is the radio of the residual stress at $y = c$ to the plane strain yield stress $2Y/\sqrt{3}$. For small elastic/plastic bending, the residual curvature is comparable to the amount of elastic spring-back.

8.4 Cylindrical Bars under Pure Torsion

We begin with a solid cylindrical bar of radius a, subjected to a twisting moment T. So long as the bar is elastic, the shear stress acting over any cross section is proportional to the radial distance r from the central axis. The applied torque T is the resultant moment of the stress distribution about this axis. If the angle of twist per unit length of the bar is denoted by θ, the elastic shear stress may be written as

$$\tau = Gr\theta = \frac{2Tr}{\pi a^4}$$

Since the shear stress has its greatest value at $r = a$, the bar begins to yield at this radius when the torque is increased to T_e, the corresponding twist being θ_e. Setting $\tau = k$ at $r = a$, we get

$$T_e = \frac{1}{2}\pi k a^3; \quad \theta_e = \frac{k}{Ga}$$

If the torque is increased further, a plastic annulus forms near the boundary, leaving a central zone elastic material within a radius c (Figure 8-4). The stress distribution in the elastic region is linear, with the shear stress reaching the value k at $r = c$. For a non-hardening material, the shear stress has the constant value k throughout the plastic region, and the stress distribution becomes

$$\tau = k\frac{r}{c} \quad (0 \leqslant r \leqslant c)$$

$$\tau = k \quad (c \leqslant r \leqslant a)$$

Since the shear stress within the elastic zone is also equal to $Gr\theta$, we have $\theta = k/(Gc)$. The twisting moment is

$$T = 2\pi \int_0^a \tau r^2 \mathrm{d}r = \frac{2}{3}\pi k\left(a^3 - \frac{1}{4}c^3\right) = \frac{1}{3}T_\mathrm{e}\left[4 - \left(\frac{\theta_\mathrm{e}}{\theta}\right)^3\right] \tag{8-12}$$

As the elastic/plastic torsion continues, the torque rapidly approaches the fully plastic value $(2/3)k\pi a^3$. Since θ tends to infinity as c tends to zero, an elastic core of material must exist for all finite values of the angle of twist.

In the case of an annealed material, there is no well-defined yield point, and the elastic/plastic boundary is therefore absent. Since the engineering shear strain at any radius r is $\gamma = r\theta$, the torque may be expressed as

$$T = 2\pi \int_0^a \tau r^2 \mathrm{d}r = \frac{2\pi}{\theta^3} \int_0^{a\theta} \tau \gamma^2 \mathrm{d}\gamma$$

When the shear stress-strain curve of the material is given, the torque can be calculated from above, using the known (τ, γ) relationship. Conversely, if the torque-twist relationship for a solid bar has been experimentally determined, the shear stress-strain curve can be easily derived from it. The differentiation of the above equation with respect to θ gives

$$\frac{\mathrm{d}}{\mathrm{d}\theta}(T\theta^3) = 2\pi a^3\theta^2\tau_0$$

where τ_0 is the value of τ at $r = a$ where the shear strain is $\gamma_0 = a\theta$. The relationship between τ_0 and γ_0 is therefore given by

$$\tau_0 = \frac{1}{2\pi a^3}\left(\theta\frac{\mathrm{d}T}{\mathrm{d}\theta} + 3T\right) \quad (\gamma_0 = a\theta) \tag{8-13}$$

The geometrical significance of the first term in the bracket is indicated in Figure 8-5. Since $\mathrm{d}T/\mathrm{d}\theta$ must be obtained numerically or graphically from the measured (T, θ) curve, the computation based on Equation (8-13) is not very accurate for the initial part of the curve. The accuracy may, however, be improved by rewriting the shear stress as

$$\tau_0 = \frac{1}{2\pi a^3}\left[\theta^2\frac{\mathrm{d}}{\mathrm{d}\theta}\left(\frac{T}{\theta}\right) + 4T\right]$$

The ratio T/θ is constant in the elastic range, and decreases slowly over the initial part of the plastic range. The contribution of the first term in the bracket is therefore small, the Equation (8-13) should give more satisfactory results.

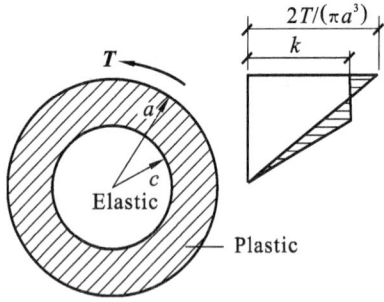

Figure 8-4 Plastic annulus and stress distribution for H=0

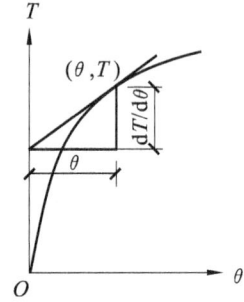

Figure 8-5 Nadia's construction for an annealed bar

Suppose, now, that a bar of radius a has a concentric circular hole of radius b. The material is assumed to strain-harden, the uniaxial stress-plastic strain curve being represented by a straight line of slope H over the relevant range. If the bar is subjected to pure torsion, yielding begins at $r=a$ when the torque and the specific angle of twist become

$$T_e = \frac{1}{2}\pi k a^3\left(1-\frac{b^4}{a^4}\right); \quad \theta_e = \frac{k}{Ga}$$

During the elastic/plastic torsion, the shear stress increases with the radius in both elastic and plastic regions. Since the total equivalent strain in any plastic element is equal to $(1/\sqrt{3})(r\theta - \tau/G)$, the assumed strain-hardening law gives

$$\tau = k + \frac{H}{3}\left(r\theta - \frac{\tau}{G}\right); \quad \theta = \frac{k}{Gc}$$

where c is the radius to the elastic/plastic boundary.

Substituting for θ, the shear stress in the plastic region is obtained as

$$\tau = k\left(\frac{1+\dfrac{Hr}{3Gc}}{1+\dfrac{H}{3G}}\right) \quad (c \leqslant r \leqslant a)$$

The stress in the elastic region $b \leqslant r \leqslant c$ is $\tau = kr/c$ as before. The stress distribution over the entire cross section furnishes the applied torque T. A straightforward integration results in

$$\left(1+\frac{H}{3G}\right)T = \frac{2}{3}\pi k a^3\left\{1+\frac{1}{4}\left[-\frac{c^3}{a^3}\left(1+\frac{3b^4}{c^4}\right)+\frac{Ha}{Gc}\left(1-\frac{b^4}{a^4}\right)\right]\right\} \tag{8-14}$$

which reduces to Equation (8-12) when $H=0$ and $b=0$. The fully plastic torque T_0 for a strain-hardening hollow bar is given by

$$\left(1+\frac{H}{3G}\right)T_0 = \frac{2}{3}\pi k a^3\left[1-\frac{b^3}{a^3}+\frac{H}{4G}\left(\frac{a}{b}-\frac{b^3}{a^3}\right)\right]$$

The variation of T/T_e with θ/θ_e for $H=0$ and $H=0.3G$ is shown in Figure 8-6. The fully plastic angle of twist per unit length has the finite value $\theta_0 = k/(Gb)$, which is independent of the rate of hardening of the material.

The residual stress left in the bar on unloading from an elastic/plastic state can be determined as in the case of bending. It is only necessary to superpose an elastic distribution of stress produced by an opposite torque equal in magnitude to that which is released. For a completely unloaded bar, we therefore have to subtract the quantity $2Tr/[\pi(a^4 - b^4)]$ from the stress existing at the moment of unloading. Using Equation (8-14) for T, the residual stress distribution for $H=0$ may be written as

$$\begin{cases} \dfrac{\tau}{k} = \dfrac{r}{c} - \dfrac{r}{3a} \dfrac{4 - \dfrac{c^3}{a^3}\left(1 + \dfrac{3b^4}{c^4}\right)}{1 - \dfrac{b^4}{a^4}} & (b \leqslant r \leqslant c) \\[4ex] \dfrac{\tau}{k} = 1 - \dfrac{r}{3a} \dfrac{4 - \dfrac{c^3}{a^3}\left(1 + \dfrac{3b^4}{c^4}\right)}{1 - \dfrac{b^4}{a^4}} & (c \leqslant r \leqslant a) \end{cases} \tag{8-15}$$

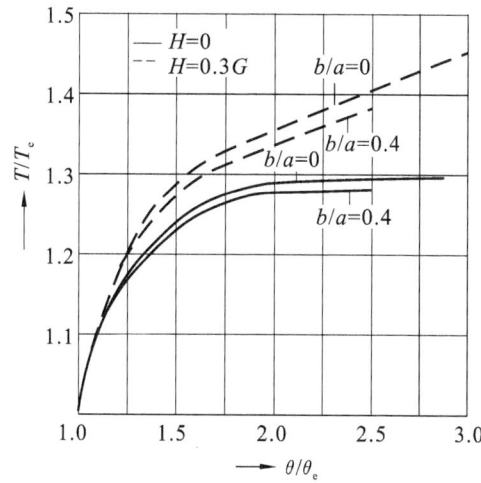

Figure 8-6   Torque-twist relationship in pure torsion of cylindrical bars in the elastic/plastic range

The residual stress is negative in an outer part of the plastic annulus and positive over the remainder of the cross section. When $b=0$, the numerically greatest residual stress occurs at $r=c$ for $c/a \leqslant 0.576$ and at $r=a$ for $c/a > 0.576$.

The angle of elastic untwist per unit length on complete unloading is equal to $2T/[\pi G(a^4 - b^4)]$, where T is given by Equation (8-14). Assuming $H=0$, the residual angle of twist per unit length may be expressed as

$$\theta = \frac{k}{Gc}\left[1 - \frac{c}{3a} \frac{4 - \dfrac{c^3}{a^3}\left(1 + \dfrac{3b^4}{c^4}\right)}{1 - \dfrac{b^4}{a^4}}\right] \tag{8-16}$$

The factor outside the square bracket is the value of θ at the moment of unloading, while the expression inside the square bracket is the residual value of τ/k at the elastic/plastic boundary. For a given elastic/plastic twist, the residual angle of twist decreases as the rate of hardening increases.

8.5 Pure Bending of a Prismatic Beam

Consider a uniform prismatic beam bent by two equal and opposite couples M applies at its ends (Figure 8-7). The cross section of the beam has an axis of symmetry Oy, and the axis of the bending couple is parallel to Oz, where O is taken on neutral plane. The plane of bending then coincides with the xy plane, the neutral fiber Ox being bent to a circular arc of radius R. During the elastic bending, O is situated at the centroid of the cross section, and the only nonzero stress $\sigma_x = \sigma$ is given by

$$\sigma = \frac{Ey}{R} = \frac{My}{I_z}$$

where E is Young's modulus for the material, and I_z the moment of inertia of the cross section about the neutral axis Oz.

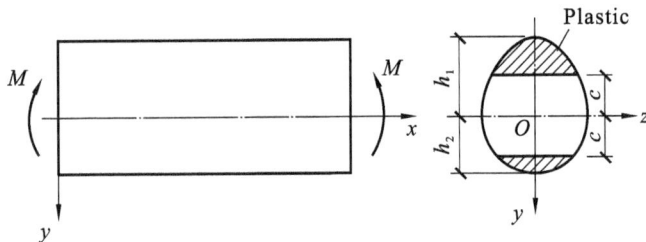

Figure 8-7 Bending of a prismatic beam under terminal couples

The longitudinal strain in the elastic beam is $\varepsilon_{xx} = y/R$, and this is accompanied by the transverse strains $\varepsilon_{yy} = \varepsilon_{zz} = -\nu y/R$, where ν is Poisson's ratio. If the components of the displacement are denoted by u, v and w with respect to the coordinate axes, then

$$\frac{\partial u}{\partial x} = \frac{y}{R}; \quad \frac{\partial v}{\partial y} = \frac{\partial w}{\partial z} = -\frac{\nu y}{R}$$

$$\frac{\partial u}{\partial y} + \frac{\partial v}{\partial x} = \frac{\partial u}{\partial z} + \frac{\partial w}{\partial x} = \frac{\partial v}{\partial z} + \frac{\partial w}{\partial y} = 0$$

Assuming an element of the x axis and an element of the yz plane to be fixed in space at $x = y = z = 0$, the solution is obtained from the conditions $u = v = w = 0$ and $\partial v/\partial x = \partial w/\partial x = \partial w/\partial y = 0$ at the origin of coordinates. The result is

$$\begin{cases} u = \dfrac{xy}{R} \\[2ex] v = -\dfrac{x^2 + \nu(y^2 - z^2)}{2R} \\[2ex] w = -\dfrac{\nu yz}{R} \end{cases} \tag{8-17}$$

The deformation is such that transverse planes remain plane during the bending. The neutral plane xz, and every parallel plane, is deformed into an anticlastic surface having a transverse curvature ν/R with an upward convexity.

Yielding first occurs in the fiber that is farthest from the neutral surface, when the longitudinal stress becomes numerically equal to Y. If the cross section is not symmetrical about the neutral axis Oz, the plastic zone spreads inward from this side before the other side begins to yield. The subsequent bending of the beam involves two separate plastic zones, with the elastic/plastic boundaries situated at equal distances $c = (Y/E)/R$ from the neutral surface. The position of the neutral surface varies with amount of bending, and is determined from the condition of zero resultant longitudinal force across any transverse section, namely

$$\int \sigma b(y)\,\mathrm{d}y = 0$$

where b is the width of the cross section at any distance y from Oz. If Oz is an axis of symmetry of the cross section, the neutral axis coincides with the centroidal axis in both elastic and plastic ranges of bending.

It is customary to assume that the state of stress is uniaxial even when the beam is partly plastic. This, however, is not strictly correct, as may be seen by considering the deformation of the beam. During a small incremental distortion, the anticlastic curvature changes by the amount $\nu\mathrm{d}(1/R)$ in the elastic region, and by the amount $\eta\mathrm{d}(1/R)$ in the plastic region, where η denotes the contraction ratio for the material beyond the yield point. The elements would not therefore fit together at the elastic/plastic interface except in the special case $\eta = \nu = 1/2$. It follows that the preceding theory of elastic/plastic bending would be strictly valid only if the material is incompressible. For $\nu < 1/2$, the necessary continuity restriction cannot be maintained without introducing transverse stresses, which affect the shape of the plastic boundaries. The problem then becomes extremely complicated. According to the simplified treatment, the displacement in both elastic and plastic region is given by Equation (8-17) with appropriate value $\nu = 1/2$, so long as the deformation is sufficiently small.

The stress σ in the elastic region varies linearly from zero on the neutral axis to a magnitude Y on the elastic/plastic boundary. In a plastic fibre, the stress has the local yield value in tension or compression, and is a given function of the strain $|y/R|$. The bending moment at any stage can be calculated from the expression

$$M = \int \sigma y b(y)\,\mathrm{d}y$$

For an annealed material, the elastic/plastic interface disappears, but the integral can still be evaluated from a given stress-strain law holding over the entire cross section. For a non-hardening material, the ratio of the fully plastic moment to the initial yield moment of a given cross section is called shape factor.

8.6   Pure Bending of a Rectangular and Circular Cross Section Beam

As a first example, consider the bending of a beam whose cross section is a rectangle of depth $2h$ and width b, the bending couple being applied in the vertical plane [Figure 8-8(a)]. In view of the symmetry of the cross section, the neutral axis always passes through its centroid, the moment of inertia about this axis being $I_z = \dfrac{2}{3}bh^3$. Plastic yielding begins at $y = \pm h$ when the bending moment and the radius of curvature become

$$M_e = \frac{2}{3}bh^2 Y; \quad R_e = \frac{Eh}{Y}$$

The radius of curvature at any stage during the elastic/plastic bending is $R = Ec/Y$, where c is the half depth of the elastic core. It is supposed that the material strain-hardens according to the law

$$\frac{\sigma}{Y} = \left(\frac{E\varepsilon}{Y}\right)^n \quad \left(\varepsilon \geqslant \frac{Y}{E}\right)$$

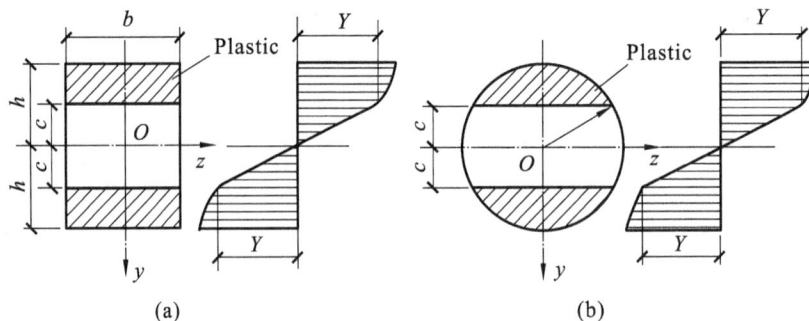

Figure 8-8   Geometry and stress distribution for the elastic/plastic bending of beams of rectangular
and circular cross sections for strain-hardening materials

where $0 \leqslant n < 1$. Evidently, σ and ε are equal to the magnitudes of the longitudinal stress and strain in the plastic regions. Since $\varepsilon = |y/R|$, the stress distribution on the tension side of the cross section may be written as

$$\begin{cases} \sigma = Y\left(\dfrac{y}{c}\right) & (0 \leqslant y \leqslant c) \\ \sigma = Y\left(\dfrac{y}{c}\right)^{n} & (c \leqslant y \leqslant h) \end{cases} \tag{8-18}$$

In view of the symmetry of the cross section, the bending moment at any stage is given by

$$M = 2b \int_{0}^{h} \sigma y \, \mathrm{d}y$$

Substituting from Equation (8-18) and integrating, the relationship between the bending moment and the curvature may be expressed as

$$\frac{M}{M_{e}} = \frac{1}{2+n}\left[3\left(\frac{R_{e}}{R}\right)^{n} - (1-n)\left(\frac{R}{R_{e}}\right)^{2}\right] \tag{8-19}$$

For a non-hardening material $(n=0)$, the moment-curvature relationship reduces to that given by Equation (8-9). The variation of M/M_{e} with R_{e}/R is shown graphically in Figure 8-9 for several values of n. The bending moment increases steadily with the curvature, except when $n=0$, for which there is a limiting moment equal to $1.5M_{e}$. The shape factor for a beam of rectangular cross section is therefore 1.5.

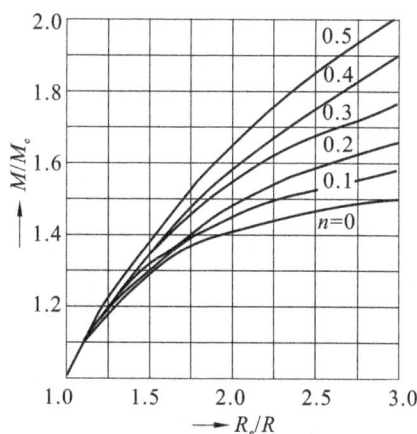

Figure 8-9   Moment-curvature relationship for strain-hardening beams of rectangular cross section in pure bending

If the bending moment of an elastic-plastic beam is released by an amount M', a purely elastic stress equal to $-M'y/I_{z}$ is superposed on Equation (8-18). The residual curvature of the beam is obtained by subtracting the spring-back curvature $M'/(EI_{z})$ from the elastic/plastic curvature $1/R$. If M' is continued to increase in the opposite sense, yielding will occur in compression at $y=h$ and tension at $y=-h$ when

$$\frac{M'}{M_e} = 1 + \left(\frac{h}{c}\right)^n = 1 + \left(\frac{R_e}{R}\right)^n \tag{8-20}$$

provided there is no Bauschinger effect. Thus, for a non-hardening material $(n=0)$, the elastic range of bending moment is always equal to $2M_e$, as shown in Figure 8-10. As the bending is continued in the negative sense, the resultant bending moment of the non-hardening beam approaches the value $-1.5M_e$ in an asymptotic manner. When the magnitude of the negative curvature becomes equal to or greater than that at the instant of unloading, the bending moment is identical to that required to bend the beam monotonically to this curvature from the unstrained state.

The stress-strain curve of a material in uniaxial tension or compression may be derived from an experimentally determined relationship between the bending moment M and the angle of bend α measured over a length l. We begin by writing the moment in the form

$$M = 2bR^2 \int_0^{\varepsilon_0} \sigma\varepsilon\,\mathrm{d}\varepsilon \quad \left(\varepsilon_0 = \frac{h}{R}\right)$$

Multiplying the expression for M by α^2, and using the fact that $R^2\alpha^2 = l^2$, we obtain the derivative

$$\frac{\mathrm{d}}{\mathrm{d}\alpha}(M\alpha^2) = 2bh^2\alpha\sigma_0$$

where σ_0 is the tensile stress corresponding to the strain ε_0 occurring at the boundary $y=h$.

The relationship between σ_0 and ε_0 may therefore be written as

$$\begin{cases} \sigma_0 = \dfrac{1}{2bh^2}\left(\alpha\dfrac{\mathrm{d}M}{\mathrm{d}\alpha} + 2M\right) \\[2mm] \varepsilon_0 = \dfrac{h\alpha}{l} \end{cases} \tag{8-21}$$

If a tangent is drawn at any point P on the M-α curve to meet the M axis at Q, then the projection of PQ parallel to the M axis represents the first term in the parenthesis of Equation (8-21), as indicated in Figure 8-11. A graphical construction of a σ_0-ε_0 curve on the basis of a given M-α curve is therefore possible.

Consider now a beam of circular cross section subjected to pure bending in the vertical diametral plane. The moment of inertia of the cross section about the neutral axis, which coincides with the horizontal diameter, is equal to $\pi a^4/4$, where a is the radius of the circular boundary. Yielding first occurs at extremities of the vertical diameter when M and R attain the values

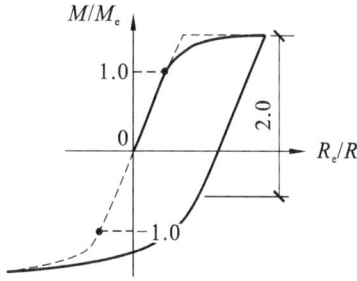

Figure 8-10　Effects of unloading and reversed loading without hardening;

the broken lines at the top represent an idealized behavior

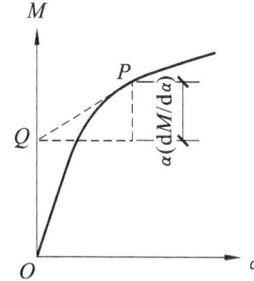

Figure 8-11　Nadia's construction for

an annealed beam

$$M_e = \frac{\pi}{4} a^3 Y; \quad R_e = \frac{Ea}{Y}$$

A typical elastic/plastic stage is specified by the distance c of the plastic boundary from the neutral axis. The elastic/plastic bending moment is given by

$$M = 2 \int_0^a \sigma y b(y) \, \mathrm{d}y = 4 \int_0^a \sigma y \sqrt{a^2 - y^2} \, \mathrm{d}y$$

where $\sigma = (y/c)Y$ for $y \leqslant c$, and $\sigma = Y$ for $y \geqslant c$, strain-hardening being neglected. Carrying out the integration, the result may be expressed as

$$\frac{M}{M_e} = \frac{2}{\pi} \left[\frac{1}{3} \left(5 - \frac{2c^2}{a^2} \right) \sqrt{1 - \frac{c^2}{a^2}} + \frac{a}{c} \arcsin \frac{c}{a} \right] \tag{8-22}$$

The radius of curvature of the elastic/plastic beam is c/a times that at the initial yielding. As the cross section in rendered increasingly plastic, the ratio M/M_e rapidly approaches the asymptotic value $16/(3\pi) \approx 1.698$, which is the shape factor for a circle cross section.

8.7　Closure

We concerned with problems of plastic flow begining at a stage in a deformable body subjected to external loads of gradually increasing magnitude in which the yield criterion is first satisfied in some stress element. We illustrated spreading of the plastic zone separated from the elastic material by an elastic/plastic boundary when loads further increase. In this chapter, we mainly introduced some examples of elastoplastic analysis such as plane strain compression of a block, plane strain bending of a beam, a cylindrical bar under pure torsion, pure bending of a prismatic beam and pure bending of a rectangular and circular cross section beam.

Problems

8-1 A solid cylindrical bar made of a strain-hardening material with an initial shear yield stress k is subjected to pure torsion until the torque has the fully plastic value corresponding to no strain-hardening. Assuming a constant plastic modulus $H = G/3$, find the associated twist ratio θ/θ_e. If the bar is fully unloaded from the partly plastic state, compute the residual shear stress at the external boundary in terms of the yield stress k.

8-2 A hollow cylindrical bar of external radius a and internal radius $0.5a$ is twisted to fully plasticity and the applied torque is subsequently released. The material is non-hardening and obeys the von Mises yield criterion. If the unloaded bar is subjected to a sufficiently large axial tension, show that yielding will restart at the inner radius.

8-3 A solid circular cylinder of radius a is rendered partially plastic to a radius b in pure torsion. An increasing axial tension is then applied to the bar while angle of twist is held constant. Assuming an incompressible and non-hardening Prandtl-Reuss material, show that the stress distribution in the region $b \leqslant r \leqslant a$ is given by

$$\frac{\sigma}{Y} = \tan\left(\frac{3G}{Y}\varepsilon\right); \quad \frac{\sqrt{3}\tau}{Y} = \sec\left(\frac{3G}{Y}\varepsilon\right)$$

where ε is the longitudinal strain. Show that the axial stress in the additional plastic region is given by Equation $\dfrac{\sigma}{Y} = \tan\left(\dfrac{3G}{Y}\varepsilon - \sqrt{1 - \dfrac{r^2}{b^2}} + \arctan\sqrt{1 - \dfrac{r^2}{b^2}}\right)$.

8-4 A non-hardening beam of rectangular cross section is bent about an axis of symmetry to an elastic/plastic curvature to κ_0. The beam is then unloaded and reloaded in the opposite sense until plastic deformation again occurs. Show that the new elastic/plastic phases involves a curvature $\kappa \leqslant \kappa_0 - 2\kappa_e$, where κ_e corresponds to the initial yielding, and that the moment-curvature relationship becomes

$$\frac{M}{M_e} = -\frac{1}{2}\left[3 + \left(\frac{\kappa_e}{\kappa_0}\right)^2\right] + \left(\frac{2\kappa_e}{\kappa_0 - \kappa}\right)^2$$

Verify that for $\kappa \leqslant \kappa_0$, the beam behaves as though the initial positive loading had never taken place.

8-5 A beam of rectangular cross section having a width b and depth $2h$ is bent about an axis parallel to the width. The material is linearly strain-hardening with an initial yield stress Y and a tangent modulus T. Show that the moment-curvature relationship during an elastic/plastic bending may be written as

$$\frac{M}{M_e}=\frac{1}{2}\left(1-\frac{T}{E}\right)\left[3-\left(\frac{R}{R_e}\right)^2\right]+\frac{TR_e}{ER}$$

If the beam is completely unloaded from a state in which half the cross section is rendered plastic, show that the residual curvature is $5(E-T)/(8E)$ times the curvature of the beam at the initial yielding.

8-6 A prismatic beam of square cross section, made of an ideally plastic material, is bent about a diagonal in the elastic/plastic range. Show that the initial yield moment is $M_e=2\sqrt{2}a^3Y/3$, where $2a$ is the length of each side of the square, and that the moment-curvature relationship for the partially plastic beam is

$$\frac{M}{M_e}=2-2\left(\frac{R}{R_e}\right)^2+\left(\frac{R}{R_e}\right)^3$$

If the applied moment I releases after half the area of the cross section has become plastic, find the limits between which a reloading moment M can vary without causing further plastic flow.

中　文　篇

1 力　系

1.1　引　言

在刚体力学的研究中,我们所解决的问题可以忽略物体的变形。你可以回想一下,对于刚体力学问题,我们只需要利用牛顿定律就可以计算出平衡状态下作用在物体上的未知力。然而在静力学的研究中,虽然从物理角度可以将物体看作"刚性"的,但仅用牛顿定律是不能解决问题的,这类问题被称为超静定问题。这些问题中物体的变形尽管很小,但对作用力的确定是十分重要的。举个例子说明,参见图 1-1(a)所示的简支梁以及图 1-1(b)所示的自由体受力图。假设施加在梁上的外荷载的位置几乎不变,运用刚体力学的方法,我们就可以很容易地求解出支座反力 A、B_x 和 B_y,可以实现这一点是因为我们知道自由体上合力为零。通过自由体上合力为零并利用未变形体的几何不变性,我们就可以很容易求解出这三个未知量。接下来假设梁的支座由两个变为三个,如图 1-2 所示。很明显,我们可以预测出该体系下的变形会比两支座支撑体系的变形更小,因此求解由几何变化带来的问题并不困难。两支座的支撑体系符合刚体力学的要求。而在第二个问题中,我们无法确定支座反力 A、C、B_x 和 B_y 的唯一值,因为会有无穷多个数值组合能满足刚体力学要求下的精确合力。为了选出正确的数值组合,即使梁的变形很小,我们也需要考虑这个变形。可以看出,运用刚体力学方法分析问题是求解支座反力系合力的必要条件,而由变形分析提供的附加条件可以确定各个支座反力的数值。与上述所讨论问题类似的超静定问题将是我们需要着手解决的一类问题。

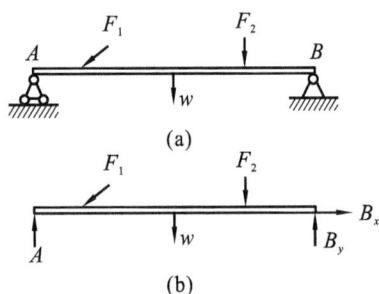

图 1-1　简支梁及其受力图　　　　图 1-2　三支撑梁及其受力图

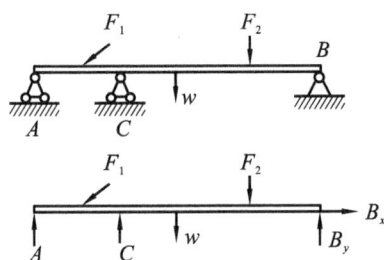

接着,选择如图 1-3 所示的处于平衡状态的任意固体。假设假想平面 M 穿过物体,我们想求出通过该平面从物体 A 部分传递到物体 B 部分力的分布情况,可将 B 部分视为一个自由体(图 1-4),运用刚体力学的方法我们可以在截面的某个位置得到一个力和一个力偶,这就是正确的合力系统分布。当然,前提是施加的力并没有因变形而明显改变其初始方向。但像求解超静定梁的支座反力问题一样,这个合力系统可以有无穷多种分布情况,我们必须研究物体的整体变形以得到足够多的补充条件,从而确定唯一的力的分布。

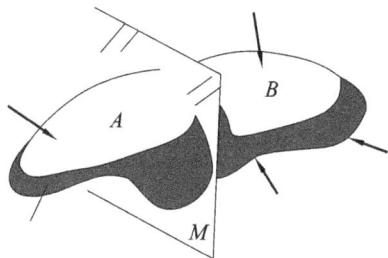

图 1-3　处于平衡状态的任意固体　　　　　图 1-4　任意固体的自由体图

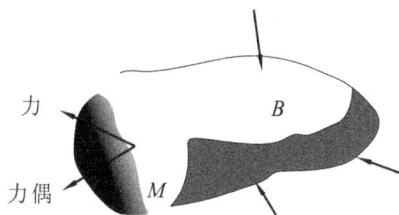

　　了解固体中力的分布对于大多数系统的设计是至关重要的。需要指出的是,近几年,单级化学火箭的射程有较大的提高,从德国 V-2 的 200 英里射程增加到中程弹道导弹的 1000 英里以上射程,很大程度上是由结构的改进而不是推力的提高造成的。

　　因此,我们将对固体中的内力分布以及超静定结构问题(如梁的问题)中某些离散力的计算感兴趣。

　　目前,对于需要考虑物体变形的问题,我们的重点一直放在某些离散力和力的分布计算上。很明显,有时物体自身的挠度才是最重要的因素,并非力的分布。因为我们解决的问题仅限于小变形问题,所以,首先运用刚体力学知识计算支座反力,然后考虑变形,我们就能确定双支撑静定梁的挠度。在三支撑超静定梁中,我们将同时计算出支座反力以及挠度,因为变形与二者有本质的联系。

　　与考虑静力学和动力学问题中的连续刚体不同,我们在研究变形体时主要考虑静力学问题,然而,有一类重要的非静力学问题可以采用本书中的公式。假设给定一个力系,物体发生运动且形状几乎未发生变化,以至我们可以运用刚体力学知识,利用物体未变形时的形状计算物体在任何时刻的运动,那么我们就可以采用本书中的方法计算由给定的组合力系以及惯性力而引起的物体的变形。但有一个重要的前提条件是上述力必须很缓慢地随时间变化。施加该条件的原因是,我们计算物体在 t 时刻的变形,是根据物体的刚体运动计算得到 t 时刻施加力系统和惯性力分布,它类似于作用在物体上的静荷载。显然,这种方法对于一个随时间变化很快的力的分布问题是不适用的。旋转圆盘作为这类问题的一个典例,可以运用上述方法来解决。

　　变形体动力学留待更高级的课程研究,其中你需要考虑梁、板、壳的振动,或波在固体中传播,结构的冲击荷载。其他课程还可能研究喷气发动机的应力或推进系统引起的火箭振动,这些都是超出本书内容的一些有趣的问题。研究这些问题需要我们先掌握本书中的基础知识。

　　接下来我们很多时候都要研究各种力,所以对力进行准确的分类是很有必要的。

1.2　力的分布类型

　　现在我们对力的分布进行分类。作用在整个物体的力场,如影响整个物体质量分布的力场,我们称其为体力分布,重力和惯性力就是这种分布的主要例子。体力分布可以表示为直接影响物体单元每单位质量或每单位体积所受到的力。因此,若 $\boldsymbol{B}(x,y,z,t)$ 是单位质量的体力分布,那么质量元 $\mathrm{d}m$ 所受的力为

$$\mathrm{d}\boldsymbol{F} = \boldsymbol{B}(x,y,z,t)\mathrm{d}m \tag{1-1}$$

　　除体力分布外还有面力分布,这些力分布作用于被研究物体的边界上。面力分布可以表示为物体边界上每单位面积所受的力,它包括图 1-3 所示的由一个作用在物体外表面的力的分布,或者由图中假想平面 M 所剖开的外露内表面上的力的分布,而图 1-4 所示的力和力偶就是面力分布的合力系统。有时我们想要区分实际边界上的面力分布与假想外露内表面(如假想平面 M)边界上的面力分布,在这种情况

下,我们将作用在实际边界上的面力分布称为表面约束。在本书后续的学习中我们将有机会体会到二者的区别。

1.3 结 语

在本章中,我们尝试指出早期刚体力学与本书所研究内容之间的关系。此外,为了便于学习变形固体,我们还对力的分布提出了一些定义。下面我们将接着考虑利用自由体受力图研究在假想外露内表面上面力的分布问题。

2 应 力

2.1 引 言

在第 1 章中,我们讨论了两种类型力的分布,分别是体力分布和面力分布。现在,我们将着重考虑面力分布。你可以回想一下,面力分布可分为实际边界面上的面力分布(我们称之为表面约束),以及通过自由体法得到的外露物体内表面上的面力分布。显然,通过自由体法我们可以得到一个物体任意内表面上的面力分布情况。由此,我们可以认为整个物体内都存在面力分布。实际上,根据这个观点我们也可以定量地描述出外力是如何在物体内传递的。

在平衡体的假想界面上取一个面积微单元 δA,如图 2-1 所示。注意整个界面上的受力可表示为刚体合力 F_R 和合力偶 M_R。对于面积微单元,同样可以用图 2-2 所示的合力 δF 和合力偶 δM 来表示。

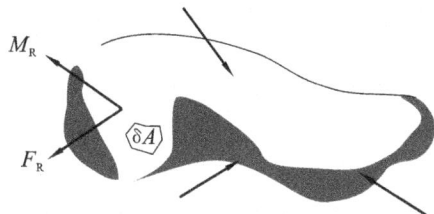

图 2-1 平衡体

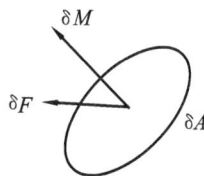

图 2-2 物体的面积微单元

如果这个面积微单元尺寸接近无穷小,那么力偶可以被忽略,因为这个区域上力的分布情况接近均匀平行分布,以至于可以像刚体力学中那样受力仅用一个合力代替。由于限定了尺寸无穷小,下面我们将忽略力偶 δM。我们将很容易把 δF 分解成一组相互垂直的力,如图 2-3 所示,δF_n 是一个垂直于该面积微单元所在平面的分量,而其他分量 δF_{s_1} 和 δF_{s_2} 则是与面积微单元所在平面相切的分量。

现在,我们可以通过对下面公式求极限,从而对正应力 σ 和剪应力 τ_s 做出定义:

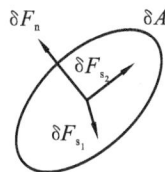

图 2-3 物体面积微单元上的一组正交分量

$$\begin{cases} \sigma = \lim\limits_{\delta A \to 0} \dfrac{\delta F_n}{\delta A} = \dfrac{\mathrm{d}F_n}{\mathrm{d}A} \\[2mm] \tau_{s_1} = \lim\limits_{\delta A \to 0} \dfrac{\delta F_{s_1}}{\delta A} = \dfrac{\mathrm{d}F_{s_1}}{\mathrm{d}A} \\[2mm] \tau_{s_2} = \lim\limits_{\delta A \to 0} \dfrac{\delta F_{s_2}}{\delta A} = \dfrac{\mathrm{d}F_{s_2}}{\mathrm{d}A} \end{cases} \tag{2-1}$$

我们发现正应力和剪应力是单位面积上力分量的强度,注意它们是标量。通过用正应力和剪应力来描述内表面上力的分布情况,我们可以知道整个物体的内力分布。

2.2 应力符号

在 2.1 节中,我们展示了如何描述几何体内某内表面上力的分布情况。通过这种方法我们可以得到在某一点处任意内表面上的正应力和剪应力,因此可以描述出外力在几何体内是如何传递的。由于我们在某一点处的每个界面上都规定了三个应力分量,因此采取有效的符号来标注它们是十分必要的。为此,我们在变形体中考虑每个面均平行于笛卡儿坐标系的微小单元体,如图 2-4 所示,每个面组合形成一个微小平行矩形六面体。三个面上的应力都采用双下标表示法进行标注。第一个下标表示应力所在平面的法线方向,第二个下标表示应力自身方向。由于正应力方向和平面法线方向共线,因此正应力仅用一个下标表示,剪应力则用两个下标表示。比如,τ_{yx} 表示应力作用的平面平行于 xz 平面,其平面法线方向为 y 方向,应力方向为 x 方向。

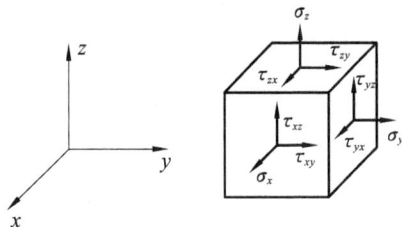

图 2-4 各面均平行于笛卡儿坐标系的微小单元体

对于应力符号规定,我们遵循以下规则:如果应力作用的面积微单元外法线方向和任何坐标轴的正方向一致,且应力方向也和任何坐标轴的正方向一致,那么应力符号为正(面积向量和应力对应的坐标轴不一定相同)。此外,如果面积向量和应力都同时指向相同或不同坐标轴的负方向,则应力符号也为正。因此,你会发现在图 2-4 中的应力符号均为正。如果面积向量和应力不是同时指向坐标轴正向或负向,则应力符号为负。

在 2.3 节中,我们将学习当给定某一点处三个正交面上的应力情况,如何通过应力转换方程确定在该点处任意面上的应力。我们已知坐标面上的应力分布就等同于已知整个物体的应力分布情况。这时,我们提出的符号标注将被证明是十分有用的。

我们应清楚地认识到应力不只存在于固体中,在本节及之后的章节中所得到的结论还适用于任何具有刚体黏性的连续介质。

接下来,我们将学习应力转换方程。

2.3 应力转换方程

我们选择一个连续的微小四面体,如图 2-5 所示。四面体正交边长分别记为 $\Delta x, \Delta y, \Delta z$,并在三个坐标面上标注出正的剪应力及正应力,在外法线为 n 的斜面上,我们只给出正应力 σ_n 及总的剪应力 τ_{ns},并能很容易地给出 n 与 x 轴,y 轴,z 轴间的方向余弦,分别记为 l, m, n。现在使用上述标注,用斜面 ABC 的面积来表示四面体三个坐标面的面积,如式(2-2)所示。

$$\begin{cases} AOC=ABCm \\ BOA=ABCn \\ COB=ABCl \end{cases} \tag{2-2}$$

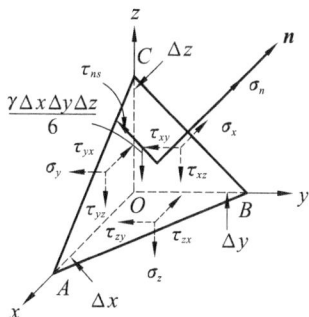

图 2-5 连续介质的微小四面体

接下来,写出在 n 方向上满足牛顿定律的方程。

$$\sigma_n ABC - \sigma_x COBl - \tau_{xz}COBn - \tau_{xy}COBm - \sigma_y AOCm - \tau_{yz}AOCn - \tau_{yx}AOCl -$$

$$\sigma_z BOAn - \tau_{zx}BOAl - \tau_{zy}BOAm - \gamma \frac{\Delta x \Delta y \Delta z}{6} n = \rho \frac{\Delta x \Delta y \Delta z}{6} a_n \quad (2\text{-}3)$$

式中,a_n 表示在 n 方向上的加速度;γ 表示物体的比重;ρ 表示密度。

现在我们将方程(2-2)代入方程(2-3)替换 AOC、BOA 和 COB,再将方程(2-3)两边同时除以 ABC,最后取极限,令 Δx,Δy 和 Δz 均趋于 0。显然,由于 $\dfrac{\Delta x \Delta y \Delta z}{ABC}$ 是一个无穷小量,我们取极限后趋于 0,因此公式最后两项可以忽略,于是可以得到任意点处的 σ_n。

$$\sigma_n = \sigma_x l^2 + \sigma_y m^2 + \sigma_z n^2 + \tau_{xy}lm + \tau_{xz}ln + \tau_{yz}mn + \tau_{yx}ml + \tau_{zx}nl + \tau_{zy}nm \quad (2\text{-}4)$$

我们会发现,某一点处任何平面上的正应力仅取决于该点处三个正交平面上的应力,以及该平面法线在上述正交参考系中对应的方向余弦。

现在我们继续计算该四面体斜面 ABC 上的剪应力,其计算方法和正应力类似。四面体斜面上剪应力 τ_{ns} 如图 2-6 所示,同样,我们给出方向余弦 l',m',n'。由于正应力和剪应力互相垂直,因此两组方向余弦必定满足下列公式:

$$ll' + mm' + nn' = 0 \quad (2\text{-}5)$$

如果沿剪应力方向(即图 2-6 中 s 方向)写出对应的牛顿定律的方程,再对方程(2-4)进行类似的处理,我们将得到以下公式:

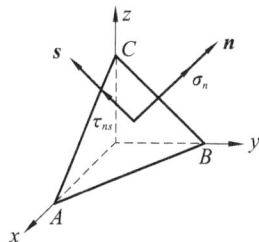

图 2-6 斜面上存在剪应力的四面体

$$\tau_{ns} = ll'\sigma_x + mm'\sigma_y + nn'\sigma_z + lm'\tau_{xy} + ml'\tau_{yx} + ln'\tau_{xz} + nl'\tau_{zx} + mn'\tau_{yz} + nm'\tau_{zy} \quad (2\text{-}6)$$

当已知某点一组正交面上的九个应力,我们就可以利用转换方程(2-4)和方程(2-6)计算出该点处的所有应力。你可以回想一下,一点处空间矢量只需要用三个分量表示,显然,应力的表示更为复杂。采用方程(2-4)和方程(2-6)得到的某一点上的变换量称为二阶张量。除应力张量外,在下面章节中我们将发现应变也是一种张量,在刚体动力学的研究中会涉及惯性张量。当改变某一点的坐标时,这些量会以某种方式变换,所以它们具有某些不同于其他量的特性,我们将继续研究其中一些性质,并采用一种更有效的符号来表示这些量。应力张量通常表示成下列形式:

$$\begin{pmatrix} \sigma_x & \tau_{xy} & \tau_{xz} \\ \tau_{yx} & \sigma_y & \tau_{yz} \\ \tau_{zx} & \tau_{zy} & \sigma_z \end{pmatrix} \quad (2\text{-}7)$$

注意,第一个下标用来表明数组的行,第二个下标用来表明数组的列。此外,还要注意从左上到右下矩阵的对角线称为主对角线,仅由正应力组成,主对角线上的所有项之和称为张量的迹。

为了进一步简化标注,先介绍下标标注法或笛卡儿标注方法。矢量 V 的分量 V_x,V_y,V_z 可记作 V_i,

其中 i 假设可取 x,y,z 所有值。因此使用下标 i 相当于表示三个标量分量,从而能够有效地注明一个矢量。对于应力张量,我们仅需给出 τ_{ij} 和 σ_i 就可以代替式(2-7)中的数组,并默认 i,j 均可取 x,y,z 可能排列中的所有值。以后我们将运用下标标注法建立一定的方程式,设计得到关于向量和张量之间非常有用的运算方式。

2.4 应力张量的对称性

我们现在证明一点处下标相反的剪应力彼此相等,例如 $\tau_{ij}=\tau_{ji}$,这说明式(2-7)中应力矩阵主对角线两侧的剪应力相等。我们将简单地证明一下该结论:先在平衡体中取一个微小单元体,如图 2-7(a)所示,如果我们只对 $O—x$ 边线取力矩,仅图 2-7(b)中所示的应力被涉及,并忽略高阶项,就很容易得出以下结论:

$$\tau_{yx} = \tau_{xy} \tag{2-8}$$

同样,我们也能证明,即使单元体运动,该结论也成立。

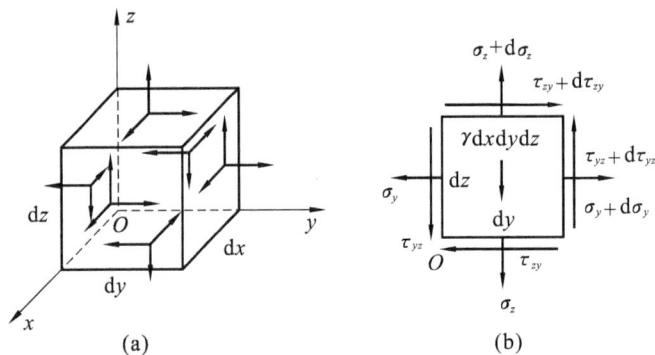

图 2-7 平衡状态下的微小单元体

一点处的剪应力 τ_{xy} 和 τ_{yx} 不仅相等,而且两者的方向是相关联的,可能的两种情况如图 2-8 所示。简言之,你可以发现剪应力都指向一角或远离一角。通过观察图 2-8,考虑单元体各角处的力矩,只有图中的标注方向能够满足牛顿定律。

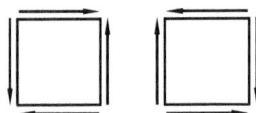

图 2-8 剪应力方向

对单元体其他坐标面进行类似考虑,我们可以对三维应力状态进行总结归纳。因此,用笛卡儿张量表示有:

$$\tau_{ij} = \tau_{ji} \tag{2-9}$$

此外,我们已在图 2-4 和图 2-5 中确定了剪应力相对于彼此的方向。

工程师处理的许多张量都是对称张量。这些张量包括应变张量(这将在下一章讨论)、刚体力学的惯性张量和电磁理论的四极张量。

利用应力张量的对称性,我们可以将转换方程(2-4)改写为以下形式:

$$\sigma_n=\sigma_x l^2+\sigma_y m^2+\sigma_z n^2+2(\tau_{xy}lm+\tau_{xz}ln+\tau_{yz}mn) \tag{2-10}$$

同样,转换方程(2-6)也可以类似地改写为如下形式:

$$\tau_{ns}=ll'\sigma_x+mm'\sigma_y+nn'\sigma_z+(lm'+ml')\tau_{xy}+(ln'+nl')\tau_{xz}+(mn'+nm')\tau_{yz} \tag{2-11}$$

2.5 主应力计算及张量不变量

我们选择一个无穷小的四面体,如图 2-9 所示,并在它的三个参考面上标出已知的应力。假设斜面 ABC 是主平面,我们将使用无下标的字母 σ 作为该平面上的主应力。主应力 σ 的方向是斜面 ABC 的法线方向,由单位向量 $\boldsymbol{n}$ 表示,其方向余弦记为 l,m,n。

为了确定主应力 σ,我们分别沿坐标轴 x,y 和 z 方向运用牛顿定律。首先沿 z 方向,我们得到公式如下:

$$\sigma ABCn - \sigma_z AOB - \tau_{yz} AOC - \tau_{xz} BOC - \gamma \frac{\Delta x \Delta y \Delta z}{6} = \rho \frac{\Delta x \Delta y \Delta z}{6} a_z \tag{2-12}$$

忽略重力和惯性力这两项高阶项,将方程(2-2)代入式(2-12),两边再同时除以 ABC 并重新整理后,利用剪应力的互等定理,我们得出如下方程:

图 2-9 微小四面体各面上的应力情况

$$\tau_{xz} l + \tau_{zy} m + (\sigma_z - \sigma) n = 0$$

在另外两个坐标轴方向上使用类似的方法,我们能得到另外两个类似于前面的方程。所得方程组如下:

$$(\sigma_x - \sigma) l + \tau_{xy} m + \tau_{xz} n = 0 \tag{2-13a}$$

$$\tau_{yx} l + (\sigma_y - \sigma) m + \tau_{yz} n = 0 \tag{2-13b}$$

$$\tau_{zx} l + \tau_{zy} m + (\sigma_z - \sigma) n = 0 \tag{2-13c}$$

现在我们想要确定主应力 σ 及其对应的方向余弦 l,m 和 n。首先,我们认为方向余弦是前面方程组中的未知数,并且需要根据应力来求解。因此,使用克拉默法则,我们得到 l 的行列式:

$$l = \frac{\begin{vmatrix} 0 & \tau_{xy} & \tau_{xz} \\ 0 & \sigma_y - \sigma & \tau_{yz} \\ 0 & \tau_{zy} & \sigma_z - \sigma \end{vmatrix}}{\begin{vmatrix} \sigma_x - \sigma & \tau_{xy} & \tau_{xz} \\ \tau_{yx} & \sigma_y - \sigma & \tau_{yz} \\ \tau_{zx} & \tau_{zy} & \sigma_z - \sigma \end{vmatrix}} \tag{2-14}$$

很明显 l 值为零,那么其他方向余弦的值也将为零,只有当方程(2-14)中分母为零时,余弦 l 才能有不确定解。由于方向余弦满足方程(2-15),故方向余弦值不能都是零。

$$l^2 + m^2 + n^2 = 1 \tag{2-15}$$

因此,解决该问题所需的一个必要前提是

$$\begin{vmatrix} \sigma_x - \sigma & \tau_{xy} & \tau_{xz} \\ \tau_{yx} & \sigma_y - \sigma & \tau_{yz} \\ \tau_{zx} & \tau_{zy} & \sigma_z - \sigma \end{vmatrix} = 0 \tag{2-16}$$

通过展开行列式,我们得到关于未知数 σ 的三次方程,即

$$\sigma^3 - (\sigma_x + \sigma_y + \sigma_z) \sigma^2 + (\sigma_x \sigma_y + \sigma_y \sigma_z + \sigma_z \sigma_x - \tau_{xy}^2 - \tau_{yz}^2 - \tau_{zx}^2) \sigma -$$
$$(\sigma_x \sigma_y \sigma_z - \sigma_x \tau_{yz}^2 - \sigma_y \tau_{xz}^2 - \sigma_z \tau_{xy}^2 + 2\tau_{xy} \tau_{yz} \tau_{xz}) = 0 \tag{2-17}$$

我们可以证明方程(2-17)总有三个实根 σ_1,σ_2 和 σ_3,这些实根就是我们前面所提到的主应力。

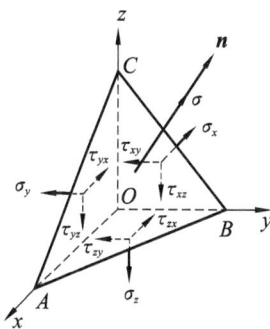

为了得到每个主应力 σ_i 对应的每组方向余弦值,我们只需将 σ_i 代入方程(2-13),联立方程(2-15)求解即可。

方程(2-17)不仅可以帮助我们求得某点处的主应力,还让我们对某点处的应力状态得出一些影响深远的结论。我们已经指出,已知某一点的应力状态,即存在唯一的二次曲面,这说明主应力仅取决于某一点的应力状态,而不取决于在该点处选择的坐标系。如果坐标系原点不变,我们将原参考系 xyz 旋转得到新参考系 $x''y''z''$,则在新参考系下方程(2-17)的实根与原实根 σ_1,σ_2 和 σ_3 相同。因此我们知道对于坐标轴的任意转动,方程(2-17)中的 σ 的幂指数的系数一定是恒定不变的,即

$$\sigma_x + \sigma_y + \sigma_z = I_1 \tag{2-18a}$$

$$\sigma_x\sigma_y + \sigma_y\sigma_z + \sigma_z\sigma_x - \tau_{xy}^2 - \tau_{yz}^2 - \tau_{zx}^2 = I_2 \tag{2-18b}$$

$$\sigma_x\sigma_y\sigma_z - \sigma_x\tau_{yz}^2 - \sigma_y\tau_{zz}^2 - \sigma_z\tau_{xy}^2 + 2\tau_{xy}\tau_{yz}\tau_{zx} = I_3 \tag{2-18c}$$

其中,给定一点的坐标 x,y,z 时,I_1,I_2 和 I_3 为常量。当然,当我们移动到另一点时,I_1,I_2 和 I_3 的值就可能发生变化。因此,上述每个量对应的值都是关于位置的函数,所以每一组量在受力体中都会形成一个标量场。这些量我们分别称为第一、第二和第三应力张量不变量。

你会发现第一应力张量不变量是应力张量中主对角线项的和,即应力张量的迹。I_1 值的三分之一是某点处的平均正应力,我们通常称它为体积应力 σ_m:

$$\sigma_m = \frac{1}{3}(\sigma_x + \sigma_y + \sigma_z) \tag{2-19}$$

在流动状态下,减去平均正应力,σ_m 就可以被称为压力。

第二应力张量不变量很容易被证明是由应力张量矩阵形成的三个行列式之和得到的。

因此,如下列布置,两个行列式已经被标明,第三个行列式由带圆圈的项组成。可以发现,这些行列式是构成主对角线中各项的子矩阵,因此,I_2 是子矩阵之和。

$$\tag{2-20}$$

第三应力张量不变量即应力张量的整个矩阵表示的行列式的值。

2.6 平面应力

现在我们提出一种简化的应力分布情况,即平面应力,它可以用来表示许多实际问题中的应力状态。我们将沿某一方向所有应力分量均为零的应力分布定义为平面应力,该方向通常被取为 z 方向。因此,平面应力要求

$$\tau_{xz} = \tau_{yz} = \sigma_z = 0 \tag{2-21}$$

在对称面内受到荷载作用的薄板,通常被视为处于平面应力状态,并取板面的法线方向为 z 方向,如图 2-10 所示。显然,如果垂直于板面的荷载为 0,板上下表面上的 σ_z 将均为 0,考虑板足够薄,我们认为板内 σ_z 也为 0。同样,由于外力平行于板面,我们认为 τ_{xz} 和 τ_{yz} 也始终为 0。

对于本节中这种特定但实用的应力分布,我们将对前文推导的一般公式进行简化。这将使我们对一般公式及其含义有更深的了解,也为我们提供了有用的简化计算公式。

同样,我们考虑一个平面应力状态下的微小棱柱单元体,如图 2-11 所示。其正应力 σ 可以通过将应

力 σ_x, σ_y 和 τ_{xy} 代入方程(2-10)得到,计算公式如下:

$$\sigma = \sigma_x l^2 + \sigma_y m^2 + 2\tau_{xy} lm \tag{2-22}$$

剪应力 τ 可通过对方程(2-11)采取类似方法得到,计算公式如下:

$$\tau = ll'\sigma_x + mm'\sigma_y + (lm' + ml')\tau_{xy} \tag{2-23}$$

对于平面应力问题,引入角度 θ 表示方向余弦会更加方便,如图 2-11 所示,具体如下:

$$l = \cos\theta \quad l' = -\sin\theta$$
$$m = \sin\theta \quad m' = \cos\theta$$

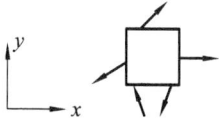

图 2-10 平面应力状态

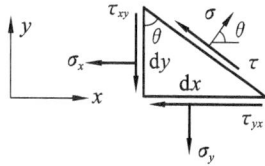

图 2-11 在平面应力状态下的微小棱柱单元体

因此,我们得到下列一对方程

$$\begin{cases} \sigma = \sigma_x \cos^2\theta + \sigma_y \sin^2\theta + 2\tau_{xy} \sin\theta\cos\theta \\ \tau = -\sigma_x \cos\theta\sin\theta + \sigma_y \sin\theta\cos\theta + (\cos^2\theta - \sin^2\theta)\tau_{xy} \end{cases} \tag{2-24}$$

注意以下三角恒等式

$$\begin{cases} \cos^2\theta = \dfrac{1}{2}(1 + \cos2\theta) \\ \sin^2\theta = \dfrac{1}{2}(1 - \cos2\theta) \\ 2\sin\theta\cos\theta = \sin2\theta \end{cases} \tag{2-25}$$

我们可以将方程(2-24)重新改写成如下形式:

$$\sigma = \frac{\sigma_x + \sigma_y}{2} + \frac{\sigma_x - \sigma_y}{2}\cos2\theta + \tau_{xy}\sin2\theta \tag{2-26a}$$

$$\tau = \frac{\sigma_y - \sigma_x}{2}\sin2\theta + \tau_{xy}\cos2\theta \tag{2-26b}$$

接下来,我们计算当前特殊情况下的主应力。方程(2-17)简化形式如下

$$\sigma^3 - (\sigma_x + \sigma_y)\sigma^2 + (\sigma_x\sigma_y - \tau_{xy}^2)\sigma = 0 \tag{2-27}$$

方程中可消去一个 σ 意味着其中一个根始终为零,很明显,该主应力一定与 z 方向一致。其他两个根可以通过解剩下的二次方程来确定,我们可以得到

$$\sigma_1, \sigma_2 = \frac{\sigma_x + \sigma_y}{2} \pm \sqrt{\left(\frac{\sigma_x - \sigma_y}{2}\right)^2 + \tau_{xy}^2} \tag{2-28}$$

其中,σ_1 和 σ_2 为剩余的两个主应力。

为了得到主应力轴的方向,我们只需要在方程(2-26b)中令 $\tau = 0$,然后求解 2θ。用 β 替换 θ,我们得到如下结果:

$$\tan2\beta = \frac{2\tau_{xy}}{\sigma_x - \sigma_y} \tag{2-29}$$

对于给定的一组应力,有两个相差 $180°$ 的 2β 值,可满足方程(2-29)。可以从上述方程中得到两个相差 $90°$ 的 β 值,它们即为该点处主应力的方向角。

很容易证明最大剪应力 τ_{max} 的方向角是

$$\tan2\beta' = -\frac{\sigma_x - \sigma_y}{2\tau_{xy}} \tag{2-30}$$

式中将 β' 取代 θ。同样,存在两个相差 90°的 β' 满足方程(2-30)。此外,根据方程(2-29)和方程(2-30)的右侧互为负倒数,我们可以得出最大剪应力平面与主应力平面夹角成 45°的结论,如图 2-12 所示。

在一般三向应力状态下,一点处最大剪应力平面也和主应力平面成 45°。在后面的章节中,为了更方便地处理这些问题,我们将采用图形法来求解。

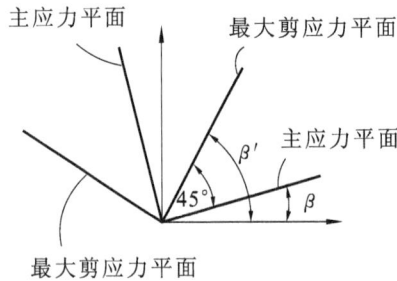

图 2-12 主轴方向

2.7 莫 尔 圆

"莫尔圆"作为一种简便的图形表示法,常用来表示一点处的平面应力状态。为了引入莫尔圆,和前面一样,我们先制定符号标记。选择如图 2-13 所示的微小矩形单元体,并标出不含一阶增量的正的剪应力和正应力。为了处理莫尔圆,我们将对剪应力制定以下规则:单元体面上的剪应力,当它关于平面中心点 O 产生顺时针弯矩时,该剪应力符号为正。关于平面中心点 O 产生逆时针弯矩的剪应力在莫尔圆中将被视为负值。因此,图 2-13 中的 τ_{xy} 被视为负值,τ_{yx} 被视为正值。

现在我们介绍应力坐标系,如图 2-14 所示,其中正应力 σ 是横坐标,剪应力 τ 是纵坐标。为了在该平面坐标系上绘制莫尔圆,我们采用前段中制定的剪应力符号规则,先绘制出单元体两个相邻正交面的应力,如图 2-13 中单元体的 a 面和 b 面。

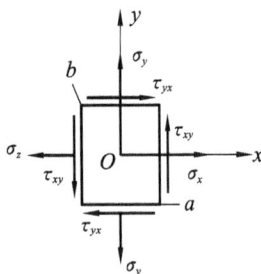

图 2-13 微小矩形单元体

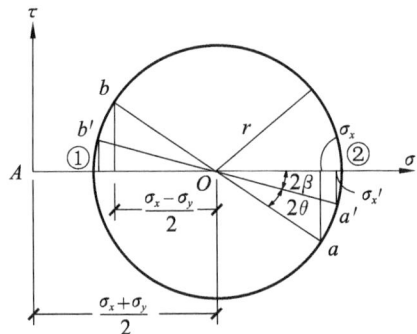

图 2-14 莫尔圆的应力坐标系

a 面和 b 面在应力图中的映射点分别表示为 a 点和 b 点。现在用一条直线连接这两点,从而得到其与 σ 轴的交点 O。如图 2-14 所示,以 O 为圆心,通过点 a 和点 b,我们可以绘制出一个圆,这就是著名的莫尔圆。

如何利用莫尔圆求解问题呢?若参考系 xy 旋转 θ 后变为参考系 $x'y'$,如图 2-15 所示,假设我们想知道单元体在新参考系下各面的应力,为了找到 a' 面对应应力坐标系中的 a' 点,我们在应力坐标系中画一条经过 O 点并从 Oa 轴旋转 2θ 的直线,其旋转方向与参考系旋转方向相同,则该直线与莫尔圆交点 a' 的坐标就是 a' 面对应的剪应力和正应力。再延长 Oa',在莫尔圆上我们可以得到点 b'。由于 Ob' 由 Oa'

旋转 180°所得,所以由 a' 面旋转 90°所得 b' 面上的应力状态对应图中 b' 点。因此,我们分别得到了对应于 x' 轴和 y' 轴的 a' 面和 b' 面上的应力状态。

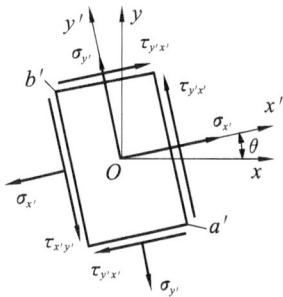

图 2-15 物理参考系 xy 旋转得到的物理参考系 $x'y'$

到目前为止,我们已经展示了如何构造莫尔圆以及如何使用它。下一步我们将证明该方法的有效性。为此,我们必须证明由莫尔圆推导出的应力是满足方程(2-26)的。观察图 2-14 中的莫尔圆,可得

$$OA = \frac{\sigma_x + \sigma_y}{2} \tag{2-31}$$

$$r = \sqrt{\left(\frac{\sigma_x - \sigma_y}{2}\right)^2 + \tau_{xy}^2} \tag{2-32}$$

上述参量已经在图 2-14 中标出。对莫尔圆进行几何推理,可以得到应力图中 a' 点对应的应力 σ:

$$\sigma = OA + r\cos(2\beta - 2\theta) \tag{2-33}$$

将方程(2-31)代入方程(2-33)并展开余弦项,可得:

$$\sigma = \frac{\sigma_x + \sigma_y}{2} + r(\cos 2\beta \cos 2\theta + \sin 2\beta \sin 2\theta) \tag{2-34}$$

从应力图我们可以清楚地看出

$$\cos 2\beta = \frac{\sigma_x - \sigma_y}{2r} \tag{2-35a}$$

$$\sin 2\beta = \frac{\tau_{xy}}{r} \tag{2-35b}$$

将这些结果代入方程(2-34),可得

$$\sigma = \frac{\sigma_x + \sigma_y}{2} + \frac{\sigma_x - \sigma_y}{2}\cos 2\theta + \tau_{xy}\sin 2\theta \tag{2-36}$$

这和我们之前推导得到的方程式一致。

用类似的方法,首先得到剪应力 τ:

$$\tau = r\sin(2\beta - 2\theta) = r(\sin 2\beta \cos 2\theta - \cos 2\beta \sin 2\theta) \tag{2-37}$$

将方程(2-35)代入方程(2-37),得到

$$\tau = \tau_{xy}\cos 2\theta - \frac{\sigma_x - \sigma_y}{2}\sin 2\theta \tag{2-38}$$

方程(2-38)验证了方程(2-26b)。因此,我们已经充分证明了莫尔圆建构的正确性,并建议使用莫尔圆。

在莫尔圆中,由于主应力一定位于应力图中的点①和点②(图 2-14),所以我们很容易确定主应力大小。如果图 2-15 中参考系 xy 逆时针旋转 β,那么应力图主应力平面将旋转 2β。

最后,从莫尔圆中我们得出最大剪应力值 $\tau_{\max}$ 为

$$\tau_{\max} = \sqrt{\left(\frac{\sigma_x - \sigma_y}{2}\right)^2 + \tau_{xy}^2} \tag{2-39}$$

该最大剪应力应位于主应力轴旋转 45°所在的平面。

虽然我们可以通过莫尔圆得到确切应力的值,但实际上我们一般不采用该方法。相反,我们会大致按比例绘制莫尔圆,以便获得某点处平面应力变化的可视图,此时,莫尔圆是最有价值的。

2.8 三维莫尔圆

本节我们将阐述广义莫尔圆,它有助于研究某一点的三维应力情况。

在受力体内取一点 O,使该点处的 x, y, z 轴和主应力方向一致。如图 2-16 所示,在 O 点处有一个微小四面体。斜面 ABC 的单位法线记为 $\boldsymbol{n}$,方向余弦为 l, m 和 n。这些方向余弦一定满足:

$$1 = l^2 + m^2 + n^2 \tag{2-40}$$

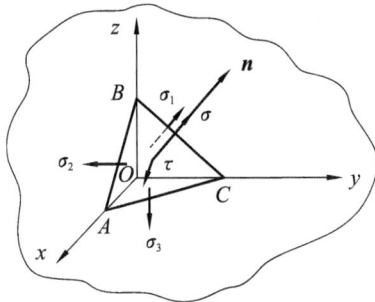

图 2-16 几何体内的微小四面体

同样通过方程(2-4),我们可以得到正应力 σ

$$\sigma = \sigma_1 l^2 + \sigma_2 m^2 + \sigma_3 n^2 \tag{2-41}$$

此外,我们也能得到 ABC 面上的最大剪应力,记为 τ。根据剪应力 τ,正应力 σ 以及主应力,利用平衡条件,即面 ABC 上的合力一定与四面体坐标平面上的合力相等且方向相反,由此可得到关于这些力的平方式。

$$(\tau^2 + \sigma^2)(ABC)^2 = (\sigma_1 BOC)^2 + (\sigma_2 AOB)^2 + (\sigma_3 AOC)^2 \tag{2-42}$$

同时除以 $(ABC)^2$ 并进行下列替换:

$$\begin{cases} \left(\dfrac{BOC}{ABC}\right)^2 = l^2 \\ \left(\dfrac{AOB}{ABC}\right)^2 = m^2 \\ \left(\dfrac{AOC}{ABC}\right)^2 = n^2 \end{cases} \tag{2-43}$$

我们得到方程

$$\tau^2 + \sigma^2 = \sigma_1^2 l^2 + \sigma_2^2 m^2 + \sigma_3^2 n^2 \tag{2-44}$$

观察方程(2-40)、方程(2-41)和方程(2-44)得出,三个方程均涉及方向余弦 l, m 和 n,以及主应力 σ_i,正应力 σ 和剪应力 τ。现在用克拉默法则求方向余弦,l 可以写成

$$l^2 = \frac{\begin{vmatrix} 1 & 1 & 1 \\ \sigma & \sigma_2 & \sigma_3 \\ \tau^2 + \sigma^2 & \sigma_2^2 & \sigma_3^2 \end{vmatrix}}{\begin{vmatrix} 1 & 1 & 1 \\ \sigma_1 & \sigma_2 & \sigma_3 \\ \sigma_1^2 & \sigma_2^2 & \sigma_3^2 \end{vmatrix}} \tag{2-45}$$

假设主应力均不相同,即没有重根(我们将在后面研究有重根的情况),此外,我们默认 $\sigma_1 > \sigma_2 > \sigma_3$。展开行列式,得到以下结果

$$l^2(\sigma_2\sigma_3^2+\sigma_1\sigma_2^2+\sigma_3\sigma_1^2-\sigma_2\sigma_1^2-\sigma_3\sigma_2^2-\sigma_1\sigma_3^2)=\sigma_2\sigma_3(\sigma_3-\sigma_2)+\tau^2(\sigma_3-\sigma_2)+\sigma^2(\sigma_3-\sigma_2)-\sigma(\sigma_3^2-\sigma_2^2)$$

两边同时除以 $(\sigma_3-\sigma_2)$,重新整理得

$$\sigma^2+\tau^2-\sigma(\sigma_3+\sigma_2)=-\sigma_2\sigma_3+\frac{l^2}{\sigma_3-\sigma_2}(\sigma_2\sigma_3^2+\sigma_1\sigma_2^2+\sigma_3\sigma_1^2-\sigma_2\sigma_1^2-\sigma_3\sigma_2^2-\sigma_1\sigma_3^2) \qquad (2\text{-}46)$$

可以很容易证明方程(2-46)右侧括号内的量可以转化为 $(\sigma_1-\sigma_2)(\sigma_1-\sigma_3)(\sigma_3-\sigma_2)$。然后,我们将方程(2-45)改写为

$$\sigma^2+\tau^2-\sigma(\sigma_3+\sigma_2)=-\sigma_3\sigma_2+l^2(\sigma_1-\sigma_2)(\sigma_1-\sigma_3) \qquad (2\text{-}47)$$

为了处理方程(2-47)中的平方项,进行如下操作:

$$\tau^2+\left(\sigma-\frac{\sigma_3+\sigma_2}{2}\right)^2=\left(\frac{\sigma_3+\sigma_2}{2}\right)^2-\sigma_3\sigma_2+l^2(\sigma_1-\sigma_2)(\sigma_1-\sigma_3) \qquad (2\text{-}48)$$

其中,$[(\sigma_3+\sigma_2)/2]^2-\sigma_3\sigma_2$ 可用 $[(\sigma_3-\sigma_2)/2]^2$ 代替,因此有

$$\tau^2+\left(\sigma-\frac{\sigma_3+\sigma_2}{2}\right)^2=\left(\frac{\sigma_3-\sigma_2}{2}\right)^2+l^2(\sigma_1-\sigma_2)(\sigma_1-\sigma_3) \qquad (2\text{-}49)$$

如果方向余弦 l 为定值,对于给定的一组主应力 σ_i,根据方程(2-49),我们就可以在平面 $\tau\sigma$ 上画出一个圆。该圆表示平面 ABC 绕 x 轴转动过程中,平面 ABC 上所有可能出现的正应力和最大剪应力组合(余弦值 l 为定值)。很明显,该圆的圆心沿 σ 轴偏移了 $(\sigma_3+\sigma_2)/2$。

我们考虑 $l=0$ 和 $l=1$ 两种极值情况,然后得到以下方程

当 $l=0$ 时:

$$\tau^2+\left(\sigma-\frac{\sigma_2+\sigma_3}{2}\right)^2=\left(\frac{\sigma_2-\sigma_3}{2}\right)^2 \qquad (2\text{-}50)$$

当 $l=1$ 时:

$$\tau^2+\left(\sigma-\frac{\sigma_2+\sigma_3}{2}\right)^2=\left(\frac{\sigma_2-\sigma_3}{2}\right)^2+(\sigma_1-\sigma_2)(\sigma_1-\sigma_3)=\left(\sigma_1-\frac{\sigma_2+\sigma_3}{2}\right)^2 \qquad (2\text{-}51)$$

现在我们得到两个圆,第一个圆半径是 $|(\sigma_2-\sigma_3)/2|$,第二个圆半径是 $|\sigma_1-(\sigma_2+\sigma_3)/2|$,对应的两个圆如图 2-17 所示。我们可以推断出任何一对应力 τ,σ 必须位于两个圆之间的区域内,该区域在图中用剖面线标出。对于剪应力 τ 我们不再考虑符号正负,所以我们只需要考虑上半部分应力圆。

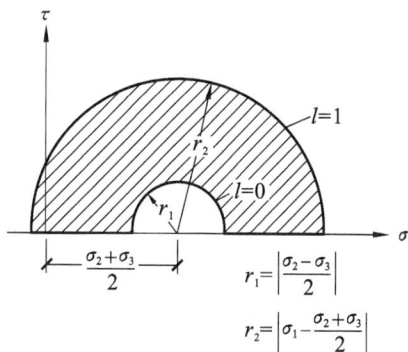

图 2-17 位于 $l=0$ 和 $l=1$ 对应的两条曲线之间的应力

和上述步骤相同,我们可以建立另外两组类似于方程(2-50)和方程(2-51)的方程,分别涉及方向余弦 m 和 n。这两组方程仅通过改变方程(2-50)和方程(2-51)的下标就可以很容易得到。而且,我们可以画出每个方程中当方向余弦取极值时对应的两个圆。图 2-18 为这类圆系,并对每个方向余弦以不同方法绘制对应的每对应力圆,在方向余弦为 0 的情况下编号为①,②,③,在方向余弦为 1 的情况下编号为 ①,② 和 ③。

像图 2-17 中讨论单组应力圆时所作的解释一样,任何一对应力 τ,σ 都必须位于各组应力圆之间的区域内。因此很明显,当我们考虑三组圆时,应力必定位于图 2-18 所示的剖面线区域内。该区域以圆①,②,③为边界。此外,应注意,这三个圆在 σ 轴的 $\sigma_1,\sigma_2,\sigma_3$ 三点处两两相切,因此当已知这三个应力时,我们可以非常容易地绘制这三个应力圆。为了简化讨论,我们在图 2-19 中画出这三个应力圆,这些就是所谓的三维应力莫尔圆。

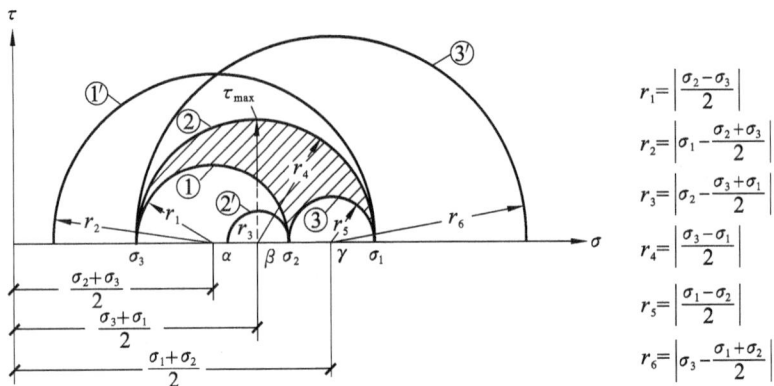

$$r_1=\left|\frac{\sigma_2-\sigma_3}{2}\right|$$

$$r_2=\left|\sigma_1-\frac{\sigma_2+\sigma_3}{2}\right|$$

$$r_3=\left|\sigma_2-\frac{\sigma_3+\sigma_1}{2}\right|$$

$$r_4=\left|\frac{\sigma_3-\sigma_1}{2}\right|$$

$$r_5=\left|\frac{\sigma_1-\sigma_2}{2}\right|$$

$$r_6=\left|\sigma_3-\frac{\sigma_1+\sigma_2}{2}\right|$$

图 2-18　三维应力莫尔圆

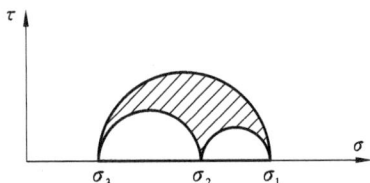

图 2-19　简化后的三维应力莫尔圆

当已知某平面的方向余弦 l,m,n,我们如何使用莫尔圆来确定任何平面上的一组应力 σ 和 τ 呢?我们知道给定方向余弦 l 时,对应的所有应力圆都是以在 σ 轴上且距原点 $(\sigma_2+\sigma_3)/2$ 处的点为圆心绘制得到的,圆心即为图 2-18 中的点 α。对于给定方向余弦 l 对应的应力圆半径可通过方程(2-52)得到:

$$r=\left[\left(\frac{\sigma_3-\sigma_2}{2}\right)^2+l^2(\sigma_1-\sigma_2)(\sigma_1-\sigma_3)\right]^{\frac{1}{2}} \tag{2-52}$$

关于方向余弦 m 的系列圆,我们用 β 作为应力圆圆心。当 m 为定值时,使用前面的方程,交换下标,即可得到对应的应力圆半径。最后,对于方向余弦 n 的系列圆用 γ 作为应力圆圆心,半径由方程(2-52)交换下标确定。三个圆的公共交点即为所求的应力 τ,σ。

我们画出如图 2-20 所示的一组圆。当给定一组主应力 σ_1,σ_2 和 σ_3 以及与给定坐标轴对应的方向余弦 l,m,n(例如 σ_1,σ_2 和 σ_3 分别平行于 x,y,z 轴),我们就能以图形的方式确定给定的一组方向余弦所对应平面上的一组应力 σ,τ。

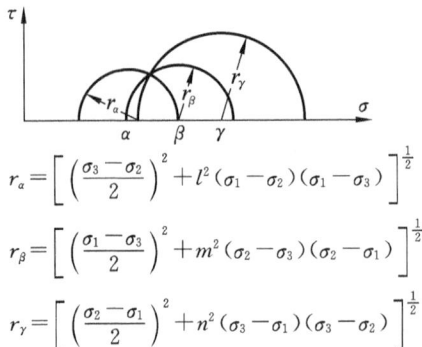

$$r_\alpha=\left[\left(\frac{\sigma_3-\sigma_2}{2}\right)^2+l^2(\sigma_1-\sigma_2)(\sigma_1-\sigma_3)\right]^{\frac{1}{2}}$$

$$r_\beta=\left[\left(\frac{\sigma_1-\sigma_3}{2}\right)^2+m^2(\sigma_2-\sigma_3)(\sigma_2-\sigma_1)\right]^{\frac{1}{2}}$$

$$r_\gamma=\left[\left(\frac{\sigma_2-\sigma_1}{2}\right)^2+n^2(\sigma_3-\sigma_1)(\sigma_3-\sigma_2)\right]^{\frac{1}{2}}$$

图 2-20 给定的一组方向余弦对应的一组应力

正如我们在讨论平面应力的莫尔圆时所指出的,我们通常不建议将莫尔圆用于计算某一点的应力。然而,通过对莫尔圆的简单观察,我们很容易得出有用的结论。如图 2-18 所示,该情况下一点的最大剪应力出现在圆②处,其中 $m=0$。该情况下最大剪应力为

$$\tau_{\max}=r_4=\frac{\sigma_1-\sigma_3}{2} \tag{2-53}$$

也就是说,最大剪应力等于最大正应力和最小正应力之差的一半。因此,如果 $\sigma_1=100$ Pa 且 $\sigma_3=-50$ Pa,则最大剪应力为 75 Pa。由于在该最大剪应力条件下 $m=0$,我们可以得出最大剪切平面与 y 轴相切的结论。根据图 2-18,用 $(\sigma_1-\sigma_3)/2$ 代替方程(2-47)中的 τ,用 $(\sigma_3+\sigma_1)/2$ 代替 σ,我们可以解出方向余弦 l。可以很容易地证明

$$l=\pm\frac{1}{\sqrt{2}} \tag{2-54}$$

我们可以通过类似的步骤得到

$$n=\pm\frac{1}{\sqrt{2}} \tag{2-55}$$

因此,最大剪应力平面将最大正应力平面和最小正应力平面形成的角度平分,如图 2-21 所示。

到目前为止,我们只考虑了主应力不同的情况。当两个主应力相等时会发生什么呢?我们先考虑 $\sigma_1=\sigma_2>\sigma_3$,如图 2-22 所示。观察涉及 l 的方程(2-52)和对应的涉及 m 的方程,我们可以发现对于这些方向余弦的所有值,r 恒为常量。

$$r=\left|\frac{\sigma_3-\sigma_2}{2}\right|=\left|\frac{\sigma_3-\sigma_1}{2}\right| \tag{2-56}$$

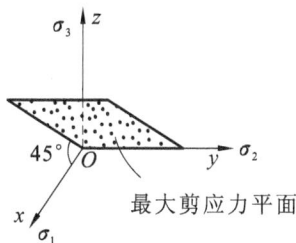

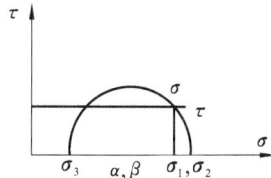

图 2-21 最大剪应力平面　　　　**图 2-22 $\sigma_1=\sigma_2>\sigma_3$ 状态下的莫尔圆**

对于已知方向余弦 n,我们得到 r

$$r=n\left[(\sigma_3-\sigma_1)(\sigma_3-\sigma_2)\right]^{\frac{1}{2}}=n(\sigma_1-\sigma_3) \tag{2-57}$$

因此,我们看到两个莫尔圆重合,图 2-19 中的阴影区域缩小为一条半圆弧。该圆弧与 l,m 确定的莫尔圆相对应。图中与 n 对应的莫尔圆也是一条圆弧,且和前面的半圆相交,交点就是对应平面的应力点。

应注意,因为图中绘制的半圆包括这两个方向余弦的所有可能值,所以 l 和 m 的值对结果没有影响。因此,当应力 σ_1 和 σ_2 相等时,每对 τ 和 σ 沿 z 轴对称分布。

2.9 结 语

在本章中,我们研究了力如何在固体内传递的定量计算方法。介绍了任一点处内表面上正应力和剪应力的概念,并指出,若已知一点处三个正交面上的应力,我们就可以确定该点任何内表面上的应力。虽然我们只在固体中得到这些结论,但必须指出,这些结论适用于静态或动态条件下的任何连续介质。因

此,我们能够通过一个参考系,有效描述在任何连续介质中在任何时间的力的分布情况。

接下来,我们通过研究某一点的应力转换方程,得出许多非常有价值的结论。由此,我们学习了某点处主应力轴、主应力以及一点处的三个应力不变量。还学习许多其他的量可以用来表示某点绕轴转动得到的转换方程,以及前文提到的主应力轴、不变量和其他特性。虽然张量的概念确实来自我们对固体的研究,但对张量的讨论绝不只局限于变形体。这些结论将推广到其他工程科学、物理学和数学中。

在第 3 章中,我们将考虑固体的几何变形。在讨论过程中,我们会证明应变也是一个二阶张量。一旦完成了该证明,我们立即能够将第 2 章中针对应力得到的所有结论应用到应变中。

习题

2-1 在图 2-23 中标注微小长方体的相应应力。

2-2 在图 2-24 中标注出应力及其正确的符号。

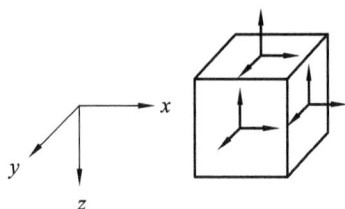

图 2-23 习题 2-1 图　　　　图 2-24 习题 2-2 图

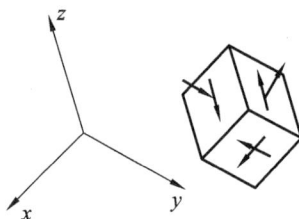

2-3 已知正交坐标系 xyz 中一点的应力分量为:

$$\sigma_x = 1000 \text{ Pa} \qquad \tau_{xy} = 200 \text{ Pa}$$
$$\sigma_y = -600 \text{ Pa} \qquad \tau_{xz} = 0 \text{ Pa}$$
$$\sigma_z = 0 \text{ Pa} \qquad \tau_{yz} = -400 \text{ Pa}$$

假设下标互换后剪应力相等($\tau_{ij} = \tau_{ji}$),那么在 $\boldsymbol{\varepsilon}$ 方向上的正应力是多大?

$$\boldsymbol{\varepsilon} = 0.11\boldsymbol{i} + 0.35\boldsymbol{j} + 0.93\boldsymbol{k}$$

2-4 推导方程(2-6)。

2-5 下述矩阵对应着某点在 xyz 参考系中的应力状态:

$$\begin{bmatrix} 200 & 100 & 0 \\ 100 & 0 & 0 \\ 0 & 0 & 500 \end{bmatrix}$$

如果新的坐标轴 $x'y'z'$ 由 xyz 绕 z 轴旋转 $60°$ 生成,那么,在 $x'y'z'$ 坐标系中,该点的应力状态是什么么?(计算过程用张量表示)

2-6 给定平面应力:

$$\sigma_x = 500 \text{ Pa}$$
$$\sigma_y = 1000 \text{ Pa}$$
$$\tau_{xy} = 500 \text{ Pa}$$

将 xy 轴旋转 $45°$ 得到 $x'y'$ 轴,则在新坐标系下剪应力和正应力是多少?

2-7 计算习题 2-6 中的主应力大小及方向。

2-8 给定以下平面应力状态:

$$\sigma_x = -1000 \text{ Pa}$$
$$\sigma_y = 500 \text{ Pa}$$
$$\tau_{xy} = 1000 \text{ Pa}$$

那么,该点处的三个应力张量不变量是什么?

2-9 计算习题 2-8 中的主应力大小及方向。

2-10 绘制该应力状态下的莫尔圆:

$$\sigma_x = -1000 \text{ Pa}$$

$$\sigma_y = 2000 \text{ Pa}$$

$$\tau_{xy} = -500 \text{ Pa}$$

2-11 在习题 2-10 绘制的莫尔圆中,近似画出从 x 轴向 y 轴旋转 30°时界面的应力状态。

2-12 证明表示平面应力的另一种方法是具有主应力 $\sigma_z = 0$。

2-13 画出 $\tau_{xy} = 500$ Pa 以及 $\sigma_x = \sigma_y = 0$ 状态下对应的莫尔圆。

3 应 变

3.1 引 言

本章我们将研究表示变形体位移的方法。回顾我们之前所学的力学课程,刚体运动可以被描述为该刚体上任一点的平移运动,再加上绕该点的旋转运动,这就是著名的沙勒定理。本章我们将讨论一般情况下的位移,除了考虑刚体平移和刚体转动产生的位移外,还要考虑其他可能产生的变形。

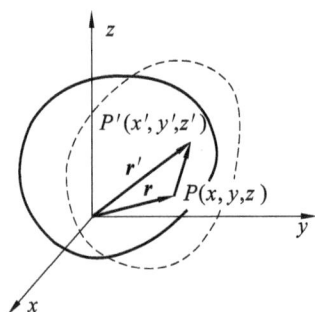

图 3-1 任意固体的变形几何体

我们现在取一个任意固体中的未变形体,建立如图 3-1 所示的固定参考系 xyz。在未变形体中任意一点 P 的位置向量 $\boldsymbol{r}$ 都可以用坐标 x,y,z 表示。其中,变形体用虚线表示。未变形体中的 P 点变形后为点 P',用坐标 x',y',z' 表示。我们用 $\boldsymbol{u}$ 表示点 P 到点 P' 的位移,并称它为位移向量。在笛卡儿分量中,位移向量通常表示为

$$\boldsymbol{u}=u_x\boldsymbol{i}+u_y\boldsymbol{j}+u_z\boldsymbol{k} \tag{3-1}$$

很明显,位移向量 $\boldsymbol{u}$ 在各点连续,因此它形成了一个向量场。我们称之为位移场,通常采用未变形体的坐标函数表示,如 $\boldsymbol{u}(x,y,z)$。

使用笛卡儿张量表示法时,可以将坐标轴 x,y,z 分别用 x_1,x_2,x_3 表示,单位向量 $\boldsymbol{i},\boldsymbol{j},\boldsymbol{k}$ 分别用 $\boldsymbol{\varepsilon}_1,\boldsymbol{\varepsilon}_2,\boldsymbol{\varepsilon}_3$ 表示。因此方程(3-1)可以写成:

$$\boldsymbol{u}=u_i\boldsymbol{\varepsilon}_i \tag{3-2}$$

其中,i 取 $1,2,3$。

3.2 小区域观点

考虑如图 3-2 所示发生变形的几何体,我们选择几何体中的任意两点(点 P 和点 Q),并用有向线段连接两点得到向量 $\boldsymbol{A}$。点 P 和点 Q 在变形后移动到点 P' 和点 Q',同样我们用有向线段连接这两点,得到第二个向量 $\boldsymbol{A}'$,如图 3-2所示。

我们想要计算 $(\boldsymbol{A}'-\boldsymbol{A})$ 的值,将其记作 $\delta\boldsymbol{A}$。为此,在图 3-2 中分别在点 P 和点 Q 处插入位移矢量 $\boldsymbol{u}_P$ 和 $\boldsymbol{u}_Q$,从而形成一个矢量多边形。可得到

$$\boldsymbol{u}_P+\boldsymbol{A}'=\boldsymbol{A}+\boldsymbol{u}_Q \tag{3-3}$$

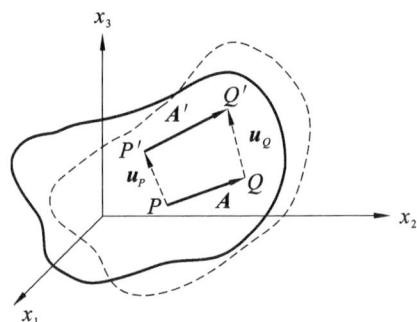

图 3-2 发生变形的物体

重新整理方程(3-3),可以得到

$$A'-A=\delta A=u_Q-u_P \tag{3-4}$$

下面我们讨论的问题中位移场 u 始终是解析函数,因此,我们可以在点 P 处对位移 u 进行泰勒级数展开,用 u_P 来表示 u_Q,即

$$u_Q=u_P+\left(\frac{\partial u}{\partial x_1}\right)_P\Delta x_1+\left(\frac{\partial u}{\partial x_2}\right)_P\Delta x_2+\left(\frac{\partial u}{\partial x_3}\right)_P\Delta x_3+\cdots \tag{3-5}$$

式中,Δx_1,Δx_2 和 Δx_3 分别是 A_1,A_2 和 A_3 的分量,可用 A 和下标表示,故方程(3-5)可以重写为

$$u_Q=u_P+\left(\frac{\partial u}{\partial x_j}\right)_P A_j+\cdots \tag{3-6}$$

如果向量 A 很小,我们只考虑了点 P 周围很小区域内的变形,那么在前面泰勒级数展开中,就可以忽略高阶项,得到

$$u_Q=u_P+\left(\frac{\partial u}{\partial x_j}\right)_P A_j \tag{3-7}$$

将方程(3-7)代入方程(3-4)中,有

$$\delta A=\left(\frac{\partial u}{\partial x_j}\right)_P A_j \tag{3-8}$$

由于点 P 是几何体中的任意一点,所以前面表达式中的下标 P 可以删去。因此在无限小的区域中,关于任意向量 A 的变化,我们表示为

$$\delta A_i=\frac{\partial u_i}{\partial x_1}A_1+\frac{\partial u_i}{\partial x_2}A_2+\frac{\partial u_i}{\partial x_3}A_3 \tag{3-9}$$

或者

$$\delta A_i=\frac{\partial u_i}{\partial x_j}A_j \tag{3-10}$$

当我们考虑几何体微小单元的变形时,方程(3-10)将十分有用。下一节中我们将介绍小变形约束,小变形约束不应与本节中介绍的小区域观点混淆。

以下例题将说明应该如何使用本节所推导的公式。

【例 3-1】 已知如下位移场

$$u=(xy\boldsymbol{i}+3x^2z\boldsymbol{j}+4\boldsymbol{k})\times10^{-2}$$

以及微小线段 Δs,该线段变形前的方向余弦如下

$$l=0.200, \quad m=0.800, \quad n=0.555$$

该线段的起点坐标为 $(2,1,3)$。在施加位移场后,所得到的新矢量 $\Delta s'$ 是什么?

【解】 我们首先计算 $\partial u_i/\partial x_j$,则

$$\frac{\partial u_1}{\partial x}=0.01y \qquad \frac{\partial u_1}{\partial y}=0.01x \qquad \frac{\partial u_1}{\partial z}=0$$

$$\frac{\partial u_2}{\partial x}=0.06xz \qquad \frac{\partial u_2}{\partial y}=0 \qquad \frac{\partial u_2}{\partial z}=0.03x^2$$

$$\frac{\partial u_3}{\partial x}=0 \qquad \frac{\partial u_3}{\partial y}=0 \qquad \frac{\partial u_3}{\partial z}=0$$

从而得到

$$\frac{\partial u_i}{\partial x_j}=\begin{bmatrix} 0.01y & 0.01x & 0 \\ 0.06xz & 0 & 0.03x^2 \\ 0 & 0 & 0 \end{bmatrix}$$

在点 $(2,1,3)$ 附近的小区域中,有

$$\left(\frac{\partial u_i}{\partial x_j}\right)_{(2,1,3)}=\begin{pmatrix}0.01 & 0.02 & 0\\0.36 & 0 & 0.12\\0 & 0 & 0\end{pmatrix}$$

利用方程(3-10)可得到

$$
\begin{aligned}
\left[\delta(\Delta s)\right]_1 &= \left(\frac{\partial u_1}{\partial x_j}\right)_P(\Delta s)_j\\
&= \left(\frac{\partial u_1}{\partial x}\right)_P(\Delta s)l+\left(\frac{\partial u_1}{\partial y}\right)_P(\Delta s)m+\left(\frac{\partial u_1}{\partial z}\right)_P(\Delta s)n\\
&= \Delta s(0.002+0.016)=0.018\Delta s
\end{aligned}
$$

$$
\begin{aligned}
\left[\delta(\Delta s)\right]_2 &= \left(\frac{\partial u_2}{\partial x_j}\right)_P(\Delta s)_j\\
&= \left(\frac{\partial u_2}{\partial x}\right)_P(\Delta s)l+\left(\frac{\partial u_2}{\partial y}\right)_P(\Delta s)m+\left(\frac{\partial u_2}{\partial z}\right)_P(\Delta s)n\\
&= \Delta s(0.072+0.0666)=0.1386\Delta s
\end{aligned}
$$

$$
\begin{aligned}
\left[\delta(\Delta s)\right]_3 &= \left(\frac{\partial u_3}{\partial x_j}\right)_P(\Delta s)_j\\
&= \left(\frac{\partial u_3}{\partial x}\right)_P(\Delta s)l+\left(\frac{\partial u_3}{\partial y}\right)_P(\Delta s)m+\left(\frac{\partial u_3}{\partial z}\right)_P(\Delta s)n\\
&= \Delta s(0)=0
\end{aligned}
$$

矢量 Δs 的变量为

$$\delta(\Delta s)=(0.018\boldsymbol{i}+0.1386\boldsymbol{j})\Delta s$$

那么新矢量 $\Delta s'$ 为

$$
\begin{aligned}
\Delta s' &= \Delta s+\delta(\Delta s)\\
&= (0.200\boldsymbol{i}+0.800\boldsymbol{j}+0.555\boldsymbol{k})\Delta s+(0.018\boldsymbol{i}+0.1386\boldsymbol{j})\Delta s\\
&= (0.218\boldsymbol{i}+0.9386\boldsymbol{j}+0.555\boldsymbol{k})\Delta s
\end{aligned}
$$

【例 3-2】 已知以下位移场

$$
\begin{aligned}
u_x &= (x^2+2y^2z+yz)\times10^{-2}\\
u_y &= [(y+z)x+3x^2z]\times10^{-2}\\
u_z &= (4y^3+2z^2)\times10^{-2}
\end{aligned}
$$

起点坐标为(1,1,1)且沿 x 轴方向的向量 Δs，在位移场作用下，增量 $\delta(\Delta s)$ 是什么？

【解】 $\partial u_i/\partial x_j$ 矩阵为

$$\frac{\partial u_i}{\partial x_j}=\begin{pmatrix}2x & (4yz+z) & (2y^2+y)\\(y+z+6xz) & x & (x+3x^2)\\0 & 12y^2 & 4z\end{pmatrix}\times10^{-2}$$

一方面，可以通过使用前面的方程来确定质点在(1,1,1)处的位移，则

$$
\begin{aligned}
(\boldsymbol{u})_{(1,1,1)} &= [(1+2+1)\boldsymbol{i}+(2+3)\boldsymbol{j}+(4+2)\boldsymbol{k}]\times10^{-2}\\
&= (4\boldsymbol{i}+5\boldsymbol{j}+6\boldsymbol{k})\times10^{-2}
\end{aligned}
$$

另一方面，为了找到位于(1,1,1)且沿 x 轴方向的向量 Δs 变化的长度，我们采用下面的公式

$$\left[\delta(\Delta s)\right]_i=\frac{\partial u_i}{\partial x}(\Delta x)+\frac{\partial u_i}{\partial y}(0)+\frac{\partial u_i}{\partial z}(0)$$

因此

$$\left[\delta(\Delta s)\right]_1=\frac{\partial u_x}{\partial x}(\Delta x)+\frac{\partial u_x}{\partial y}(0)+\frac{\partial u_x}{\partial z}(0)=2x\Delta x\times10^{-2}$$

$$[\delta(\Delta s)]_2 = \frac{\partial u_y}{\partial x}(\Delta x) + \frac{\partial u_y}{\partial y}(0) + \frac{\partial u_y}{\partial z}(0) = (y+z+6xz)\Delta x \times 10^{-2}$$

$$[\delta(\Delta s)]_3 = \frac{\partial u_z}{\partial x}(\Delta x) + \frac{\partial u_z}{\partial y}(0) + \frac{\partial u_z}{\partial z}(0) = 0$$

故我们得到 $\delta(\Delta s)$

$$[\delta(\Delta s)]_P = [2x\boldsymbol{i} + (y+z+6xz)\boldsymbol{j}](\Delta x) \times 10^{-2}$$

对于给定具体位置 $(1,1,1)$，我们得到

$$[\delta(\Delta s)]_{(1,1,1)} = (2\boldsymbol{i} + 8\boldsymbol{j})(\Delta x) \times 10^{-2}$$

3.3 小变形约束

考虑两个不同的位移场 $\boldsymbol{u}^{(1)}$ 和 $\boldsymbol{u}^{(2)}$。在任何小区域中，在第一个位移场 $\boldsymbol{u}^{(1)}$ 作用下，矢量 $\boldsymbol{A}$ 的增量为

$$\delta A_i = \left(\frac{\partial u_i^{(1)}}{\partial x_j}\right)_a A_j \tag{3-11}$$

其中，$()_a$ 表示括号中的项是根据未变形几何体在点 a 处坐标计算得到的，如图 3-3 所示。我们很容易将前面公式中 δA_i 用 $(A_i' - A_i)$ 表示，其中有一撇的字母对应的几何体表示第一次变形几何体。然后，我们得到 A_i'

$$A_i' = A_i + \left(\frac{\partial u_i^{(1)}}{\partial x_j}\right)_a A_j \tag{3-12}$$

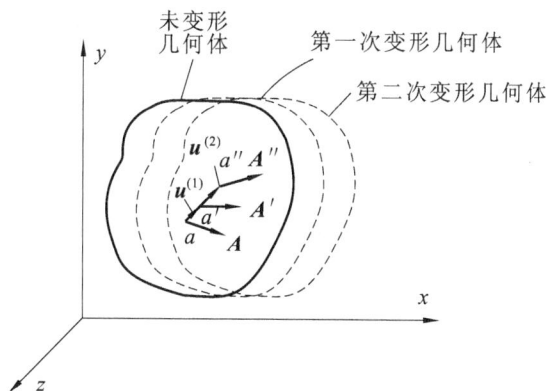

图 3-3 从未变形到两次变形的几何体的图解

现在推测第二次变形几何体是由第一次变形几何体经过变形得到的，如图 3-3 所示。也就是说，第一次变形几何体经变形后得到的变形几何体可认为是发生第二次变形的未变形几何体。那么 A_i'' 表示为

$$A_i'' = A_i' + \left(\frac{\partial u_i^{(2)}}{\partial x_k}\right)_{a'} A_k' \tag{3-13}$$

其中 $()_{a'}$ 表示括号中的项是根据第一次变形几何体在点 a' 处坐标计算得到的。注意，这里我们使用 k 而不是 j 作为名义下标。这在后面的计算步骤中是十分有用的（名义下标改变不影响结果）。为此，我们把方程 (3-12) 代入方程 (3-13)，来替换方程 (3-13) 中的 A_i' 和 A_k'，从而得到

$$A_i'' = A_i + \left(\frac{\partial u_i^{(1)}}{\partial x_j}\right)_a A_j + \left(\frac{\partial u_i^{(2)}}{\partial x_k}\right)_{a'}\left[A_k + \left(\frac{\partial u_k^{(1)}}{\partial x_j}\right)_a A_j\right] \tag{3-14}$$

该方程也可表示为

$$A''_i = A_i + \left(\frac{\partial u_i^{(1)}}{\partial x_j}\right)_a A_j + \left(\frac{\partial u_i^{(2)}}{\partial x_k}\right)_{a'} A_k + \left(\frac{\partial u_i^{(2)}}{\partial x_k}\right)_{a'} \left(\frac{\partial u_k^{(1)}}{\partial x_j}\right)_a A_j \tag{3-15}$$

将等式(3-15)右侧第三项中的名义下标 k 改为 j,合并同类项后得到

$$A''_i = A_i + \left[\left(\frac{\partial u_i^{(1)}}{\partial x_j}\right)_a + \left(\frac{\partial u_i^{(2)}}{\partial x_j}\right)_{a'}\right] A_j + \left(\frac{\partial u_i^{(2)}}{\partial x_k}\right)_{a'} \left(\frac{\partial u_k^{(1)}}{\partial x_j}\right)_a A_j \tag{3-16}$$

接下来,我们在点 a 处进行泰勒级数展开,计算点 a' 处 $(\partial u_i^{(2)}/\partial x_j)_{a'}$ 的值。

$$\left(\frac{\partial u_i^{(2)}}{\partial x_j}\right)_{a'} = \left(\frac{\partial u_i^{(2)}}{\partial x_j}\right)_a + \left[\frac{\partial}{\partial x_k}\left(\frac{\partial u_i^{(2)}}{\partial x_j}\right)\right]_a u_k^{(1)} + \left[\frac{\partial^2}{\partial x_k \partial x_l}\left(\frac{\partial u_i^{(2)}}{\partial x_j}\right)\right]_a \frac{u_k^{(1)} u_l^{(1)}}{2} + \cdots \tag{3-17}$$

现在我们施加小变形约束,即 $u_i^{(1)}$,$u_i^{(2)}$,$(\partial u_i^{(2)}/\partial x_j)_a$,$(\partial u_i^{(2)}/\partial x_j)_{a'}$ 都非常小。这意味着我们只用保留方程(3-17)等号右边的第一项,也就是说,方程可变为

$$\left(\frac{\partial u_i^{(2)}}{\partial x_j}\right)_{a'} = \left(\frac{\partial u_i^{(2)}}{\partial x_j}\right)_a \tag{3-18}$$

然后,我们可以使用未变形的几何体来计算连续变形后的结果,此时前文公式中的下标 a 和 a' 也变得不重要了。此外,在方程(3-16)中,和偏导数本身相比,偏导数的乘积可以忽略。简而言之,对于小变形来说,我们可以将方程(3-16)改写为

$$A''_i - A_i = [\delta A_i]_{\text{total}} = \left(\frac{\partial u_i^{(1)}}{\partial x_j} + \frac{\partial u_i^{(2)}}{\partial x_j}\right)_a A_j \tag{3-19}$$

施加小变形约束,将出现如下基本简化:

(1) 初始几何体经过一系列无穷小位移后得到的总变形,只要将每次按初始几何体计算得到的位移进行叠加就能得到。这个可以从方程(3-19)中明显看出,它满足叠加原理。

(2) 施加无穷小位移的顺序对总变形没有影响。这满足交换定律,同样也可以从方程(3-19)中看出。

我们的研究都局限于小变形情况。缺少简化的大变形理论将会使研究变得非常困难,也超出了本书的范围。幸运的是,大多数工程问题都可以用小变形理论来处理。

总的来说,我们通常利用 3.2 节中介绍的小区域观点,以便使用方程(3-10)。要明白,小区域观点与大变形或小变形无关,它适用于这两种情况。此外,我们将运用小变形约束。因此,使用小区域观点和小变形约束就意味着我们将考虑单元体在几何体发生小变形情况下产生的变形。

3.4 刚体转动和单元体的纯变形

在 3.2 节中,我们推导了方程(3-9)和方程(3-10),通过这些方程,我们可以确定发生变形的微小单元体中线段长度和方向的变化。显然,知道单元体内线段的变形就相当于知道单元体是如何变形的。因此,在此类研究中 $\partial u_i / \partial x_j$ 项是关键量。本节我们将指出如何用这些量来表示刚体转动,以及所谓的纯变形。

作为第一步,我们先将 $\partial u_i / \partial x_j$ 表示为

$$\frac{\partial u_i}{\partial x_j} = \frac{1}{2}\left(\frac{\partial u_i}{\partial x_j} + \frac{\partial u_j}{\partial x_i}\right) + \frac{1}{2}\left(\frac{\partial u_i}{\partial x_j} - \frac{\partial u_j}{\partial x_i}\right) \tag{3-20}$$

可以认为在这个方程中,矩阵 $\partial u_i / \partial x_j$ 是由矩阵 ε_{ij} 和矩阵 ω_{ij} 组合得到的。因此

$$\frac{\partial u_i}{\partial x_j} = \varepsilon_{ij} + \omega_{ij} \tag{3-21}$$

其中

$$\varepsilon_{ij} = \frac{1}{2}\left(\frac{\partial u_i}{\partial x_j} + \frac{\partial u_j}{\partial x_i}\right) \tag{3-22}$$

$$\omega_{ij} = \frac{1}{2}\left(\frac{\partial u_i}{\partial x_j} - \frac{\partial u_j}{\partial x_i}\right) \tag{3-23}$$

现在我们将方程(3-10)表示成如下形式：

$$\delta A_i = (\varepsilon_{ij} + \omega_{ij})A_j \tag{3-24}$$

我们将在本节和后续章节中仔细研究量 ε_{ij} 和 ω_{ij}。

我们首先研究矩阵 ω_{ij}。先计算出方程(3-23)中该矩阵的所有项，有

$$\omega_{11} = \frac{1}{2}\left(\frac{\partial u_1}{\partial x_1} - \frac{\partial u_1}{\partial x_1}\right) = 0 \quad \omega_{12} = \frac{1}{2}\left(\frac{\partial u_1}{\partial x_2} - \frac{\partial u_2}{\partial x_1}\right) \quad \omega_{13} = \frac{1}{2}\left(\frac{\partial u_1}{\partial x_3} - \frac{\partial u_3}{\partial x_1}\right)$$

$$\omega_{21} = \frac{1}{2}\left(\frac{\partial u_2}{\partial x_1} - \frac{\partial u_1}{\partial x_2}\right) \quad \omega_{22} = \frac{1}{2}\left(\frac{\partial u_2}{\partial x_2} - \frac{\partial u_2}{\partial x_2}\right) = 0 \quad \omega_{23} = \frac{1}{2}\left(\frac{\partial u_2}{\partial x_3} - \frac{\partial u_3}{\partial x_2}\right) \tag{3-25}$$

$$\omega_{31} = \frac{1}{2}\left(\frac{\partial u_3}{\partial x_1} - \frac{\partial u_1}{\partial x_3}\right) \quad \omega_{32} = \frac{1}{2}\left(\frac{\partial u_3}{\partial x_2} - \frac{\partial u_2}{\partial x_3}\right) \quad \omega_{33} = \frac{1}{2}\left(\frac{\partial u_3}{\partial x_3} - \frac{\partial u_3}{\partial x_3}\right) = 0$$

我们有意将这些方程作为项放置在一个矩阵中。很明显可以看出，矩阵 ω_{ij} 的主对角线项为零，非对角线上对应的两项互为相反数。因此，我们可以得到

$$\omega_{ij} = -\omega_{ji} \tag{3-26}$$

显然，当 $i=j$ 时，ω_{ij} 项必须为零才能满足方程。这样的矩阵称为螺旋对称矩阵或反对称矩阵。

我们首先证明非对角线上的项是单元体的刚体旋转分量。如图 3-4 所示，我们考虑一个在点 P 处发生无穷小变形的单元体。由于变形，点 P 处的单元体移动到点 P'，因此，单元体的位移用 $\mathbf{u}(P)$ 表示。此外，我们可以想象该单元体发生的刚体转动以及形状的变化。现在假设，当单元体从图 3-4 中字母不加撇的位置变形到字母加一撇的位置时，只发生了平移和刚体转动。研究图 3-4 中的线段 Δy。很显然，我们依据该线段绕 x 轴（或任何平行于 x 轴的轴）的旋转量将得到该单元体绕 x 轴旋转的旋转分量。在图 3-5 中，我们展示了在变形体中该线段的放大图，将其记为 $\Delta y'$，并标出线段 Δy 端点在 z 轴方向上的位移分量。通过这些位移分量，我们可以用下述方式表示线段 Δy 绕 x 轴转动。

$$(\delta\boldsymbol{\phi})_x = \frac{\left(u_z + \frac{\partial u_z}{\partial y}\Delta y\right) - u_z}{(\Delta y')_{y'}} \tag{3-27}$$

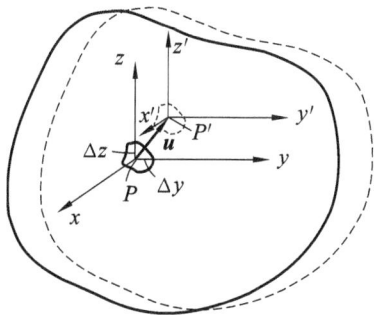

图 3-4　在几何体内部点 P 处发生无穷小变形的单元体

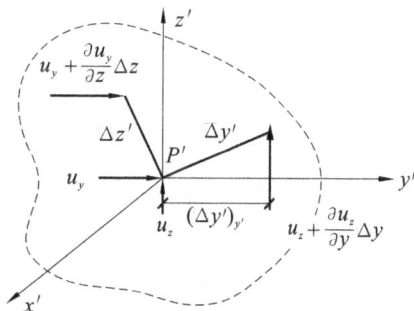

图 3-5　变形体中线段的放大图

其中，$(\Delta y')_{y'}$ 是图 3-5 中线段 $\Delta y'$ 在 y' 轴的投影。对于小变形，我们可以假设方程(3-27)中有 $(\Delta y')_{y'} = \Delta y$，因此，$(\delta\boldsymbol{\phi})_x$ 变成

$$(\delta\boldsymbol{\phi})_x = \frac{\partial u_z}{\partial y} \tag{3-28}$$

我们也可以对图 3-4 中的线段 Δz 段进行类似研究。根据图 3-5，可以得到

$$(\delta\boldsymbol{\phi})_x = \frac{-\left(u_y + \frac{\partial u_y}{\partial z}\Delta z - u_y\right)}{\Delta z} \tag{3-29}$$

注意:这些项的符号是通过右手螺旋法则确定的。化简方程(3-29)可变为

$$(\delta\boldsymbol{\phi})_x = -\frac{\partial u_y}{\partial z} \tag{3-30}$$

将方程(3-28)和方程(3-30)相加,可得$(\delta\boldsymbol{\phi})_x$表示如下

$$(\delta\boldsymbol{\phi})_x = \frac{1}{2}\left(\frac{\partial u_z}{\partial y} - \frac{\partial u_y}{\partial z}\right) \tag{3-31}$$

而等式右边的项就是ω_{zy}。通过类似的论证,我们可以证明$(\delta\boldsymbol{\phi})_y$和$(\delta\boldsymbol{\phi})_z$分别等于$\omega_{xz}$和$\omega_{yx}$。从而得到

$$(\delta\boldsymbol{\phi})_x = \omega_{zy}$$
$$(\delta\boldsymbol{\phi})_y = \omega_{xz} \tag{3-32}$$
$$(\delta\boldsymbol{\phi})_z = \omega_{yx}$$

因此,非对角项可以表示刚体转动的旋转分量。旋转矢量$\delta\boldsymbol{\phi}$可以表示如下:

$$\delta\boldsymbol{\phi} = \omega_{zy}\boldsymbol{i} + \omega_{xz}\boldsymbol{j} + \omega_{yx}\boldsymbol{k} \tag{3-33}$$

现在我们可以很容易指出,方程(3-24)中ω_{ij}对δA_i的贡献就是刚体转动的结果。回顾刚体力学中,对于刚体中的一个固定向量$\boldsymbol{A}$,它的时间变化率是

$$\frac{\mathrm{d}\boldsymbol{A}}{\mathrm{d}t} = \boldsymbol{\omega} \times \boldsymbol{A} \tag{3-34}$$

式中,$\boldsymbol{\omega}$是刚体的角速度。

方程(3-34)可以写成

$$\mathrm{d}\boldsymbol{A} = \boldsymbol{\omega}\mathrm{d}t \times \boldsymbol{A} = \mathrm{d}\boldsymbol{\phi} \times \boldsymbol{A} \tag{3-35}$$

式中,$\mathrm{d}\boldsymbol{A}$表示矢量$\boldsymbol{A}$由于转动$\mathrm{d}\boldsymbol{\phi}$而发生的变化。取方程(3-35)中的第$i$个分量,用$\delta$代替$\mathrm{d}$,可以得到

$$(\delta\boldsymbol{A})_i = (\delta\boldsymbol{\phi} \times \boldsymbol{A})_i \tag{3-36}$$

简单地验算一下$i=1$时的情况。我们得到向量叉乘

$$(\delta\boldsymbol{A})_1 = (\delta\boldsymbol{\phi})_2 A_3 - (\delta\boldsymbol{\phi})_3 A_2 \tag{3-37}$$

用方程(3-32)替换$(\delta\boldsymbol{\phi})_2$和$(\delta\boldsymbol{\phi})_3$,我们得到

$$(\delta\boldsymbol{A})_1 = \omega_{13} A_3 - \omega_{21} A_2 \tag{3-38}$$

由于$\omega_{21} = -\omega_{12}$且$\omega_{11} = 0$,方程(3-36)可表示如下:

$$(\delta\boldsymbol{A})_1 = \omega_{11} A_1 + \omega_{12} A_2 + \omega_{13} A_3 = \omega_{1j} A_j \tag{3-39}$$

同样,考虑方程(3-37)中$\delta\boldsymbol{A}$的其他分量,我们可以得出以下结论:

$$(\delta\boldsymbol{A})_i = \delta A_i = \omega_{ij} A_j \tag{3-40}$$

方程(3-40)与方程(3-24)相同,但仅出现矩阵ω_{ij}。因此,我们可以推断方程(3-40)中的ω_{ij}表示发生无穷小变形情况下刚体转动对单元体变形所做的贡献。现在,我们已经确定了ω_{ij}的物理含义,以及ω_{ij}对单元体变形做出的贡献。正如预期那样,ω_{ij}被称为旋转矩阵。

接下来,我们像确定矩阵ω_{ij}中各项一样计算ε_{ij}中的项。

$$\varepsilon_{11} = \frac{\partial u_1}{\partial x_1} \qquad \varepsilon_{12} = \frac{1}{2}\left(\frac{\partial u_1}{\partial x_2} + \frac{\partial u_2}{\partial x_1}\right) \quad \varepsilon_{13} = \frac{1}{2}\left(\frac{\partial u_1}{\partial x_3} + \frac{\partial u_3}{\partial x_1}\right)$$

$$\varepsilon_{21} = \frac{1}{2}\left(\frac{\partial u_2}{\partial x_1} + \frac{\partial u_1}{\partial x_2}\right) \quad \varepsilon_{22} = \frac{\partial u_2}{\partial x_2} \qquad \varepsilon_{23} = \frac{1}{2}\left(\frac{\partial u_2}{\partial x_3} + \frac{\partial u_3}{\partial x_2}\right) \tag{3-41}$$

$$\varepsilon_{31} = \frac{1}{2}\left(\frac{\partial u_3}{\partial x_1} + \frac{\partial u_1}{\partial x_2}\right) \quad \varepsilon_{32} = \frac{1}{2}\left(\frac{\partial u_3}{\partial x_2} + \frac{\partial u_2}{\partial x_3}\right) \quad \varepsilon_{33} = \frac{\partial u_3}{\partial x_3}$$

观察该矩阵,可以看出它是一个对称矩阵。我们已经建立了方程(3-25)中的反对称矩阵ω_{ij}以及方程(3-41)中的对称矩阵ε_{ij}。既然矩阵ω_{ij}表示单元体的刚体转动,我们接着必须得出矩阵ε_{ij}表示单元体纯变形的结论。因此,我们称矩阵ε_{ij}为应变矩阵。在展示了矩阵ε_{ij}的张量的性质后,我们将称之为应变张量。下一节中,我们将仔细研究应变矩阵ε_{ij}中各项的性质。

【**例 3-3**】 某物体按照如下变形场发生变形：

$$u_1 = 0.003x_1 + 0.002x_2$$
$$u_2 = -0.001x_1 + 0.0005x_3$$
$$u_3 = 0.0006x_1 + 0.003x_2 - 0.003x_3$$

那么旋转矩阵 ω_{ij} 和应变矩阵 ε_{ij} 是什么？

【**解**】 通过前面方程可以很容易确定矩阵 $\partial u_i / \partial x_j$

$$\frac{\partial u_i}{\partial x_j} = \begin{pmatrix} 0.003 & 0.002 & 0 \\ -0.001 & 0 & 0.0005 \\ 0.0006 & 0.003 & -0.003 \end{pmatrix}$$

因此

$$\omega_{11} = 0 \quad \omega_{12} = \frac{1}{2} \times (0.002 + 0.001) = 0.0015 \quad \omega_{13} = \frac{1}{2} \times (0 - 0.0006) = -0.0003$$

$$\omega_{21} = \frac{1}{2} \times (-0.001 - 0.002) = -0.0015 \quad \omega_{22} = 0 \quad \omega_{23} = \frac{1}{2} \times (0.0005 - 0.003) = -0.00125$$

$$\omega_{31} = \frac{1}{2} \times (0.0006 - 0) = 0.0003 \quad \omega_{32} = \frac{1}{2} \times (0.003 - 0.0005) = 0.00125 \quad \omega_{33} = 0$$

根据方程(3-32)得到旋转分量 $\delta\boldsymbol{\phi}$，那么

$$(\delta\boldsymbol{\phi})_1 = \omega_{32} = 0.00125 \text{rad}$$
$$(\delta\boldsymbol{\phi})_2 = \omega_{13} = -0.0003 \text{rad}$$
$$(\delta\boldsymbol{\phi})_3 = \omega_{21} = -0.0015 \text{rad}$$

最后，该位移的应变矩阵为

$$\varepsilon_{11} = 0.003 \qquad \varepsilon_{12} = \frac{1}{2} \times (0.002 - 0.001) = 0.0005 \qquad \varepsilon_{13} = \frac{1}{2} \times (0 + 0.0006) = 0.0003$$

$$\varepsilon_{21} = \frac{1}{2} \times (-0.001 + 0.002) = 0.0005 \quad \varepsilon_{22} = 0 \qquad \varepsilon_{23} = \frac{1}{2} \times (0.0005 + 0.003) = 0.00175$$

$$\varepsilon_{31} = \frac{1}{2} \times (0.0006 + 0) = 0.0003 \qquad \varepsilon_{32} = \frac{1}{2} \times (0.003 + 0.0005) = 0.00175 \quad \varepsilon_{33} = -0.003$$

例 3-3 是一种仿射变形情况。我们注意到应变矩阵和旋转矩阵由常数组成，这意味着物体内的每个小单元体都具有相同的转动和纯变形，这种变形称为均匀变形。

【**例 3-4**】 已知位移场如下

$$u_x = (3x^2 y + 6) \times 10^{-2}$$
$$u_y = (y^2 + 6xz) \times 10^{-2}$$
$$u_z = (6z^2 + 2yz + 10) \times 10^{-2}$$

在 $x = 1, y = 0, z = 2$ 处，单元体的转动变形是什么？

【**解**】 根据位移场，我们可以先确定矩阵 $\partial u_i / \partial x_j$。那么

$$\frac{\partial u_x}{\partial x} = 0.06xy \quad \frac{\partial u_x}{\partial y} = 0.03x^2 \quad \frac{\partial u_x}{\partial z} = 0$$

$$\frac{\partial u_y}{\partial x} = 0.06z \quad \frac{\partial u_y}{\partial y} = 0.02y \quad \frac{\partial u_y}{\partial z} = 0.06x$$

$$\frac{\partial u_z}{\partial x} = 0 \quad \frac{\partial u_z}{\partial y} = 0.02z \quad \frac{\partial u_z}{\partial z} = 0.12z + 0.02y$$

因此

$$\frac{\partial u_i}{\partial x_j} = \begin{bmatrix} 0.06xy & 0.03x^2 & 0 \\ 0.06z & 0.02y & 0.06x \\ 0 & 0.02z & 0.12z+0.02y \end{bmatrix}$$

接下来,从该变形场中我们能得到旋转矩阵 ω_{ij}。通过计算我们得到

$$\omega_{ij} = \begin{bmatrix} 0 & \frac{1}{2}(3x^2-6z) & 0 \\ -\frac{1}{2}(3x^2-6z) & 0 & \frac{1}{2}(6x-2z) \\ 0 & -\frac{1}{2}(6x-2z) & 0 \end{bmatrix} \times 10^{-2}$$

根据单元体信息,我们有

$$\omega_{ij} = \begin{bmatrix} 0 & -0.045 & 0 \\ 0.045 & 0 & 0.01 \\ 0 & -0.01 & 0 \end{bmatrix}$$

因此我们得到旋转分量

$$(\delta\pmb{\phi})_1 = \omega_{32} = -0.01\text{rad}$$
$$(\delta\pmb{\phi})_2 = \omega_{13} = 0\text{rad}$$
$$(\delta\pmb{\phi})_3 = \omega_{21} = 0.045\text{rad}$$

3.5 应变项的物理解释

到目前为止,知道小变形中的位移场 $\pmb{u}$,我们就可以确定矩阵 ω_{ij} 和 ε_{ij},计算出该点处微小单元体的刚体转动和纯变形情况。我们已经将旋转矩阵 ω_{ij} 的项与旋转矢量 $\delta\pmb{\phi}$ 联系起来。下一步,我们将通过对某点计算得到的矩阵 ε_{ij} 中的项与一些具有物理含义的几何解释联系起来,这将有助于我们后面的学习。

首先,我们沿 x 轴连接点 P 和点 Q 形成一条线段 Δx,如图 3-6 所示。变形后,点 P 移动到点 P' 且点 Q 移动到点 Q'。我们将线段 $\overline{P'Q'}$ 在 x 轴方向上的投影记作 $(\overline{P'Q'})_x$,它可以通过原长 Δx 以及点 P 和点 Q 在 x 轴方向上的位移按以下方式计算得到

$$(\overline{P'Q'})_x = \Delta x + (u_x)_Q - (u_x)_P \tag{3-42}$$

接下来,我们采用 3.2 节中同样的方法,将方程(3-42)中的 $(u_x)_Q$ 在点 P 处进行泰勒级数展开,可以得到

$$(\overline{P'Q'})_x = \Delta x + \left[(u_x)_P + \left(\frac{\partial u_x}{\partial x}\right)_P \Delta x + \cdots\right] - (u_x)_P \tag{3-43}$$

那么,线段 Δx 在 x 方向上的伸长量为

$$(\overline{P'Q'})_x - \Delta x = \left(\frac{\partial u_x}{\partial x}\right)_P \Delta x + \cdots$$

等式两边同时除以 Δx,并对每一项求极限,令 $\Delta x \to 0$,那么

$$\lim_{\Delta x \to 0} \frac{(\overline{P'Q'})_x - \Delta x}{\Delta x} = \left(\frac{\partial u_x}{\partial x}\right)_P \tag{3-44}$$

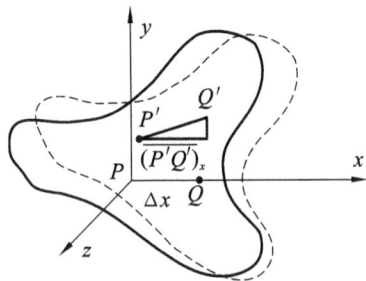

图 3-6 沿 x 轴方向上长度为 Δx 的微小线段 PQ

注意,方程(3-44)中等号右侧表示点 P 处的应变分量 ε_{xx}。因此,等式左侧就是 ε_{xx},即原本沿 x 轴方向的无穷小线段在变形后沿 x 轴方向的单位长度拉伸量。用这种方法我们可以对 ε_{yy} 和 ε_{zz} 给出类似的解释。这些量都是应变矩阵的对角项,我们称之为正应变。一般来说,我们认为某点处的 ε_{pp} 表示原本沿 p 轴方

向的无穷小线段在变形后沿 p 轴方向的单位长度拉伸量。由于只讨论小变形情况，所以我们可以用方程(3-44)中的 $\overline{P'Q'}$ 来代替其分量 $(\overline{P'Q'})_x$，这对结果的影响可以忽略不计。那么，ε_{pp} 也可以解释为原本沿 p 轴方向上线段单位长度的变化。

现在我们选择一条沿 x 轴方向长为 Δx 的线段 $\overline{PQ}$ 以及一条沿 y 轴方向长为 Δy 的线段 $\overline{PR}$，如图 3-7 所示。变形后，点 P、点 Q 以及点 R 分别移动到点 P'、点 Q' 和点 R'。我们将研究线段 $\overline{P'Q'}$ 和线段 $\overline{P'R'}$ 在 xy 平面上的投影，因此在图 3-8 中绘制出该投影。α 是线段 $\overline{P'R'}$ 的平面投影与 y 轴之间的夹角，β 是线段 $\overline{P'Q'}$ 的平面投影与 x 轴之间的夹角。点 P 在 x 轴方向上的位移记作 u_x，点 R 沿 x 轴方向上的位移通过对泰勒级数展开式 $[u_x+(\partial u_x/\partial y)\Delta y+\cdots]$ 计算得到(注意展开式中 $\Delta x=\Delta z=0$)。最后注意，线段 $\overline{P'R'}$ 沿 y 轴方向上的投影长度分量的数学表达式可以写为

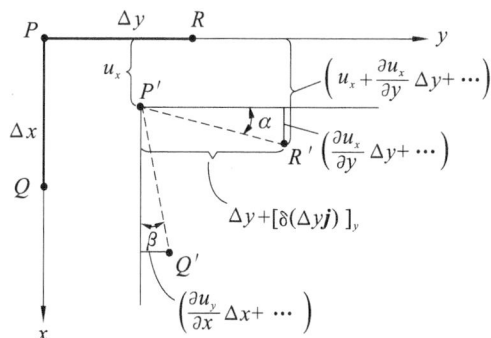

图 3-7 线段 $\overline{PQ}$ 和线段 $\overline{PR}$ 的变形情况

$$\overline{(P'R')}_y=\Delta y+[\delta(\Delta y\boldsymbol{j})]_y \tag{3-45}$$

根据图 3-8，很容易得到 $\tan\alpha$ 的值，有

$$\tan\alpha=\frac{\dfrac{\partial u_x}{\partial y}\Delta y+\cdots}{\Delta y+[\delta(\Delta y\boldsymbol{j})]_y} \tag{3-46}$$

现在令 $\Delta x\to 0$ 求极限，并忽略高阶项，最终得到的简化关系为

$$\tan\alpha=\frac{\partial u_x}{\partial y} \tag{3-47}$$

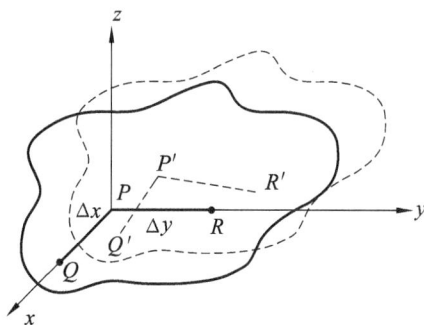

图 3-8 线段 $\overline{P'Q'}$ 和 $\overline{P'R'}$ 在 xy 平面上的投影

对于小变形来说，$\tan\alpha=\alpha$，可以得到

$$\alpha=\frac{\partial u_x}{\partial y} \tag{3-48}$$

通过类似方法，对于角度 β 有

$$\beta=\frac{\partial u_y}{\partial x} \tag{3-49}$$

两角之和 $(\alpha+\beta)$ 表示当我们将变形几何体投影到由未变形几何体中的线段形成的平面上时，在点 P 处由两条互相垂直的微小线段产生的直角减少量。因为存在小变形约束，变形几何体中微小线段之间的直角改变量可以通过将变形几何体投影在 xy 平面上得到的角度改变量来代替。根据方程(3-48)和方程(3-49)，可以得到

$$\alpha+\beta=\frac{\partial u_x}{\partial y}+\frac{\partial u_y}{\partial x}=2\varepsilon_{xy} \tag{3-50}$$

因此,$2\varepsilon_{ij}$ 表示在某点由两条微小正交线段 dx_i 和 dx_j 构成的直角因变形造成的角度减少量。那么,我们可以对应变矩阵的非对角项进行物理解释。这种应变称为剪应变。通常使用剪切角 γ_{ij} 表示线段 dx_i 和 dx_j 构成的直角的总减少量。因此

$$\gamma_{ij} = 2\varepsilon_{ij} \tag{3-51}$$

在后面章节中,我们将大量地使用剪切角 γ_{ij}。

如图 3-9 所示,对于属于微小三维平行矩形六面体的应变单元,现在我们将对它进行进一步的物理说明。首先我们考虑,该单元体的应变矩阵 ε_{ij} 中只有正应变为非零。通过讨论,我们可以推断出平行矩形六面体在变形过程中仍然是平行矩形六面体。但我们也应该指出,该单元体可能存在刚体转动,那么该平行矩形六面体的各边在变形后可能不再平行于坐标轴(这种情况已在图中表示)。

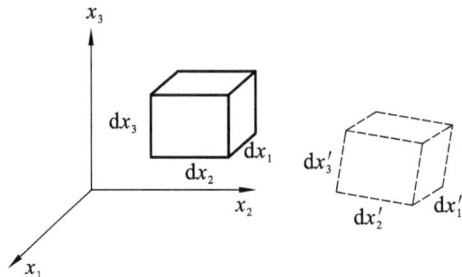

图 3-9 微小三维平行矩形六面体

利用我们对正应变的几何解释,平行矩形六面体各边的新长度为

$$\begin{cases} dx_1' = dx_1(1+\varepsilon_{11}) \\ dx_2' = dx_2(1+\varepsilon_{22}) \\ dx_3' = dx_3(1+\varepsilon_{33}) \end{cases} \tag{3-52}$$

计算该单元体的体积变化将很有意义。我们现在来表示单元体的体积变化:

$$dx_1'dx_2'dx_3' - dx_1dx_2dx_3 = dx_1dx_2dx_3(1+\varepsilon_{11})(1+\varepsilon_{22})(1+\varepsilon_{33}) - dx_1dx_2dx_3 \tag{3-53}$$

对于小变形,可以进行以下近似处理

$$(1+\varepsilon_{11})(1+\varepsilon_{22})(1+\varepsilon_{33}) = 1 + (\varepsilon_{11}+\varepsilon_{22}+\varepsilon_{33}) \tag{3-54}$$

在式(3-54)中我们忽略了应变的乘积。将式(3-54)代入方程(3-53),我们得到

$$dx_1'dx_2'dx_3' - dx_1dx_2dx_3 = (\varepsilon_{11}+\varepsilon_{22}+\varepsilon_{33})dx_1dx_2dx_3 \tag{3-55}$$

那么有

$$\frac{dx_1'dx_2'dx_3' - dx_1dx_2dx_3}{dx_1dx_2dx_3} = \varepsilon_{11}+\varepsilon_{22}+\varepsilon_{33} \tag{3-56}$$

我们可以将方程(3-56)解释为在某点处单位体积的变化量,并称之为立方体膨胀,它由正应变相加得到。

接下来我们考虑只有剪应变非零时一点处的应变状态。很明显,前面讨论的平行矩形六面体在变形中会改变矩形形状。

那么,对于未变形体中的一个微小平行矩形六面体,我们总结得到:一方面,各边界方向上的正应变可能引起膨胀,即单元体的体积发生变化,但不影响边界的正交性。另一方面,剪应变会破坏边界的正交性,从而影响单元体的基本形状,但不改变单元体体积。

现在我们解释了应变矩阵对角项为正应变,应变矩阵非对角项即为剪应变。立方体膨胀的概念进一步解释了应变矩阵的迹。

下面,我们将证明应变分量 ε_{ij} 是二阶张量。当完成这个证明后,我们将获得关于某点应变性质的更多信息。

3.6 应变转换方程

利用 3.5 节中对应变项的几何解释,我们将证明应变项可以形成二阶张量场。为此,我们必须证明当某点坐标轴转动时,该点的应变项将按照方程(2-10)和方程(2-11)进行转换。

我们先计算点 P 在如图 3-10 所示单位向量 $\boldsymbol{n}$ 方向上产生的正应变。在图中我们选择一条长度为 Δn 的线段 $\overline{PQ}$,并且选用有一撇的字母表示变形体。将点 P 在 $\boldsymbol{n}$ 方向上的位移投影到坐标轴上,我们可以得到下述公式。

$$(u_n)_P = (u_x)_P l + (u_y)_P m + (u_z)_P n \tag{3-57}$$

此外,点 Q 在 $\boldsymbol{n}$ 方向上的位移可表示为在点 P 处的泰勒级数展开,如下所示:

$$(u_n)_Q = (u_n)_P + \left(\frac{\partial u_n}{\partial x}\right)_P \Delta x + \left(\frac{\partial u_n}{\partial y}\right)_P \Delta y + \left(\frac{\partial u_n}{\partial z}\right)_P \Delta z + \cdots \tag{3-58}$$

图 3-10 线段 $\overline{PQ}$ 对应的变形几何体

求解方程(3-58)中 $(u_n)_Q - (u_n)_P$ 的值,并用方程(3-57)右侧代替 u_n,得到

$$(u_n)_Q - (u_n)_P = \left(\frac{\partial u_x}{\partial x}\right)_P \Delta x l + \left(\frac{\partial u_y}{\partial x}\right)_P \Delta x m + \left(\frac{\partial u_z}{\partial x}\right)_P \Delta x n + \left(\frac{\partial u_x}{\partial y}\right)_P \Delta y l +$$

$$\left(\frac{\partial u_y}{\partial y}\right)_P \Delta y m + \left(\frac{\partial u_z}{\partial y}\right)_P \Delta y n + \left(\frac{\partial u_x}{\partial z}\right)_P \Delta z l + \left(\frac{\partial u_y}{\partial z}\right)_P \Delta z m + \left(\frac{\partial u_z}{\partial z}\right)_P \Delta z n + \cdots \tag{3-59}$$

现在把等式(3-59)两边除以 $\overline{PQ}$ 的原长 Δn,并取极限,令 $\Delta n \to 0$,忽略高阶项,可得

$$\lim_{\Delta n \to 0} \frac{(u_n)_Q - (u_n)_P}{\Delta n} = \left(\frac{\partial u_x}{\partial x}\right)_P \frac{\Delta x}{\Delta n} l + \left(\frac{\partial u_y}{\partial x}\right)_P \frac{\Delta x}{\Delta n} m + \left(\frac{\partial u_z}{\partial x}\right)_P \frac{\Delta x}{\Delta n} n + \left(\frac{\partial u_x}{\partial y}\right)_P \frac{\Delta y}{\Delta n} l +$$

$$\left(\frac{\partial u_y}{\partial y}\right)_P \frac{\Delta y}{\Delta n} m + \left(\frac{\partial u_z}{\partial y}\right)_P \frac{\Delta y}{\Delta n} n + \left(\frac{\partial u_x}{\partial z}\right)_P \frac{\Delta z}{\Delta n} l + \left(\frac{\partial u_y}{\partial z}\right)_P \frac{\Delta z}{\Delta n} m + \left(\frac{\partial u_z}{\partial z}\right)_P \frac{\Delta z}{\Delta n} n \tag{3-60}$$

根据几何解释,方程左侧显然是正应变 ε_{nn}。在方程的右侧,我们可以用 l,m 和 n 分别替换 $\Delta x/\Delta n$,$\Delta y/\Delta n$,$\Delta z/\Delta n$。由于前面的方程在任意点 P 都有效,所以我们去掉下标 P,那么有

$$\varepsilon_{nn} = \frac{\partial u_x}{\partial x} l^2 + \frac{\partial u_y}{\partial y} m^2 + \frac{\partial u_z}{\partial z} n^2 + \left(\frac{\partial u_x}{\partial y} + \frac{\partial u_y}{\partial x}\right) lm + \left(\frac{\partial u_x}{\partial z} + \frac{\partial u_z}{\partial x}\right) ln + \left(\frac{\partial u_y}{\partial z} + \frac{\partial u_z}{\partial y}\right) mn \tag{3-61}$$

图 3-11 变形几何体中线段 $\overline{PQ}$ 和线段 $\overline{PR}$

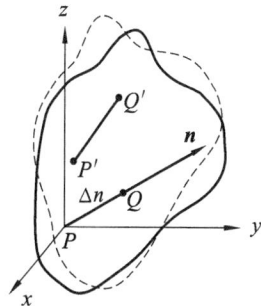

根据应变各项的定义,我们可以将上述方程表示为

$$\varepsilon_{nn} = \varepsilon_{xx} l^2 + \varepsilon_{yy} m^2 + \varepsilon_{zz} n^2 + 2(\varepsilon_{xy} lm + \varepsilon_{zx} ln + \varepsilon_{yz} nm) \tag{3-62}$$

现在我们考虑剪应变项。在图 3-11 中,我们沿 $\boldsymbol{n}$ 方向绘制一条长度为 Δn 的线段 $\overline{PQ}$ 以及沿 $\boldsymbol{s}$ 方向绘制一条长度为 Δs 的线段 $\overline{PR}$。变形后的几何体用点 P',点 Q' 和点 R' 表示。剪应变 ε_{ns} 可通过以下公式给出

$$\varepsilon_{ns} = \frac{1}{2}\left(\frac{\partial u_n}{\partial s} + \frac{\partial u_s}{\partial n}\right) \tag{3-63}$$

和前文一样,我们可以用沿各坐标轴方向的位移来表示 u_n 和 u_s,即

$$\begin{cases} u_n = u_x l + u_y m + u_z n \\ u_s = u_x l' + u_y m' + u_z n' \end{cases} \tag{3-64}$$

那么,利用前面的关系式,我们对方程(3-63)进行计算可得

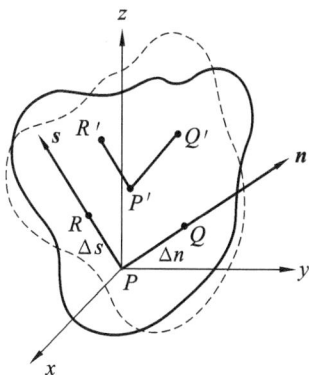

$$\varepsilon_{ns}=\frac{1}{2}\Big(\frac{\partial u_x}{\partial s}l+\frac{\partial u_y}{\partial s}m+\frac{\partial u_z}{\partial s}n+\frac{\partial u_x}{\partial n}l'+\frac{\partial u_y}{\partial n}m'+\frac{\partial u_z}{\partial n}n'\Big) \tag{3-65}$$

由于前面方程中的偏导数可以表示为

$$\begin{cases}\frac{\partial u_x}{\partial s}=\Big(\frac{\partial u_x}{\partial x}l'+\frac{\partial u_x}{\partial y}m'+\frac{\partial u_x}{\partial z}n'\Big),\frac{\partial u_y}{\partial s}=\Big(\frac{\partial u_y}{\partial x}l'+\frac{\partial u_y}{\partial y}m'+\frac{\partial u_y}{\partial z}n'\Big),\frac{\partial u_z}{\partial s}=\Big(\frac{\partial u_z}{\partial x}l'+\frac{\partial u_z}{\partial y}m'+\frac{\partial u_z}{\partial z}n'\Big)\\ \frac{\partial u_x}{\partial n}=\Big(\frac{\partial u_x}{\partial x}l+\frac{\partial u_x}{\partial y}m+\frac{\partial u_x}{\partial z}n\Big),\frac{\partial u_y}{\partial n}=\Big(\frac{\partial u_y}{\partial x}l+\frac{\partial u_y}{\partial y}m+\frac{\partial u_y}{\partial z}n\Big),\frac{\partial u_z}{\partial n}=\Big(\frac{\partial u_z}{\partial x}l+\frac{\partial u_z}{\partial y}m+\frac{\partial u_z}{\partial z}n\Big)\end{cases} \tag{3-66}$$

因此,我们得到

$$\varepsilon_{ns}=\frac{1}{2}\Big[2\times\frac{\partial u_x}{\partial x}ll'+2\times\frac{\partial u_y}{\partial y}mm'+2\times\frac{\partial u_z}{\partial z}nn'+\Big(\frac{\partial u_x}{\partial y}+\frac{\partial u_y}{\partial x}\Big)(lm'+l'm)+$$

$$\Big(\frac{\partial u_x}{\partial z}+\frac{\partial u_z}{\partial x}\Big)(ln'+l'n)+\Big(\frac{\partial u_y}{\partial z}+\frac{\partial u_z}{\partial y}\Big)(mn'+nm')\Big]$$

$$=\varepsilon_{xx}ll'+\varepsilon_{yy}mm'+\varepsilon_{zz}nn'+\varepsilon_{xy}(lm'+l'm)+\varepsilon_{zx}(ln'+l'n)+\varepsilon_{yz}(mn'+nm') \tag{3-67}$$

将前面方程中的 s 变为 n,我们可以认为上述方程表示一般情况下的应变转换方程。因此,我们可以得出结论:应变是一个二阶张量。

既然证明应变是一个二阶张量,那么我们就可以将第 2 章中关于应力张量所有结论应用到应变张量中。即

(1)对于任意一种应变状态,我们可以将其与被称为应变二次曲面的二阶张量相关联。

(2)具有三个正交方向,其中任意两个方向上的正应变为特定应变状态下的极值。此外,这些轴上的剪应变为零,这就是主应变轴。

(3)应变的三个张量不变量为

$$\varepsilon_{11}+\varepsilon_{22}+\varepsilon_{33}=\mathrm{I}_\varepsilon \tag{3-68}$$

$$\begin{vmatrix}\varepsilon_{22}&\varepsilon_{23}\\\varepsilon_{32}&\varepsilon_{33}\end{vmatrix}+\begin{vmatrix}\varepsilon_{11}&\varepsilon_{13}\\\varepsilon_{31}&\varepsilon_{33}\end{vmatrix}+\begin{vmatrix}\varepsilon_{11}&\varepsilon_{12}\\\varepsilon_{21}&\varepsilon_{22}\end{vmatrix}=\mathrm{II}_\varepsilon \tag{3-69}$$

$$\begin{vmatrix}\varepsilon_{11}&\varepsilon_{12}&\varepsilon_{13}\\\varepsilon_{21}&\varepsilon_{22}&\varepsilon_{23}\\\varepsilon_{31}&\varepsilon_{32}&\varepsilon_{33}\end{vmatrix}=\mathrm{III}_\varepsilon \tag{3-70}$$

(4)可以采用与第 2 章中求主应力方法相同的三次方程来求解主应变,即

$$\varepsilon^3-(\mathrm{I}_\varepsilon)\varepsilon^2+(\mathrm{II}_\varepsilon)\varepsilon-(\mathrm{III}_\varepsilon)=0 \tag{3-71}$$

其中 (I_ε),$(\mathrm{II}_\varepsilon)$ 和 $(\mathrm{III}_\varepsilon)$ 是前面给出的三个张量不变量。

(5)我们可以像计算平面应力一样,用公式 $\varepsilon_{zz}=\varepsilon_{yz}=\varepsilon_{zx}=0$ 表示平面应变。这种类型的应变适用于纵轴沿 z 方向,外部荷载不随 z 变化的棱柱体,且棱柱的末端必须是刚性约束的情况,图 3-12 所示的大坝就是这类例子。第 2 章中给出的所有平面应力公式都可以推广到平面应变,这也包括莫尔圆的相关内容。

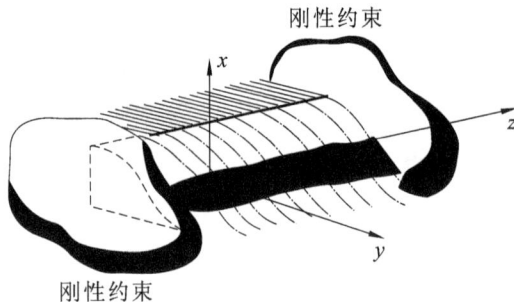

图 3-12　两端为刚性约束的大坝

在本章末尾,我们给出了一些习题,这些习题要求你运用二阶张量的一般性质来求解。

3.7 二维极坐标中的应变项

在求解圆形物体平面问题时，一般采用极坐标较为方便。在极坐标中，使用 ε_r 代表径向线段正应变，ε_θ 代表切向线段正应变，$\gamma_{r\theta}$ 代表径向与切向线段之间直角的改变量，被称为剪切角，u_r 代表径向位移，u_θ 代表切向位移。

假定仅有径向位移而没有切向位移，如图 3-13 所示。由于发生径向位移，径向线段 PA 移到 $P'A'$，切向线段 PB 移到 $P'B'$，而 P、A、B 三点的位移分别为

$$PP' = u_r \quad AA' = u_r + \frac{\partial u_r}{\partial r}\mathrm{d}r \quad BB' = u_r + \frac{\partial u_r}{\partial \theta}\mathrm{d}\theta$$

因此，径向线段 PA 的正应变为

$$\varepsilon_r = \frac{P'A' - PA}{PA} = \frac{AA' - PP'}{PA} = \frac{\left(u_r + \frac{\partial u_r}{\partial r}\mathrm{d}r\right) - u_r}{\mathrm{d}r} = \frac{\partial u_r}{\partial r} \tag{3-72}$$

切向线段 PB 的正应变为

$$\varepsilon_\theta = \frac{P'B' - PB}{PB} = \frac{(r + u_r)\mathrm{d}\theta - r\mathrm{d}\theta}{r\mathrm{d}\theta} = \frac{u_r}{r} \tag{3-73}$$

径向线段 PA 的转角为

$$\alpha = 0 \tag{3-74}$$

切向线段 PB 的转角为

$$\beta = \frac{BB' - PP'}{PB} = \frac{\left(u_r + \frac{\partial u_r}{\partial \theta}\mathrm{d}\theta\right) - u_r}{r\mathrm{d}\theta} = \frac{1}{r}\frac{\partial u_r}{\partial \theta} \tag{3-75}$$

则剪切角为

$$\gamma_{r\theta} = \alpha + \beta = \frac{1}{r}\frac{\partial u_r}{\partial \theta} \tag{3-76}$$

假定仅有切向位移而没有径向位移，如图 3-14 所示。由于发生切向位移，径向线段 PA 移到 $P''A''$，切向线段 PB 移到 $P''B''$，而 P、A、B 三点的位移分别为

$$PP'' = u_\theta \quad AA'' = u_\theta + \frac{\partial u_\theta}{\partial r}\mathrm{d}r \quad BB'' = u_\theta + \frac{\partial u_\theta}{\partial \theta}\mathrm{d}\theta$$

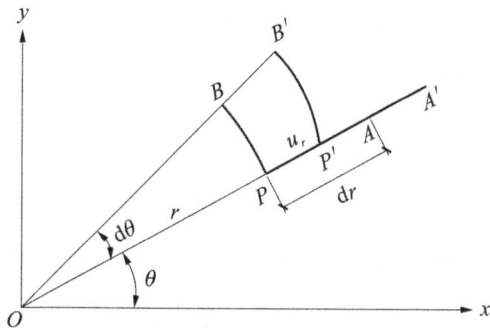

图 3-13　仅有径向位移几何关系图　　　　图 3-14　仅有切向位移几何关系图

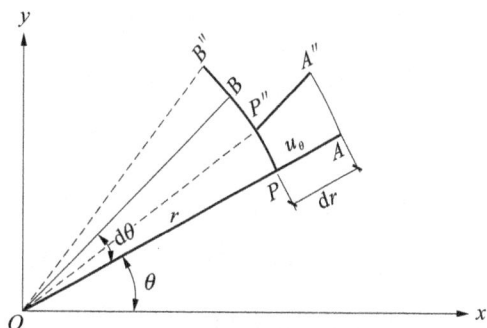

因此,径向线段 PA 的正应变为

$$\varepsilon_r = 0 \tag{3-77}$$

切向线段 PB 的正应变为

$$\varepsilon_\theta = \frac{P''B'' - PB}{PB} = \frac{BB'' - PP''}{PB} = \frac{\left(u_\theta + \frac{\partial u_\theta}{\partial \theta} d\theta\right) - u_\theta}{r d\theta} = \frac{1}{r}\frac{\partial u_\theta}{\partial \theta} \tag{3-78}$$

径向线段 PA 的转角为

$$\alpha = \frac{AA'' - PP''}{PA} = \frac{\left(u_\theta + \frac{\partial u_\theta}{\partial r} dr\right) - u_\theta}{dr} = \frac{\partial u_\theta}{\partial r} \tag{3-79}$$

切向线段 PB 的转角为

$$\beta = -\angle POP'' = -\frac{PP''}{OP} = -\frac{u_\theta}{r} \tag{3-80}$$

则剪切角为

$$\gamma_{r\theta} = \alpha + \beta = \frac{\partial u_\theta}{\partial r} - \frac{u_\theta}{r} \tag{3-81}$$

因此,如果沿径向和切向均有位移,则分别叠加可得

$$\begin{cases} \varepsilon_r = \dfrac{\partial u_r}{\partial r} \\[2mm] \varepsilon_\theta = \dfrac{u_r}{r} + \dfrac{1}{r}\dfrac{\partial u_\theta}{\partial \theta} \\[2mm] \gamma_{r\theta} = \dfrac{1}{r}\dfrac{\partial u_r}{\partial \theta} + \dfrac{\partial u_\theta}{\partial r} - \dfrac{u_\theta}{r} \end{cases} \tag{3-82}$$

这就是极坐标中的各应变项。

3.8 相 容 方 程

我们进一步考虑应变和位移间的关系:

$$\varepsilon_{ij} = \frac{1}{2}\left(\frac{\partial u_i}{\partial x_j} + \frac{\partial u_j}{\partial x_i}\right) \tag{3-83}$$

如果给定位移场,我们可以根据前面的方程对位移场求偏导数,从而很容易计算出应变张量。反过来,根据应变场求位移场就不那么简单了。这里,位移场由三个函数 u_i 组成,函数通过方程(3-83)中六个偏微分方程的积分来确定。为了保证 u_i 的解是单值且连续的,我们必须对应变函数 ε_{ij} 施加一定的限制,也就是说,我们不能取任意张量场 ε_{ij},并认为它一定与单值、连续的位移场相关联。但实际的变形是单值的,另外,我们想研究的变形也是连续的。基于这些考虑,我们对 ε_{ij} 做出的限制将适用于所有方程,我们称之为相容方程。相容方程可以将应变与前面提到的限制条件适当地联系起来。

相容方程的完整推导超出了本书的范围。我们将建立的相容方程作为满足单值、连续的位移场和应变场之间适当关系的必要条件。后面,我们将不加证明地指出这些方程何时满足充分性要求。首先,我们用完整的符号改写方程(3-83)。

$$\varepsilon_{xx} = \frac{\partial u_x}{\partial x} \tag{3-84a}$$

$$\varepsilon_{yy} = \frac{\partial u_y}{\partial y} \tag{3-84b}$$

$$\varepsilon_{zz} = \frac{\partial u_z}{\partial z} \tag{3-84c}$$

$$\gamma_{xy} = \frac{\partial u_y}{\partial x} + \frac{\partial u_x}{\partial y} \tag{3-84d}$$

$$\gamma_{xz} = \frac{\partial u_z}{\partial x} + \frac{\partial u_x}{\partial z} \tag{3-84e}$$

$$\gamma_{yz} = \frac{\partial u_z}{\partial y} + \frac{\partial u_y}{\partial z} \tag{3-84f}$$

接下来从这些方程中消除位移项。首先,将微分方程(3-84a)对 y 进行两次偏分,将微分方程(3-84b)对 x 进行两次偏分,有

$$\frac{\partial^2 \varepsilon_{xx}}{\partial y^2} = \frac{\partial^3 u_x}{\partial y^2 \partial x} \tag{3-85a}$$

$$\frac{\partial^2 \varepsilon_{yy}}{\partial x^2} = \frac{\partial^3 u_y}{\partial x^2 \partial y} \tag{3-85b}$$

两式相加,得到

$$\frac{\partial^2 \varepsilon_{xx}}{\partial y^2} + \frac{\partial^2 \varepsilon_{yy}}{\partial x^2} = \frac{\partial^3 u_x}{\partial y^2 \partial x} + \frac{\partial^3 u_y}{\partial x^2 \partial y} \tag{3-86}$$

将方程(3-84d)对 x 和 y 进行混合偏分,可得到

$$\frac{\partial^2 \gamma_{xy}}{\partial x \partial y} = \frac{\partial^3 u_x}{\partial x \partial y^2} + \frac{\partial^3 u_y}{\partial y \partial x^2} \tag{3-87}$$

假设所有的导数都是连续的,我们就可以改变偏分顺序。我们可以看出方程(3-86)和方程(3-87)的右侧相等,这两个方程左侧也将相等,可以写成下列形式:

$$\frac{\partial^2 \varepsilon_{xx}}{\partial y^2} + \frac{\partial^2 \varepsilon_{yy}}{\partial x^2} = \frac{\partial^2 \gamma_{xy}}{\partial x \partial y} \tag{3-88}$$

通过类似方法,我们可以得到另外两个方程

$$\frac{\partial^2 \varepsilon_{yy}}{\partial z^2} + \frac{\partial^2 \varepsilon_{zz}}{\partial y^2} = \frac{\partial^2 \gamma_{yz}}{\partial y \partial z} \tag{3-89a}$$

$$\frac{\partial^2 \varepsilon_{zz}}{\partial x^2} + \frac{\partial^2 \varepsilon_{xx}}{\partial z^2} = \frac{\partial^2 \gamma_{zx}}{\partial z \partial x} \tag{3-89b}$$

方程(3-88)和方程(3-89)是六个相容方程中的三个。为了得到剩下的相容方程,我们将方程(3-84a)对 z 和 y 进行混合偏分,因此

$$\frac{\partial^2 \varepsilon_{xx}}{\partial y \partial z} = \frac{\partial^3 u_x}{\partial y \partial z \partial x} \tag{3-90}$$

将方程(3-84d)对 z 和 x 进行混合偏分,得到

$$\frac{\partial^2 \gamma_{xy}}{\partial x \partial z} = \frac{\partial^3 u_x}{\partial x \partial z \partial y} + \frac{\partial^3 u_y}{\partial z \partial^2 x} \tag{3-91}$$

同样,将方程(3-84e)对 y 和 x 进行混合偏分,有

$$\frac{\partial^2 \gamma_{xz}}{\partial x \partial y} = \frac{\partial^3 u_x}{\partial x \partial y \partial z} + \frac{\partial^3 u_z}{\partial^2 x \partial y} \tag{3-92}$$

最后,将方程(3-84f)对 x 进行两次偏分,有

$$\frac{\partial^2 \gamma_{yz}}{\partial x^2} = \frac{\partial^3 u_z}{\partial x^2 \partial y} + \frac{\partial^3 u_y}{\partial x^2 \partial z} \tag{3-93}$$

将方程(3-91)和方程(3-92)相加,再减去方程(3-93),并通过改变偏分的顺序,可得到以下方程:

$$-\frac{\partial^2 \gamma_{yz}}{\partial x^2} + \frac{\partial^2 \gamma_{zx}}{\partial x \partial y} + \frac{\partial^2 \gamma_{xy}}{\partial x \partial z} = 2\frac{\partial^3 u_x}{\partial x \partial y \partial z} \tag{3-94}$$

将方程(3-90)的左侧乘2并代入方程(3-94),就可以得到另一个相容方程:

$$2\frac{\partial^2 \varepsilon_{xx}}{\partial y \partial z} = \frac{\partial}{\partial x}\left(-\frac{\partial \gamma_{yz}}{\partial x} + \frac{\partial \gamma_{xz}}{\partial y} + \frac{\partial \gamma_{xy}}{\partial z}\right) \tag{3-95}$$

通过类似方法,可以得到其他相容方程:

$$\begin{cases} 2\dfrac{\partial^2 \varepsilon_{yy}}{\partial z \partial x} = \dfrac{\partial}{\partial y}\left(-\dfrac{\partial \gamma_{zx}}{\partial y} + \dfrac{\partial \gamma_{yx}}{\partial z} + \dfrac{\partial \gamma_{yz}}{\partial x}\right) \\[3mm] 2\dfrac{\partial^2 \varepsilon_{zz}}{\partial x \partial y} = \dfrac{\partial}{\partial z}\left(-\dfrac{\partial \gamma_{xy}}{\partial z} + \dfrac{\partial \gamma_{zy}}{\partial x} + \dfrac{\partial \gamma_{zx}}{\partial y}\right) \end{cases} \tag{3-96}$$

完整的相容方程组如下:

$$\frac{\partial^2 \varepsilon_{xx}}{\partial y^2} + \frac{\partial^2 \varepsilon_{yy}}{\partial x^2} = \frac{\partial^2 \gamma_{xy}}{\partial x \partial y} \tag{3-97a}$$

$$\frac{\partial^2 \varepsilon_{yy}}{\partial z^2} + \frac{\partial^2 \varepsilon_{zz}}{\partial y^2} = \frac{\partial^2 \gamma_{yz}}{\partial y \partial z} \tag{3-97b}$$

$$\frac{\partial^2 \varepsilon_{zz}}{\partial x^2} + \frac{\partial^2 \varepsilon_{xx}}{\partial z^2} = \frac{\partial^2 \gamma_{zx}}{\partial z \partial x} \tag{3-97c}$$

$$2\frac{\partial^2 \varepsilon_{xx}}{\partial y \partial z} = \frac{\partial}{\partial x}\left(-\frac{\partial \gamma_{yz}}{\partial x} + \frac{\partial \gamma_{xz}}{\partial y} + \frac{\partial \gamma_{xy}}{\partial z}\right) \tag{3-97d}$$

$$2\frac{\partial^2 \varepsilon_{yy}}{\partial z \partial x} = \frac{\partial}{\partial y}\left(-\frac{\partial \gamma_{zx}}{\partial y} + \frac{\partial \gamma_{yx}}{\partial z} + \frac{\partial \gamma_{yz}}{\partial x}\right) \tag{3-97e}$$

$$2\frac{\partial^2 \varepsilon_{zz}}{\partial x \partial y} = \frac{\partial}{\partial z}\left(-\frac{\partial \gamma_{xy}}{\partial z} + \frac{\partial \gamma_{zy}}{\partial x} + \frac{\partial \gamma_{zx}}{\partial y}\right) \tag{3-97f}$$

如果应变场是单值且连续的,则应变张量必定满足上述方程,这样的应变才是具有物理意义的变形。

3.9 关于单连体和多连体的说明

为了说明相容方程的充分条件,我们需要区分两类物体,即单连体和多连体。

我们将物体内的每条闭合路径连续收缩到一个点且不会切割该物体边界的体定义为单连体。该定义允许我们在路径缩小到某一点的过程中以任何方式移动路径。由于任何路径 a 都可以收缩到一个点,而不与外边界 S_1 或材料内部小孔的闭合内边界 S_2 相切割,所以图 3-15 所示物体就是单连体。

多连体是指存在一条或多条路径不能按照上述的方式收缩到某一点的体。图 3-16 所示的环就是一个多连体示例,其中路径 a 很显然不能在不切割边界的情况下收缩至一点。

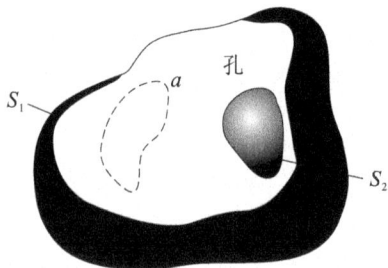

图 3-15 单连体

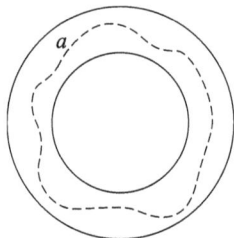

图 3-16 多连体

现在我们不加证明地指出,对于一点固定的单连体,满足相容方程既是单值又是连续位移场的充分条件,也是必要条件。如果是多连体,相容方程成立只是得到单值、连续位移场的必要条件,对于充分条件所需要的其他要求,我们在此不做讨论。

3.10 结　语

在本章中,我们研究了旋转矩阵和应变矩阵。我们能通过这些公式来描述在小变形或无穷小变形条件下微单元体的刚体转动和纯变形。本章的公式均是几何公式,因此适用于荷载作用下发生小变形的任何连续介质。在流体和黏弹性介质中,变形速率在分析中具有重要意义,因此我们更关注应变速率等。另外,在第 2 章中,我们应用牛顿定律推导某一点的应力公式,这些应力公式同样也适用于所有连续介质。

将应力和变形联系起来,我们发现连续介质的性质变得很重要。在第 4 章中,我们将对某些介质的应力和应变张量之间关系的实验数据进行分析。除了呈现某些实验结果外,我们还将对这些结果进行有用的推理和提出理想的公式。这将为我们从第 5 章开始求解应力和应变分布奠定基础。

习题

3-1　给定以下位移场:

$$u=(x^2+y)i+(3+z)j+(x^2+2y)k$$

求:在点 $(3,1,-2)$ 处变形后的位置是多少?

3-2　未变形几何体内两点坐标为 $(0,0,1)$ 和 $(2,0,-1)$,那么变形后两点间的距离是多少(假设将习题 3-1 中的位移场施加在该物体上)?

3-3　给定以下位移场:

$$u=(0.16x^2+\sin y)i+\left(0.1z+\frac{x}{y^3}\right)j+0.004k$$

已知在未变形几何体中两点的位置向量如下:

$$r_1=10i+3j$$
$$r_2=4k+3j$$

那么变形后两点间的距离增量是多少?

3-4　给定以下位移场:

$$u_i=\lambda_{ij}x_j$$

其中矩阵 λ_{ij} 由常数组成,我们称之为仿射变形。设 λ_{ij} 为

$$\lambda_{ij}=\begin{pmatrix} +0.2 & -0.05 & -0.1 \\ +0.03 & +0.1 & -0.02 \\ +0.003 & -0.2 & +0.03 \end{pmatrix}$$

已知未变形几何体中某点的位置向量 $r=i-j+3k$,那么该点位移是多少?

3-5　证明仿射变形的性质(以习题 3-4 为例)。

(1)平面截面在变形过程中始终保持平面。

(2)直线在变形过程中始终保持为直线。

3-6　在习题 3-4 中,位于未变形几何体中 xy 平面上的平面方程是什么?(提示:先在 xy 平面中找到两个位置矢量,再在未变形几何体中找出同时垂直于这两个矢量的法向量 n。所求平面中的位置向量可以用 $r'=x'i+y'j+z'k$ 表示,故 $n\cdot r'=0$]

3-7　给定以下位移场

$$u = [y^2 i + 3yz j + (4 + 6x^2) k] \times 10^{-2}$$

对于空间坐标系中任意一点附近小区域中的矩阵 $\partial u_i / \partial x_j$ 是什么? 若坐标点为 $(0,1,3)$,那么该矩阵是什么?

3-8 在坐标点 $(1,0,2)$ 处,沿 y 轴方向的微小线段在习题3-7所给的位移场作用下,其单位长度的增加量是多少?

3-9 给出以下位移场

$$u = \frac{x}{100} i + \frac{y^2}{200} j + \left(0.02 + \frac{y^2 z}{500}\right) k$$

在下述位置处的位移是多少?

在原点处的小区域中变形矩阵是什么?

$$r = 6i + 2j$$

3-10 解释为什么只有刚体的微小转动可以用矢量表示。

3-11 给定以下位移场

$$u = [x^2 i + (2y^2 + 3z) j + 10k] \times 10^{-3}$$

(1)由于变形,未变形几何体中两点 $(0,1,0)$ 和 $(2,-1,3)$ 之间的距离增加了多少?

(2)在点 $(1,3,2)$ 的小区域内,原沿 y 轴方向的微小线段在 y 轴方向上的单位长度伸长量是多少?

3-12 在位置 $(1,0,1)$ 处的微小矢量 $\Delta s = 0.002 i + 0.003 j + 0.0004 k$ 受到习题3-11位移场作用,其矢量变形后的变形量是多少? 变形几何体中的新矢量怎么表示?

3-13 假设我们在 P 点有两个连续的小变形,可用变形矩阵表示

$$\left(\frac{\partial u_i^{(1)}}{\partial x_j}\right)_P = \begin{bmatrix} 0.02 & 0.01 & 0 \\ 0 & 0.01 & -0.02 \\ 0 & 0 & -0.02 \end{bmatrix}$$

以及

$$\left(\frac{\partial u_i^{(2)}}{\partial x_j}\right)_P = \begin{bmatrix} 0.01 & 0.015 & -0.02 \\ 0 & 0 & -0.01 \\ 0 & -0.03 & 0.04 \end{bmatrix}$$

那么下列给定的矢量 Δs 在 P 点处的总变形是多少?

$$\Delta s = (6i + 10j + 2k) \times 10^{-3}$$

3-14 一个物体固定在点 O,若在该点绕 z 轴旋转一个角度 φ,那么该运动的矩阵 $\partial u_i / \partial x_j$ 是什么?(提示:依次考虑沿 x,y,z 轴线段)

3-15 单元体产生一个小转动 $\delta\boldsymbol{\phi}$,表示如下:

$$\delta\boldsymbol{\phi} = 0.0002 i + 0.0005 j - 0.0002 k \quad (\text{rad})$$

在该点的矩阵 $\partial u_i / \partial x_j$ 是什么?

3-16 给定以下位移场

$$u = [(6y + 5z) i + (-6x + 3z) j + (-5x - 3y) k] \times 10^{-3}$$

证明该位移场表示刚体转动的位移场。该物体的旋转向量 $\delta\boldsymbol{\phi}$ 是什么?

3-17 点 P 处的单元体在下述变形矩阵作用下产生变形

$$\left(\frac{\partial u_i}{\partial x_j}\right)_P = \begin{bmatrix} 0.01 & 0 & 0 \\ -0.02 & 0.03 & 0 \\ 0 & -0.02 & -0.01 \end{bmatrix}$$

那么该单元体的旋转矩阵、应变矩阵和旋转角分别是多少?

3-18 给定位移场

$$u = [(x^3 + 10) i + 3yz j + (z^2 - yx) k] \times 10^{-2}$$

那么物体在位移场作用下的刚体平移是多少？在位置$(2,1,0)$处单元体的转动量是多少？

3-19 给定以下位移场

$$\boldsymbol{u}=[(y^2+3z)\boldsymbol{i}+(x+3yz)\boldsymbol{j}+(\sin z)\boldsymbol{k}]\times 10^{-2}$$

求解在$(1,2,3)$处的单元体的应变矩阵和旋转矩阵。该单元体的刚体转动量以及位移分别是多少？

3-20 给定以下位移场

$$u_x=0.06x+0.05y-0.01z$$
$$u_y=0.01x-0.03z$$
$$u_z=-0.02x+0.01z$$

该物体中各点的正应变ε_{xx}是多少？如果在未变形几何体中有一条与x轴平行且长度为的10^{-3}m 长的线段，那么该线段变形后的长度是多少？

3-21 使用正应变的几何定义，确定习题 3-20 中xy平面方向上与x轴成$30°$方向上的ε_{kk}。

3-22 在习题 3-17 中，点P处的剪应变ε_{xy}和剪切角γ_{xy}是多少？

3-23 给定下述位移场

$$\boldsymbol{u}=[xy\boldsymbol{i}+3y\boldsymbol{j}+(6+10z)\boldsymbol{k}]\times 10^{-2}$$

点$(2,1,3)$处的正应变ε_{xx}，ε_{yy}和剪应变ε_{yz}是多少？

3-24 施加习题 3-18 中给出的位移场，物体中某位置x,y,z处分别平行于y轴和z轴的微小线段所构成的直角变化量是多少？这两条线段的转动量是多少？

3-25 求出在习题 3-23 位移场作用下，点$(3,0,-2)$处的两条分别平行于x轴和y轴的微小线段 dx和 dy所构成的直角变化量。同时，求出该点处单位体积的体积变化量。

3-26 使用应变转换方程求解习题 3-21。

3-27 在物体上某一点存在以下应变状态

$$\varepsilon_{ij}=\begin{pmatrix} 0.01 & -0.02 & 0 \\ -0.02 & 0.03 & -0.01 \\ 0 & -0.01 & 0 \end{pmatrix}$$

那么，该点在方向余弦为$l=0.6$，$m=0$ 和$n=0.8$的方向p上正应变ε_{pp}是多少？

3-28 在习题 3-27 的应变状态下，选用图 3-17 中的一组坐标轴x'，y'和z'。那么，在新参考系下该点处的应变张量是什么？

3-29 物体内某一点的主应变为$\varepsilon_{xx}=0.002$，$\varepsilon_{yy}=0.001$ 和$\varepsilon_{zz}=0$。那么，对于习题 3-28 采用的坐标轴方向，该点处的剪切角$\gamma_{x'y'}$是多少？

3-30 在习题 3-27 中，该点处的三个应变张量不变量是什么？

3-31 坐标系如图 3-18 所示，证明平面应变为

$$\varepsilon_{x'x'}=\frac{\varepsilon_{xx}+\varepsilon_{yy}}{2}+\frac{\varepsilon_{xx}-\varepsilon_{yy}}{2}\cos 2\theta+\frac{\gamma_{xy}}{2}\sin 2\theta$$

$$\gamma_{x'y'}=\gamma_{xy}\cos 2\theta-(\varepsilon_{xx}-\varepsilon_{yy})\sin 2\theta$$

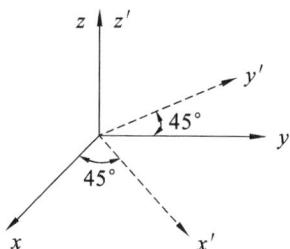

图 3-17 习题 3-28 图   **图 3-18 习题 3-31 图**

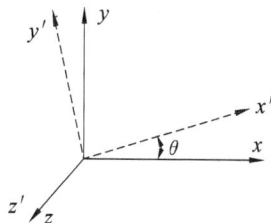

3-32 根据习题 3-31 中给出的公式,可以确定某一点在参考系 xyz 绕 z 轴旋转 $30°$ 得到新坐标系 $x'y'z'$ 中产生的应变。在该点处应变如下

$$\varepsilon_{xx} = 0.01$$
$$\varepsilon_{yy} = -0.02$$
$$\varepsilon_{xy} = 0.02$$

那么,在该点的主应变是多少? 主应变轴方向是什么?

3-33 推导出平面应变状态下一点处的最大剪应变公式,并确定最大剪应变的方向。

3-34 证明在平面应变状态下,最大剪切角 $(\gamma_{x'y'})_{\max}$ 的方向一定与 x 轴和 y 轴的夹角为 $45°$,且最大剪切角 $(\gamma_{x'y'})_{\max}$ 等于主应变之差。

3-35 习题 3-23 中最大剪切角 $(\gamma_{x'y'})_{\max}$ 是多少?

3-36 习题 3-32 中的最大剪切角 $(\gamma_{x'y'})_{\max}$ 是多少?

3-37 对于平面应变的情况,我们如何构造莫尔圆?

3-38 画出习题 3-32 中应变状态对应的莫尔圆,并求出图 3-18 所给 x',y' 轴的应变,其中 $\theta = 30°$。

3-39 写出平面应变的相容方程。

3-40 给出如下平面应变分布

$$\varepsilon_{xx} = 3x^2 y$$
$$\varepsilon_{yy} = 4y^2 x$$
$$\varepsilon_{xy} = yx + x^3$$

求证是否满足相容方程。

4 应力-应变关系

4.1 引　　言

在前几章中我们分别研究了应力和应变。很明显,应力和应变是相互联系的,在本章中我们将对这种关系进行介绍。

我们现在所掌握的所有基本知识几乎都来源于材料的宏观试验,在本章中,我们关心的正是这些试验的结果以及这些试验所产生的宏观理论。然而一直以来,固态物理学家和工程师们一直致力于研究力学性能的微观机理,即研究原子和分子的运动。即使对所涉及的知识还未达到全面的认识,在这一方向上我们也已经取得了很大的进展。今后,我们将越来越多地转向这一基本方向。现代技术是将结构置于更加复杂的环境和条件中,在这些环境和条件下进行如我们本章所述的宏观试验,将逐渐失去意义。为了更好地理解材料在高温、动荷载、辐射、温度梯度、振动等条件组合下的性能,我们需要理解力学作用与原子、分子结构之间的关系。目前,我们还不能对这些结果进行定量表示。但可以肯定的是,理解这些关系对这些材料的定性表示可能有很大帮助。

本章中,我们将首先考虑简单的单轴加载情况,然后将单轴加载的结果和概念推广应用至一般的应力状态。

4.2　拉　伸　试　验

在研究应力-应变关系中,最基础的试验就是简单的拉伸试验。如图 4-1 所示,拉伸试验机在圆柱体试件的中心线处施加一个拉力 F,并用测距仪 1 始终测量试件上两点间的距离 L,同时采用测距仪 2 测量圆柱体试件的直径 D。随着力 F 的变化,我们可以测出每个力 F 对应状态下的圆柱体试件直径 D 和两点间距离 L 的数值。因此,对于任何状态,我们都可以得到下述信息。

(1)实际应力 $(\sigma_z)_{act}$ 计算式为 F/A_{act},其中 A_{act} 为圆柱体的横截面面积,由测距仪 2 测得的实际直径 D 计算得到。

(2)工程应力 $(\sigma_z)_{eng}$ 计算式为 F/A_0,其中 A_0 为圆柱体未发生应变的初始截面面积。

(3)应变 ε_{zz} 由 $\Delta L/L_0$ 表示,其中 ΔL 由测距仪 1 测得,L_0 为未发生应变的试件长度。

对于上述试验,通常要绘制工程应力 $(\sigma_z)_{eng}$ 和应变 ε_{zz} 的关系图。因为试件的体积只发生微小的变化,且随着拉伸荷载的增加,试件的截面面积 A_{act} 会发生收缩,因此工程应力始终小于实际应力。施加小荷载时,A_{act} 不会明显小于 A_0,因此使用更简单的工程应力不会有较大的误差。然而当荷载过大时,A_{act} 与 A_0 之间会有很大的差别,这就会导致 $(\sigma_z)_{eng}$ 和 ε_{zz} 的应力-应变曲线表现得"不自然"。因为工程应力

$(\sigma_z)_{eng}$与力F成正比,比值为$1/A_0$。而且由于侧向收缩不宜准确测量,所以工程技术人员通常以工程应力$(\sigma_z)_{eng}$代替实际应力$(\sigma_z)_{act}$进行计算。

我们在图 4-2 中绘制出简单拉伸试验中得到的应力-应变曲线,这是典型的低碳钢试件的应力-应变曲线。虽然该材料的曲线与其他材料明显不同,但我们会仔细地分析这条曲线,以便给出最一般的定义。可以注意到在早期加载过程中,该曲线是一条直线,即应力、应变是呈比例关系的,那么我们可以表示成:

$$\sigma_z = E\varepsilon_{zz} \tag{4-1}$$

式中,比例系数E被称为杨氏模量,单位为 MPa,这个可以很容易验证得到。大约在 300 年前,罗伯特·胡克就提出了这个结论。他对受轴向荷载拉伸的金属杆进行试验,得出拉伸长度与力成正比的结论,这就是每个高中生都知道的胡克定律。应力-应变之间线性关系的终止点,被称为比例极限。然而,它的数值并不容易测量。

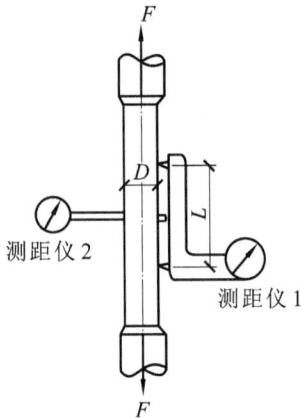

图 4-1　简单拉伸试验

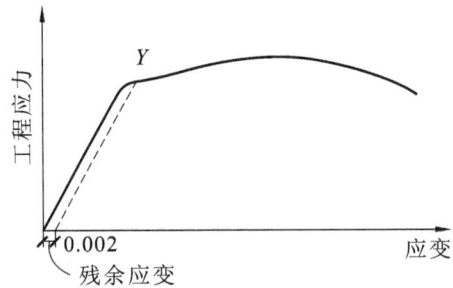

图 4-2　简单拉伸试验的应力-应变曲线

并不是所有材料的应力-应变曲线开头都有一段有限的直线,例如橡胶材料一般不具有这种性质,橡胶试件的应力-应变曲线如图 4-3 所示。虽然图 4-2 和图 4-3 中的低碳钢与橡胶的曲线形状有明显差异,但有一个相似之处是,当橡胶试件和低碳钢试件卸载至零时,试件均沿加载曲线恢复至原来的形状,但前提是作用在低碳钢上的荷载产生的应力σ_z要低于比例极限,如图 4-4 所示。因此,这两种材料都可以表现出理想的弹性性能。

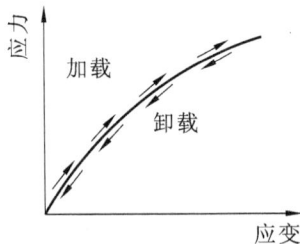

图 4-3　橡胶试件的应力-应变曲线

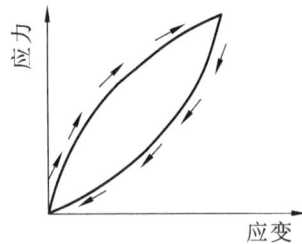

图 4-4　一种特殊试件的应力-应变曲线

低碳钢试件有一个接近比例极限的应力水平,并且当试件从高于上述应力水平的应力卸载时将不再恢复至初始形状。我们称这个应力水平为弹性极限,它是弹性状态的极限,非弹性状态的起点。对于钢材,它的弹性极限与比例极限非常接近,二者几乎很难区分。对于比例极限与弹性极限接近的材料(如钢材),我们称之为线弹性材料。

回到钢材的应力-应变图,我们需要指出的是,弹性极限和比例极限一样是很难精确测量的。因此,工程技术人员更多是将屈服应力或屈服点定义为非弹性状态的起点,其值为材料在卸载过程中产生较小残余应变(通常为 0.002)时对应的应力值。图 4-2 应力-应变图中的点Y即对应屈服点。

目前,我们对应力-应变图的讨论只涉及屈服点。在达到屈服点之前,试件的实际横截面面积与初始横截面面积相差不大,在计算中用哪一种面积都可以。这是我们一直采用 A_0 的原因。然而如前所述,在拉伸试验的任何时候,随着荷载的施加,试件的横截面面积是持续减小的(在压缩试验中,随着荷载的施加,试件的横截面面积是相应增大的),这种横向效应称为泊松效应。在屈服点之后,应变 ε_{zz} 可能会迅速增加,同时由于泊松效应,横截面面积也会迅速发生变化。这将造成工程应力与实际应力的值有明显的差异。为了说明这一点,图 4-5 为拉伸试验中采用实际应力和工程应力试件的应力-应变曲线(图中虚线是在压缩试验中使用工程应力的试件对应的曲线)。你可以发现实际应力会不断增加直至试件发生破坏。

正如前面指出的,工程应力与力 F 成正比,因此试件的极限承载能力为工程应力-应变曲线的最高点,该点的工程应力在图中用 U 表示,称为极限应力。我们可以用数学表达式 $F = \sigma_z A$ 表达极限应力状态。用符号 δ 表示伸长长度 ΔL,那么将上述表达式对 δ 求微分,可得

$$\frac{\mathrm{d}F}{\mathrm{d}\delta} = A\frac{\mathrm{d}\sigma_z}{\mathrm{d}\delta} + \sigma_z\frac{\mathrm{d}A}{\mathrm{d}\delta} \tag{4-2}$$

对于低于屈服点较小的力,在拉伸试验中 $\dfrac{\mathrm{d}A}{\mathrm{d}\delta}$ 的值始终为负数,$A\dfrac{\mathrm{d}\sigma_z}{\mathrm{d}\delta}$ 的值始终为正数,且在数量级上 $\dfrac{\mathrm{d}A}{\mathrm{d}\delta}$ 项相对于 $A\dfrac{\mathrm{d}\sigma_z}{\mathrm{d}\delta}$

图 4-5　采用实际应力和工程应力的应力-应变曲线图

项会非常小。然而,当力 F 增加直至超过屈服点时,$\dfrac{\mathrm{d}A}{\mathrm{d}\delta}$ 的大小会显著增加,直到达到极限应力状态时,方程(4-2)右边变为零。在这一点之后会发生什么呢? 由应力-应变曲线得知,试件的承载能力会随着应变的增加而降低。事实上,在试验中当加载刚超过极限应力时,试件就无法继续承受施加的荷载而发生破坏。破坏发生如此迅速以至于很难记录到应力-应变曲线上在极限应力点后面的数据。

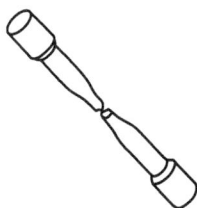

超过上述极限应力点后试件的承载能力下降不是因为试件的整个面积迅速减小,而是因为试件的某个局部区域的面积迅速减小,我们称这种现象为试件的"颈缩"。试件发生"颈缩"的位置主要取决于材料的局部缺陷。图 4-6 所示为试件加载至破坏状态的示意图。通过观察试件破坏部分,能够很容易看出颈缩现象。当在一个小的区域(如在拉伸试件的颈缩区域)内迅速产生较大的非弹性变形时,我们就认为这个区域存在塑性流动。

图 4-6　试件加载破坏图

上述过程引导我们实现了钢试件的拉伸试验,因此我们研究了一种最重要的结构材料性能。那其他材料的性质呢? 本节中,我们也研究了一些橡胶材料,可以看出其和低碳钢材料相比有很大的区别。为了使我们的讨论更有意义,在研究其他材料时我们可以用低碳钢作为比较的基础。此外,我们在研究低碳钢时提出的定义同样适用于一般讨论。图 4-7 中展示了各种钢材和铝合金材料的应力-应变图。在表 4-1 中我们列出了在本节中已经讨论的一些重要材料的参数(有关此类材料更精确、详细的信息请参考结构材料手册)。

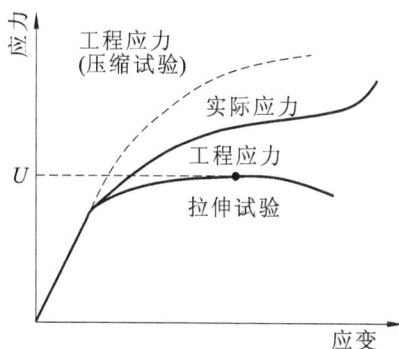

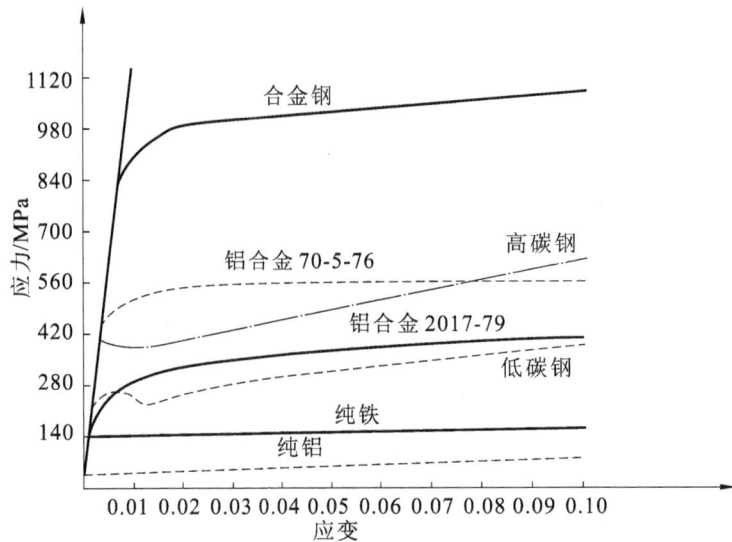

图 4-7　各种钢材和铝合金材料的应力-应变图

表 4-1　　　　　　　　　　　　常见工程材料的一些力学性能

材料	E	U	Y	G
	弹性模量/MPa	极限应力/MPa	屈服应力(0.002)/MPa	剪切模量/MPa
铝	6.90×10^4	4.14×10^2	3.11×10^2	2.76×10^4
铸黄铜	8.97×10^4	3.11×10^2	1.38×10^2	3.45×10^4
冷拉铜	1.17×10^5	3.80×10^2	2.76×10^2	4.14×10^4
铸铁	9.66×10^4	1.38×10^2	—	3.86×10^4
镁	4.49×10^4	2.42×10^2	1.59×10^2	1.66×10^4
结构钢	2.00×10^5	4.14×10^2	2.42×10^2	8.28×10^4
不锈钢	1.93×10^5	8.28×10^2	5.52×10^2	6.90×10^4

注:除了铸铁,材料的弹性模量在拉伸和压缩情况下是相同的,这里只给出了拉伸情况下的弹性模量。

4.3　应 变 硬 化

在大多数延性材料如低碳钢、铝和铜中我们可以观察到,在超过屈服点后想要发生持续变形需要不断增加实际应力。图 4-5 所示的应力-应变图就是这种情况,我们称之为应变硬化。可以用位错理论定性地解释金属的应变硬化。

在塑性范围内存在被称为应变硬化的另一种重要现象,应变硬化与具有线弹性范围的试件以及卸载开始于塑性阶段有关。4.2 节中我们讨论了当荷载在弹性范围内时线弹性材料的卸载过程,以及荷载超出弹性范围时非线性弹性材料的卸载过程。在这些情况下,当我们完全移除荷载时,试件会恢复到原来的形状,即没有永久变形。此外,在应力-应变图中卸载路径与加载路径重合。对于线弹性材料,当荷载在塑性范围内进行卸载时,将不再沿着加载路径移动,而是沿着基本上平行于原始加载路径的线弹性部分的新路径移动。如图 4-8 所示,初始加载在 A 点处停止,第一次卸载沿直线至横坐标上 B 点处。因此,我们在横坐标上用 OB 表示永久变形,显然,弹性恢复量为线段 BE。第二次加载,我们将沿着路径 BA

移动。当应力低于 A 点进行第二次卸载时,将沿路径 BA 回到 B 点,因此实际上我们有一个从 B 点到 A 点的线弹性范围。观察图 4-8 可发现,因为第一次加载到塑性范围,第二次加载时屈服点已经提高。通过这种方式提高屈服点,这一物理现象被称为应变硬化现象。超过新屈服点后,第二次加载路径将沿着 AC 进行,之后将沿着第一次没有中断加载的应力曲线移动。在 C 点处进行第二次卸载并将重复相同的过程。

需要注意的是,卸载和再加载的曲线并不完全重合。相反,它们形成了一个小的滞回曲线,图 4-9 放大了该曲线。其中,用滞回曲线的面积表示一个循环中的能量损失,这个能量损失非常小。

最后,需要指出的是,仅在初始加载方向上可以观察到应变硬化使屈服点发生变化。也就是说,在与初始加载方向垂直的方向上,材料的屈服应力没有增加。

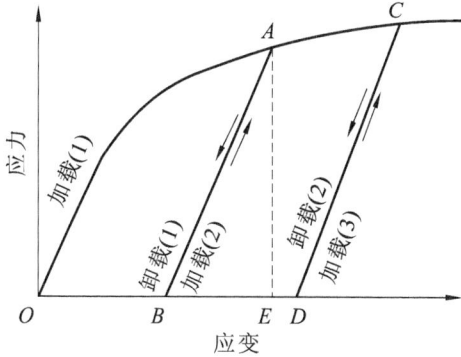

图 4-8 应变硬化现象图

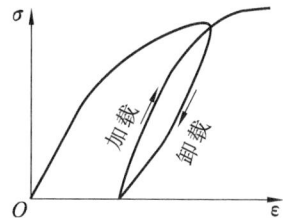

图 4-9 一个小的滞回曲线

4.4 与拉伸试验有关的其他性能

在前面描述的简单拉伸试验和简单压缩试验的基础上(两者除了荷载方向不同,基本是相同的),我们可以做一些其他有用的分类,这对描述材料的力学性能是十分有意义的。

首先,我们可以根据试件在拉伸试验中的性能将材料分为两类。在断裂前无明显塑性变形的材料,如玻璃,称为脆性材料。在断裂前表现出大量塑性变形的材料,如低碳钢,称为延性材料。对于脆性材料,拉伸试验中的应力-应变曲线与压缩试验的曲线不同。此外,通过多次试验发现,脆性材料的断裂点非常分散。相反,对于延性材料,在拉伸试验与压缩试验中具有基本相同的应力-应变曲线,并且有相同的屈服点、断裂点等,在多次试验中,它们的可重复性很高。

前面已经指出,在试件上的拉伸荷载会引起横向收缩,称为泊松效应。同样,在压缩试验中也会产生横向膨胀。在线弹性范围内,我们从这些试验中发现横向应变与纵向应变成正比,可以表示为

$$\varepsilon_{lat} = -\nu\varepsilon_{long} \tag{4-3}$$

式中,比例常数 ν 称为泊松比,对于工程材料,它的取值范围为 $0.2\sim0.5$。在 4.6 节中,我们将看到泊松比被视为表征线性、弹性、均匀性、各向同性材料一般力学性能的基本常数之一。同时通过简单的拉伸试验就可以计算出这个常数。

在前面研究的拉伸试验中,我们假设试件的温度是均匀的且接近室温。同时,我们还指出进行试验时既应避免产生动力效应,又应避免产生长时效应(徐变效应)。随着技术的不断进步,了解在温度不接近室温以及动力效应和徐变效应变得非常重要的条件下的材料性能正在变得越来越有必要。因此,我们将对这些效应进行简要讨论。

时间效应。快速加载将产生与慢速加载不同的应力-应变曲线,这对高温下的软材料或结构材料尤

为如此。在这些情况下,加载速率信息一定会作为相关信息被包含在应力-应变曲线内。本质上,屈服应力会随加载速率的增加而增大。

现在我们观察另一个极端状态下的长期效应。我们将考虑一个在长时间内受到恒定拉伸荷载的试件情况,对于一些材料,包括高温下的结构材料,随着时间变化,应变将会有小的但持续的增加,这种现象称为徐变效应。在锅炉、反应堆、燃气轮机等结构中,这种效应具有重要作用。如果荷载维持的时间足够长,即使应力最初被认为低于极限应力,试件也有可能发生破坏。出于试验的目的,徐变应变被定义为试件达到所需荷载后产生的应变。因此,徐变应变是在恒定应力的作用下发生的,通常绘制成不同应力值关于时间的函数,如图4-10所示。注意,应力越大,在任何给定的时间对应的徐变应变就越大。

温度效应。我们已经指出温度升高如何加剧徐变现象,我们也可以表述为温度降低将导致应力-应变图的线弹性部分斜率增加,这意味着弹性模量随着温度的降低而增大。当温度升高时,应力-应变图的线性部分的斜率将会减小,表明弹性模量随温度的升高而减小。

另一个重要的热效应是大多数材料因温度升高而产生的膨胀现象。如果温度场不均匀,就会产生一个被称为热应力的应力场。

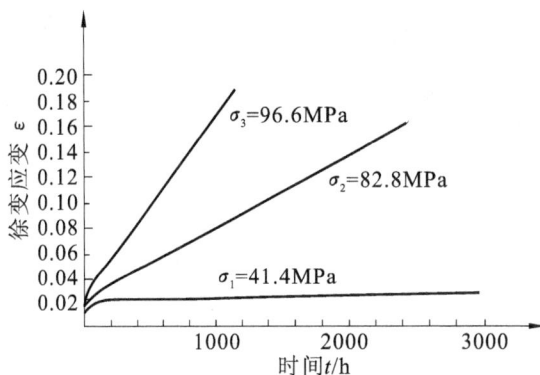

图 4-10 不同应力值的时间函数

4.5 理想的一维应力-应变定律

很明显,应力-应变关系一般来说都是很复杂的,可能会产生许多不同的结果。在前面几节中,我们介绍了部分这些复杂的关系。为了能够分析在特定条件下的材料性能,有时我们会采用理想的应力-应变关系。

最简单的理想应力-应变关系当然是刚体对应的应力-应变关系,如图4-11所示。我们已经在刚体力学课程中使用过这种模型,你可以回忆一下,在第1章中我们在计算静定结构的支反力时就采用了刚体模型。

在图4-12中,我们展示了理想线弹性材料的应力-应变曲线,该模型是本书中主要采用的模型。我们应该记得应力-应变图是从简单的一维应力状态中得到的,并且在4.6节中将这个模型推广至一般的应力状态。由此得到的公式我们称之为广义胡克定律。当应力低于屈服应力时,我们可以使用这些结果来分析常见结构材料(如钢和铝)构成的物体。

在某些情况下,可能会涉及远超出弹性变形范围的塑性变形,因此,绘制如图4-13所示的理想应力-应变图是很有用的,它展示了物体在一定应力下的刚体性能和随后呈现出的理想塑性性能。如果我们将应变硬化纳入塑性范围,虽然我们的模型将变得更复杂,但也会变得更精确。在图4-14中我们展示了具有应变硬化的理想刚塑性性能来说明这种情况。

有时,对于不能忽略弹性变形并且几乎不存在应变硬化的情况,我们可以采用如图 4-15 所示的理想弹塑性应力-应变曲线。最后,考虑到应变硬化,我们得到如图 4-16 所示的曲线,它更接近某些实际的应力-应变曲线。

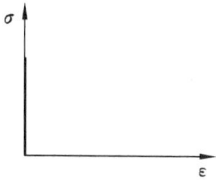

图 4-11　刚体性能

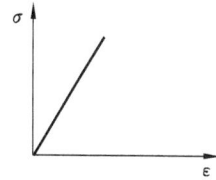

图 4-12　理想弹性性能

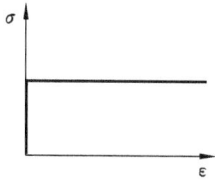

图 4-13　理想刚塑性性能

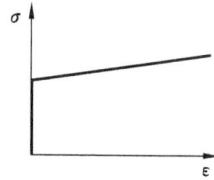

图 4-14　具有应变硬化的理想刚塑性性能

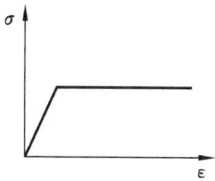

图 4-15　理想弹塑性性能

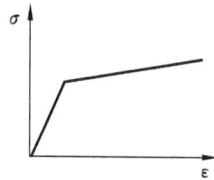

图 4-16　具有应变硬化的弹塑性性能

在本章后续小节中,我们将采用理想弹性行为的广义形式。同样,我们将有机会使用其他的理想化方法来解决塑性范围内的简单应力状态问题。

4.6　广义胡克定律

在一维试验中,我们发现许多结构材料在应力远低于屈服应力时,应力和应变之间呈线性关系,如式(4-1)所示。

现在我们假定,在一般应力状态下,符合一维应力的方程(4-1)的材料将有以下应力-应变关系

$$\begin{cases} \sigma_x = C_{11}\varepsilon_{xx} + C_{12}\varepsilon_{yy} + C_{13}\varepsilon_{zz} + C_{14}\gamma_{xy} + C_{15}\gamma_{yz} + C_{16}\gamma_{xz} \\ \sigma_y = C_{21}\varepsilon_{xx} + C_{22}\varepsilon_{yy} + C_{23}\varepsilon_{zz} + C_{24}\gamma_{xy} + C_{25}\gamma_{yz} + C_{26}\gamma_{xz} \\ \sigma_z = C_{31}\varepsilon_{xx} + C_{32}\varepsilon_{yy} + C_{33}\varepsilon_{zz} + C_{34}\gamma_{xy} + C_{35}\gamma_{yz} + C_{36}\gamma_{xz} \\ \tau_{xy} = C_{41}\varepsilon_{xx} + C_{42}\varepsilon_{yy} + C_{43}\varepsilon_{zz} + C_{44}\gamma_{xy} + C_{45}\gamma_{yz} + C_{46}\gamma_{xz} \\ \tau_{yz} = C_{51}\varepsilon_{xx} + C_{52}\varepsilon_{yy} + C_{53}\varepsilon_{zz} + C_{54}\gamma_{xy} + C_{55}\gamma_{yz} + C_{56}\gamma_{xz} \\ \tau_{xz} = C_{61}\varepsilon_{xx} + C_{62}\varepsilon_{yy} + C_{63}\varepsilon_{zz} + C_{64}\gamma_{xy} + C_{65}\gamma_{yz} + C_{66}\gamma_{xz} \end{cases} \tag{4-4}$$

其中各项 C_{ij} 构成一个 6×6 的常数矩阵,其值取决于材料本身。我们可以看出,在式(4-4)中,一点处的每个应力都与这点的所有应变是线性相关的。显然,对于沿坐标轴 xyz 某一方向上的简单单轴应力,这种关系方程(4-4)将退化为方程(4-1)所示的关系。选择 z 轴方向,注意剪应变为零,我们可以得到

$$\sigma_z = C_{31}\varepsilon_{xx} + C_{32}\varepsilon_{yy} + C_{33}\varepsilon_{zz} \tag{4-5}$$

但是我们已经证明了 ε_{yy} 和 ε_{xx} 均与 ε_{zz} 呈比例关系。

$$\sigma_z = C_{31}(-\nu\varepsilon_{zz}) + C_{32}(-\nu\varepsilon_{zz}) + C_{33}\varepsilon_{zz} \tag{4-6}$$

因此就有

$$\sigma_z = (-\nu C_{31} - \nu C_{32} + C_{33})\varepsilon_{zz} \tag{4-7}$$

可以看到，方程（4-7）与方程（4-1）具有相同的形式。我们可以依据一般应力状态下应力-应变关系包含从拉伸试验中推导出的简单应力-应变关系这一事实，验证其合理性。虽然，使用线性应力-应变定律是间接验证方法。但我们发现，在处理线弹性材料时，利用这种关系得到的分析结果与实验结果是一致的。

如果材料整体组成相同，我们可以说它是均质的。线弹性均质物体有 36 个常数或弹性模量，在比例极限下，它们必须与应力和应变相关。乍一看，我们似乎要在一个极其复杂的情况下处理问题。然而，我们在工程应用中所接触的大多数材料，其力学性能在任何方向上都是一致的。也就是说，对于新参考系 $x'y'z'$ 下广义胡克定律的常数 C_{ij} 与在参考系 xyz 下方程（4-4）的常数 C_{ij} 的数值是相同的。因此 $\sigma_{x'}$ 的表达式为

$$\sigma_{x'} = C_{11}\varepsilon_{x'x'} + C_{12}\varepsilon_{y'y'} + C_{13}\varepsilon_{z'z'} + C_{14}\gamma_{x'y'} + C_{15}\gamma_{y'z'} + C_{16}\gamma_{x'z'} \tag{4-8}$$

其中 C_{ij} 与方程（4-4）中的值相同。当一种材料在各个方向上的力学性能都相同时，我们就称这种材料是各向同性材料。现在我们将证明，对于各向同性材料，36 个弹性模量将会简化为 2 个弹性模量。

为了证明这一点，我们将坐标轴绕 x 轴，y 轴，z 轴进行一定顺序的旋转，每次都保证新方向上的应力-应变关系与原来形式相同。

如图 4-17 所示，将坐标系绕 z 轴旋转 180°。新坐标轴关于原坐标系的方向余弦如下所示。

$$\begin{array}{c|ccc} & x & y & z \\ \hline x' & -1 & 0 & 0 \\ y' & 0 & -1 & 0 \\ z' & 0 & 0 & 1 \end{array} \tag{4-9}$$

新参考系中的应力 $\sigma_{x'}$、$\sigma_{y'}$、$\sigma_{z'}$、$\tau_{x'y'}$、$\tau_{y'z'}$、$\tau_{x'z'}$ 与原来参考系中的应力 σ_x、σ_y、σ_z、τ_{xy}、τ_{yz}、τ_{zx} 有关，利用方程（4-9）中的方向余弦可以得到

$$\begin{aligned} \sigma_{x'} &= \sigma_x & \tau_{x'y'} &= \tau_{xy} \\ \sigma_{y'} &= \sigma_y & \tau_{y'z'} &= -\tau_{yz} \\ \sigma_{z'} &= \sigma_z & \tau_{x'z'} &= -\tau_{zx} \end{aligned} \tag{4-10}$$

同样，利用方程（4-9）中的方向余弦对应变进行转换，可以得到下列应变关系。

$$\begin{aligned} \varepsilon_{x'x'} &= \varepsilon_{xx} & \gamma_{x'y'} &= \gamma_{xy} \\ \varepsilon_{y'y'} &= \varepsilon_{yy} & \gamma_{y'z'} &= -\gamma_{yz} \\ \varepsilon_{z'z'} &= \varepsilon_{zz} & \gamma_{x'z'} &= -\gamma_{zx} \end{aligned} \tag{4-11}$$

将方程（4-10）和方程（4-11）中的应力和应变转换结果代入方程（4-8）中，我们可以得到

$$\sigma_x = C_{11}\varepsilon_{xx} + C_{12}\varepsilon_{yy} + C_{13}\varepsilon_{zz} + C_{14}\gamma_{xy} - C_{15}\gamma_{yz} - C_{16}\gamma_{zx} \tag{4-12}$$

将方程（4-12）与方程（4-4）中的第一个方程进行比较，我们可以看到满足各向同性的一个必要条件是

$$C_{15} = C_{16} = 0 \tag{4-13}$$

同样以这种方式计算其他的应力，对于各向同性材料，我们可以推导出下列条件必须成立。

$$\begin{array}{lllll} C_{25} = 0 & C_{36} = 0 & C_{51} = 0 & C_{54} = 0 & C_{63} = 0 \\ C_{26} = 0 & C_{45} = 0 & C_{52} = 0 & C_{61} = 0 & C_{64} = 0 \\ C_{35} = 0 & C_{46} = 0 & C_{53} = 0 & C_{62} = 0 \end{array} \tag{4-14}$$

然后将弹性模量矩阵进行简化

$$\begin{bmatrix} C_{11} & C_{12} & C_{13} & C_{14} & 0 & 0 \\ C_{21} & C_{22} & C_{23} & C_{24} & 0 & 0 \\ C_{31} & C_{32} & C_{33} & C_{34} & 0 & 0 \\ C_{41} & C_{42} & C_{43} & C_{44} & 0 & 0 \\ 0 & 0 & 0 & 0 & C_{55} & C_{56} \\ 0 & 0 & 0 & 0 & C_{65} & C_{66} \end{bmatrix} \tag{4-15}$$

观察图 4-17,可以清楚地看到,上述矩阵在 x 方向与 x' 方向上的力学性能相同,在 y 方向和 y' 方向也是这样。因此,我们可以说平面 yz 和平面 xz 是弹性对称的。为了实现平面 xy 的对称,利用各向同性条件,我们将绕 x 轴旋转 180°,如图 4-18 所示。矩阵将被简化为下列形式

$$\begin{bmatrix} C_{11} & C_{12} & C_{13} & 0 & 0 & 0 \\ C_{21} & C_{22} & C_{23} & 0 & 0 & 0 \\ C_{31} & C_{32} & C_{33} & 0 & 0 & 0 \\ 0 & 0 & 0 & C_{44} & 0 & 0 \\ 0 & 0 & 0 & 0 & C_{55} & 0 \\ 0 & 0 & 0 & 0 & 0 & C_{66} \end{bmatrix} \tag{4-16}$$

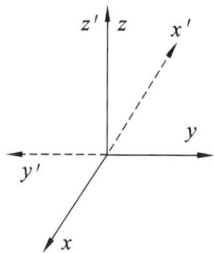

图 4-17 绕 z 轴旋转 180°

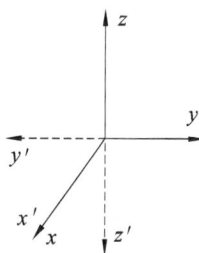

图 4-18 绕 x 轴旋转 180°

这个模量矩阵表示了三个正交平面弹性对称的情况。这就意味着当 x、y、z 方向相反时,材料的力学性能不会发生变化。具有这种性能的材料称为正交异性材料。这种有限的各向同性的情况对某些材料(如晶体)是很重要的。然而,在本书中我们只讨论各向同性材料的性能。

就像第一次绕轴转动一样,通过一系列其他的绕轴转动,我们可以将独立弹性模量的数量减少至 2 个。并且我们已经证明了广义胡克定律可以简化为下列形式

$$\begin{aligned} \sigma_x &= (2G+\lambda)\varepsilon_{xx} + \lambda(\varepsilon_{yy}+\varepsilon_{zz}) \\ \sigma_y &= (2G+\lambda)\varepsilon_{yy} + \lambda(\varepsilon_{xx}+\varepsilon_{zz}) \\ \sigma_z &= (2G+\lambda)\varepsilon_{zz} + \lambda(\varepsilon_{xx}+\varepsilon_{yy}) \\ \tau_{xy} &= G\gamma_{xy} \\ \tau_{yz} &= G\gamma_{yz} \\ \tau_{xz} &= G\gamma_{xz} \end{aligned} \tag{4-17}$$

式中,λ 是拉梅常数;G 是剪切弹性模量。

那么,杨氏模量 E 与常数 λ 和 G 有什么关系呢? 为了解决这个问题,我们可以利用上述方程中应力来求解应变。我们通过纯代数的方法可以得到

$$\varepsilon_{xx} = \frac{\lambda+G}{G(3\lambda+2G)}\sigma_x - \frac{\lambda}{2G(3\lambda+2G)}(\sigma_y+\sigma_z) \tag{4-18a}$$

$$\varepsilon_{yy} = \frac{\lambda+G}{G(3\lambda+2G)}\sigma_y - \frac{\lambda}{2G(3\lambda+2G)}(\sigma_x+\sigma_z) \tag{4-18b}$$

$$\varepsilon_{zz} = \frac{\lambda + G}{G(3\lambda + 2G)}\sigma_z - \frac{\lambda}{2G(3\lambda + 2G)}(\sigma_x + \sigma_y) \tag{4-18c}$$

$$\gamma_{xy} = \frac{1}{G}\tau_{xy} \tag{4-18d}$$

$$\gamma_{xz} = \frac{1}{G}\tau_{xz} \tag{4-18e}$$

$$\gamma_{yz} = \frac{1}{G}\tau_{yz} \tag{4-18f}$$

在简单拉伸试验中,很明显有 $\sigma_y = \sigma_x = 0$,因此方程(4-18c)可转化为:

$$\varepsilon_{zz} = \frac{\lambda + G}{G(3\lambda + 2G)}\sigma_z \tag{4-19}$$

将方程(4-19)与方程(4-1)对比,可以得到

$$E = \frac{G(3\lambda + 2G)}{\lambda + G} \tag{4-20}$$

为了建立涉及泊松比的另一种形式的胡克定律,在简单拉伸试验中有

$$\varepsilon_{xx} = -\nu\varepsilon_{zz}$$

现在用方程(4-18a)和方程(4-18c)代替上式中的 ε_{xx} 和 ε_{zz},并代入方程(4-20)中的 E 可以得到

$$\frac{\sigma_x}{E} - \frac{\lambda}{2G(3\lambda + 2G)}(\sigma_y + \sigma_z) = -\nu\left[\frac{\sigma_z}{E} - \frac{\lambda}{2G(3\lambda + 2G)}(\sigma_x + \sigma_y)\right] \tag{4-21}$$

在一维应力状态下设 $\sigma_x = \sigma_y = 0$,消去 σ_z 后,有

$$\frac{\lambda}{2G(3\lambda + 2G)} = \frac{\nu}{E} \tag{4-22}$$

将方程(4-20)和方程(4-22)代入方程(4-18),胡克定律可以表示为

$$\varepsilon_{xx} = \frac{1}{E}\left[\sigma_x - \nu(\sigma_y + \sigma_z)\right] \tag{4-23a}$$

$$\varepsilon_{yy} = \frac{1}{E}\left[\sigma_y - \nu(\sigma_x + \sigma_z)\right] \tag{4-23b}$$

$$\varepsilon_{zz} = \frac{1}{E}\left[\sigma_z - \nu(\sigma_x + \sigma_y)\right] \tag{4-23c}$$

$$\gamma_{xy} = \frac{1}{G}\tau_{xy} \tag{4-23d}$$

$$\gamma_{xz} = \frac{1}{G}\tau_{xz} \tag{4-23e}$$

$$\gamma_{yz} = \frac{1}{G}\tau_{yz} \tag{4-23f}$$

其中我们使用到了常数 ν、E 和 G。

在本书中,我们一般采用上述形式的广义胡克定律,我们需要记得只有两个常数是独立的。因此,了解上述形式的广义胡克定律中给出的常数之间的关系是有帮助的。为此,由方程(4-20)可求出 λ。

$$\lambda = \frac{2G^2 - EG}{E - 3G} \tag{4-24}$$

再由方程(4-22)求出 λ。

$$\lambda = \frac{4G^2\nu}{E - 6G\nu} \tag{4-25}$$

最后令式(4-24)和式(4-25)右边相等可以解出 G,得到想要的关系。

$$G = \frac{E}{2(1 + \nu)} \tag{4-26}$$

由于 E 和 ν 可以由简单拉伸试验测得,故通常将它们视为基本常数。根据方程(4-26)并利用这些常数可以计算 G。

可以很容易证明,方程(4-23)可以用张量符号表示为

$$\varepsilon_{ij}=\frac{1+\nu}{E}\tau_{ij}-\frac{\nu}{E}\sigma_k\delta_{ij} \tag{4-27}$$

式中,δ_{ij} 为狄拉克函数,当 $i=j$ 时为 1,当 $i \neq j$ 时为 0。并且当 $i=j$ 时 $\tau_{ij}=\sigma_i$。

4.7 结 语

在本章中,我们首先考虑简单的单轴加载情况,接着我们尝试将单轴加载的结果和概念推广到一般应力状态。在研究被称为简单拉伸试验的应力-应变关系时,我们介绍了最基本的试验过程,也说明了实际应力、工程应力、应变、比例极限、弹性极限、屈服点等概念。为了能分析在特定条件下的材料性能,我们使用了应力-应变关系的理想模型,考虑了简单的单轴加载并将这些单轴加载的结果和概念推广到一般应力状态,我们称之为广义胡克定律。

习题

4-1　一个拉伸试件的直径为 15 mm,初始长度为 50 mm,加载 4500 N 时纵向测距仪测得长度增量为 0.025 mm,横向测距仪测得直径减少量为 0.00125 mm。那么,该试件的实际应力与工程应力分别是多少？并计算弹性模量。

4-2　如图 4-19 所示的一条假想应力-应变曲线,该材料的比例极限、极限应力和弹性模量是多少？

4-3　确定习题 4-1 中的泊松比。

4-4　说明什么情况下非线性弹性材料可以用作减震器。

4-5　在如图 4-19 对应的拉伸试验中,试件从应力为 380 MPa 处卸载,应变的弹性恢复为多少？ 永久变形是多少？ 当试件重新加载时的比例极限是多少？

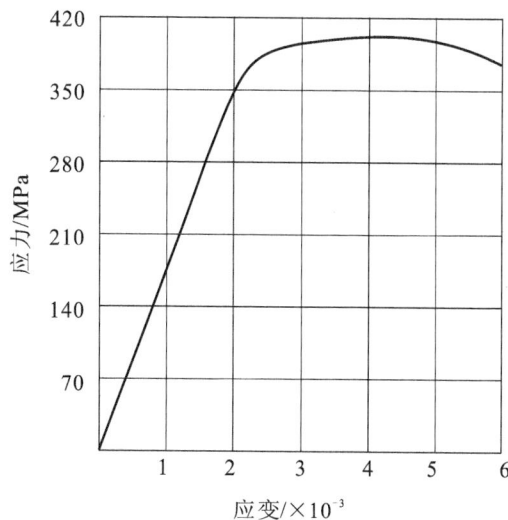

图 4-19　习题 4-2 图

4-6　固体材料的理想塑性行为与流体的理想塑性行为相同吗？ 并加以解释。

4-7　证明方程(4-13)。

4-8　对于一个平面的弹性对称问题,矩阵 C_{ij} 的非零系数是什么?

4-9　对于钢材,已知

$$E=2.1\times10^5\ \text{MPa}$$

$$G=1.1\times10^5\ \text{MPa}$$

计算拉梅常数。若该材料某一点的应变状态为

$$\varepsilon_{ij}=\begin{pmatrix} 0.001 & 0 & -0.002 \\ 0 & -0.003 & 0.0005 \\ -0.002 & 0.0005 & 0 \end{pmatrix}$$

请确定该点的应力张量。

4-10　当剪切模量为 6.9×10^4 MPa,弹性模量为 1.7×10^5 MPa 时,计算下列应力状态下的应变张量。

$$\sigma_{ij}=\begin{pmatrix} 10 & -50 & 0 \\ -50 & 5 & 5 \\ 0 & 5 & -20 \end{pmatrix}\ \text{MPa}$$

4-11　对于各向同性的线弹性材料,证明其主应力轴与主应变轴相一致(对各向异性材料不成立,对各向同性非线性材料成立)。

4-12　当泊松比为 0.30,杨氏模量为 2.1×10^5 MPa 时,计算下列应力状态下的应变张量。

$$\sigma_{ij}=\begin{pmatrix} 0 & 10 & -20 \\ 10 & 5 & -30 \\ -20 & -30 & 10 \end{pmatrix}\ \text{MPa}$$

5 弹性力学基本方程

5.1 引　　言

在前面几章中,我们将应力和应变视为相互独立的实体进行研究,几乎没有要求确定在特定物体中的应力或应变分布。在第 4 章中我们已经证明了,对于线弹性材料,一旦知道应力分布,就可以确定应变分布,反之亦然。我们现在将建立在已知力或位移的条件下确定某些物体的应力和应变场的方法。因此,我们下一项任务是针对我们提出的问题,建立必须满足的基本定律,获得具有物理意义的解。

这种情况下,我们将不会像前面几章那样依赖张量符号。在前述章节的基础推导中,通过使用张量符号,我们可以在一个高度概括的水平去处理问题并发现一些重要的数学关系,否则,这些数学关系将会是模糊不清的。接下来你会注意到,当我们尝试解决问题时,通常会倾向于使用不作删减的标注方法。

5.2　弹性理论方程

任何连续介质都必须满足两个基本定律:牛顿定律、能量守恒定律。

除了这些基本定律,还必须满足特定材料的本构关系。对于线弹性材料,我们已经提出了广义胡克定律。

我们现在来研究第一个基本定律,因为它与我们的理论有关。由于本书主要研究的是惯性空间中处于平衡状态的物体,因此物体中的每个单元都必定满足牛顿定律的平衡条件。这就如求任意一个单元上的合力都为零。然后考虑物体中的一个平行矩形六面体单元,如图 5-1 所示。物体表面的应力以及单位体积上的体力分量如图 5-1 所示。在外表面上,我们已经表达了应用泰勒展开的前两项,这是 x 方向力产生的应力贡献。通过对 x 方向的力求和,我们得到

$$-\sigma_x \mathrm{d}y\mathrm{d}z + \left(\sigma_x + \frac{\partial \sigma_x}{\partial x}\mathrm{d}x\right)\mathrm{d}y\mathrm{d}z - \tau_{yx}\mathrm{d}x\mathrm{d}z + \left(\tau_{yx} + \frac{\partial \tau_{yx}}{\partial y}\mathrm{d}y\right)\mathrm{d}x\mathrm{d}z - \tau_{zx}\mathrm{d}x\mathrm{d}y +$$

$$\left(\tau_{zx} + \frac{\partial \tau_{zx}}{\partial z}\mathrm{d}z\right)\mathrm{d}x\mathrm{d}y + B_x\mathrm{d}x\mathrm{d}y\mathrm{d}z = 0 \tag{5-1}$$

通过消项以及等式两边同时除以 $\mathrm{d}x\mathrm{d}y\mathrm{d}z$ 得到的简化方程为

$$\frac{\partial \sigma_x}{\partial x} + \frac{\partial \tau_{yx}}{\partial y} + \frac{\partial \tau_{zx}}{\partial z} + B_x = 0 \tag{5-2}$$

通过考虑另外两个坐标方向上的力,我们可以建立另外两个类似的方程。利用应力张量的对称性,我们可以将牛顿定律写成如下形式

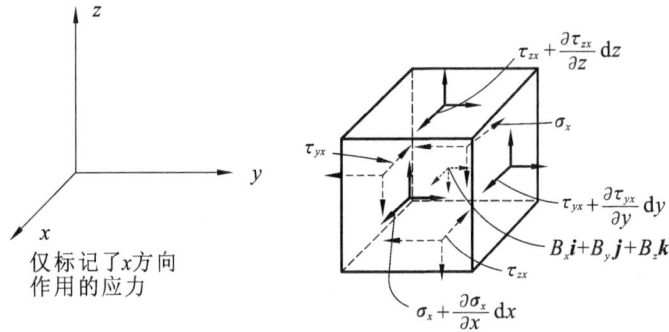

图 5-1 平行矩形六面体单元

$$\begin{cases} \dfrac{\partial \sigma_x}{\partial x} + \dfrac{\partial \tau_{xy}}{\partial y} + \dfrac{\partial \tau_{xz}}{\partial z} + B_x = 0 \\[2mm] \dfrac{\partial \tau_{yx}}{\partial x} + \dfrac{\partial \sigma_y}{\partial y} + \dfrac{\partial \tau_{yz}}{\partial z} + B_y = 0 \\[2mm] \dfrac{\partial \tau_{zx}}{\partial x} + \dfrac{\partial \tau_{zy}}{\partial y} + \dfrac{\partial \sigma_z}{\partial z} + B_z = 0 \end{cases} \tag{5-3}$$

采用笛卡儿张量标注,方程(5-3)可记为

$$\frac{\partial \tau_{ij}}{\partial x_j} + B_i = 0 \tag{5-4}$$

式中,当 $i=j$ 时,$\tau_{ij}=\sigma_i$。对于前面描述的"准静态"问题,只要把惯性力作为体力分布 B 的一部分,我们仍然可以使用前面的方程。

对于在基本理论中考虑的线弹性介质,在我们研究的问题中,既没有明显的热传递,也没有内摩擦。因此,牛顿定律的满足保证了能量守恒定律的满足。同时,就像刚体力学一样,牛顿定律和能量守恒定律提供了不同解决问题的途径。

如前所述,在该理论中起关键作用的本构定律是广义胡克定律,再次表示如下

$$\begin{cases} \varepsilon_{xx} = \dfrac{1}{E}[\sigma_x - \nu(\sigma_y + \sigma_z)] \\[2mm] \varepsilon_{yy} = \dfrac{1}{E}[\sigma_y - \nu(\sigma_x + \sigma_z)] \\[2mm] \varepsilon_{zz} = \dfrac{1}{E}[\sigma_z - \nu(\sigma_x + \sigma_y)] \\[2mm] \gamma_{xy} = \dfrac{1}{G}\tau_{xy} \\[2mm] \gamma_{xz} = \dfrac{1}{G}\tau_{xz} \\[2mm] \gamma_{yz} = \dfrac{1}{G}\tau_{yz} \end{cases} \tag{5-5}$$

上述方程的笛卡儿张量表达式为

$$\varepsilon_{ij} = \frac{1+\nu}{E}\tau_{ij} - \frac{\nu}{E}(\sigma_k)\delta_{ij} \tag{5-6}$$

根据几何关系,我们在第 3 章中说明了应变张量与位移场的关系如下

$$\begin{cases} \varepsilon_{xx} = \dfrac{\partial u_x}{\partial x} \quad \gamma_{xy} = \dfrac{\partial u_x}{\partial y} + \dfrac{\partial u_y}{\partial x} \\[2mm] \varepsilon_{yy} = \dfrac{\partial u_y}{\partial y} \quad \gamma_{yz} = \dfrac{\partial u_y}{\partial z} + \dfrac{\partial u_z}{\partial y} \\[2mm] \varepsilon_{zz} = \dfrac{\partial u_z}{\partial z} \quad \gamma_{xz} = \dfrac{\partial u_x}{\partial z} + \dfrac{\partial u_z}{\partial x} \end{cases} \tag{5-7}$$

用张量符号表示就是

$$\varepsilon_{ij}=\frac{1}{2}\left(\frac{\partial u_i}{\partial x_j}+\frac{\partial u_j}{\partial x_i}\right) \tag{5-8}$$

因此,我们总共有 15 个方程,即 3 个平衡方程、6 个胡克定律方程以及 6 个应变位移关系方程。从这些方程中,我们可以求解 6 个应力分量、6 个应变分量和 3 个位移分量,总共 15 个未知量。

在第 3 章中我们已经证明如果位移不明确,那么应变所满足的一些特定方程,即相容方程,是表示单值、连续的位移场的必要条件,即

$$\begin{cases}\dfrac{\partial^2\varepsilon_{xx}}{\partial y^2}+\dfrac{\partial^2\varepsilon_{yy}}{\partial x^2}=\dfrac{\partial^2\gamma_{xy}}{\partial x\partial y}\\[2mm]\dfrac{\partial^2\varepsilon_{yy}}{\partial z^2}+\dfrac{\partial^2\varepsilon_{zz}}{\partial y^2}=\dfrac{\partial^2\gamma_{yz}}{\partial y\partial z}\\[2mm]\dfrac{\partial^2\varepsilon_{zz}}{\partial x^2}+\dfrac{\partial^2\varepsilon_{xx}}{\partial z^2}=\dfrac{\partial^2\gamma_{zx}}{\partial z\partial x}\\[2mm]2\dfrac{\partial^2\varepsilon_{xx}}{\partial y\partial z}=\dfrac{\partial}{\partial x}\left(-\dfrac{\partial\gamma_{yz}}{\partial x}+\dfrac{\partial\gamma_{zx}}{\partial y}+\dfrac{\partial\gamma_{xy}}{\partial z}\right)\\[2mm]2\dfrac{\partial^2\varepsilon_{yy}}{\partial z\partial x}=\dfrac{\partial}{\partial y}\left(-\dfrac{\partial\gamma_{zx}}{\partial y}+\dfrac{\partial\gamma_{yx}}{\partial z}+\dfrac{\partial\gamma_{yz}}{\partial x}\right)\\[2mm]2\dfrac{\partial^2\varepsilon_{zz}}{\partial x\partial y}=\dfrac{\partial}{\partial z}\left(-\dfrac{\partial\gamma_{xy}}{\partial z}+\dfrac{\partial\gamma_{zy}}{\partial x}+\dfrac{\partial\gamma_{zx}}{\partial y}\right)\end{cases} \tag{5-9}$$

如果使用张量符号,上述方程就可用下述方程表示:

$$\frac{\partial^2\varepsilon_{ij}}{\partial x_k\partial x_l}+\frac{\partial^2\varepsilon_{kl}}{\partial x_i\partial x_j}=\frac{\partial^2\varepsilon_{ik}}{\partial x_j\partial x_l}+\frac{\partial^2\varepsilon_{jl}}{\partial x_i\partial x_k} \tag{5-10}$$

本节中所给出的方程构成了弹性力学的基本方程。

5.3　二维极坐标系下的平衡微分方程

在求解平面问题时,对于圆形物体,用极坐标求解比用直角坐标方便。在极坐标系中,平面内任一点 P 的位置,用径向坐标 r 及切向坐标 θ 来表示,如图 5-2 所示。

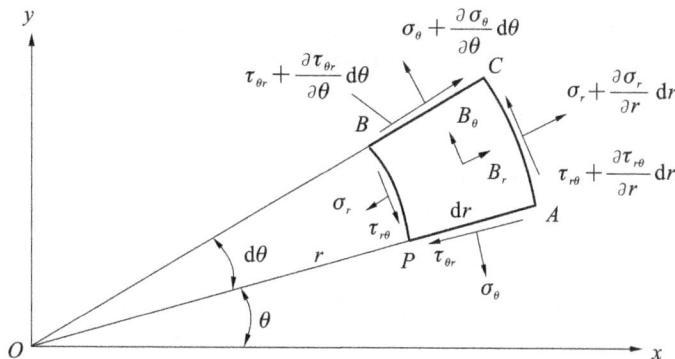

图 5-2　极坐标下微单元受力分析

为了表明极坐标中的应力分量,取出微单元体 $PACB$,如图 5-2 所示。σ_r 表示沿 r 方向的正应力,称为径向正应力,σ_θ 表示沿 θ 方向的正应力,称为切向正应力;$\tau_{r\theta}$ 及 $\tau_{\theta r}$ 表示剪应力,根据剪应力的互等定理,$\tau_{r\theta}=\tau_{\theta r}$。各应力分量的正负号规定和直角坐标系一致。图中所示的应力分量都是正的。径向及切向的体力分量分别用 B_r 及 B_θ 表示。

与直角坐标系中相似,由于应力随坐标 r 变化,设 PB 面上的径向正应力为 σ_r,则 AC 面上应力为 $\sigma_r + \frac{\partial \sigma_r}{\partial r} dr$。同样,这两个面上的剪应力分别为 $\tau_{r\theta}$ 及 $\tau_{r\theta} + \frac{\partial \tau_{r\theta}}{\partial r} dr$。$PA$ 及 BC 两个面上的切向正应力分别为 σ_θ 及 $\sigma_\theta + \frac{\partial \sigma_\theta}{\partial \theta} d\theta$,剪应力分别为 $\tau_{\theta r}$ 及 $\tau_{\theta r} + \frac{\partial \tau_{\theta r}}{\partial \theta} d\theta$。与直角坐标系不同之处有两点:一是单元体两侧弧形面积 PB 及 AC 不等;二是两切向平面 PA 及 CB 不平行。

取微单元体的厚度等于 1,于是,PB 及 AC 两面的面积分别等于 $r d\theta$ 及 $(r+dr) d\theta$,PA 及 BC 两面的面积等于 dr,微单元体的体积等于 $r d\theta dr$。由于 $d\theta$ 是微小的,可以取 $\sin \frac{d\theta}{2}$ 为 $\frac{d\theta}{2}$,取 $\cos \frac{d\theta}{2}$ 为 1。

将微单元体所受各力投影到微单元体中心的径向轴上,列出径向的平衡方程,得

$$\left(\sigma_r + \frac{\partial \sigma_r}{\partial r} dr\right)(r+dr)d\theta - \sigma_r r d\theta - \left(\sigma_\theta + \frac{\partial \sigma_\theta}{\partial \theta} d\theta\right) dr \frac{d\theta}{2} - \sigma_\theta dr \frac{d\theta}{2} +$$
$$\left(\tau_{\theta r} + \frac{\partial \tau_{\theta r}}{\partial \theta} d\theta\right) dr - \tau_{\theta r} dr + B_r r d\theta dr = 0 \tag{5-11}$$

用 $\tau_{r\theta}$ 代替 $\tau_{\theta r}$,简化以后,除以 $r d\theta dr$,再略去高阶项,得

$$\frac{\partial \sigma_r}{\partial r} + \frac{1}{r} \frac{\partial \tau_{r\theta}}{\partial \theta} + \frac{\sigma_r - \sigma_\theta}{r} + B_r = 0 \tag{5-12}$$

将所有力投影到微单元体中心的切向轴上,列出切向的平衡方程,得

$$\left(\sigma_\theta + \frac{\partial \sigma_\theta}{\partial \theta} d\theta\right) dr - \sigma_\theta dr + \left(\tau_{r\theta} + \frac{\partial \tau_{r\theta}}{\partial r} dr\right)(r+dr)d\theta - \tau_{r\theta} r d\theta +$$
$$\left(\tau_{\theta r} + \frac{\partial \tau_{\theta r}}{\partial \theta} d\theta\right) dr \frac{d\theta}{2} - \tau_{\theta r} dr \frac{d\theta}{2} + B_r r d\theta dr = 0 \tag{5-13}$$

用 $\tau_{r\theta}$ 代替 $\tau_{\theta r}$,简化以后,除以 $r d\theta dr$,再略去高阶项,得

$$\frac{1}{r} \frac{\partial \sigma_\theta}{\partial \theta} + \frac{\partial \tau_{r\theta}}{\partial r} + \frac{2\tau_{r\theta}}{r} + B_\theta = 0 \tag{5-14}$$

因此,二维极坐标系下的平衡方程如式(5-15)所示。

$$\begin{cases} \dfrac{\partial \sigma_r}{\partial r} + \dfrac{1}{r} \dfrac{\partial \tau_{r\theta}}{\partial \theta} + \dfrac{\sigma_r - \sigma_\theta}{r} + B_r = 0 \\[3mm] \dfrac{1}{r} \dfrac{\partial \sigma_\theta}{\partial \theta} + \dfrac{\partial \tau_{r\theta}}{\partial r} + \dfrac{2\tau_{r\theta}}{r} + B_\theta = 0 \end{cases} \tag{5-15}$$

5.4 边界值问题

很明显,当物体内部的单元体受到给定的体力分布,同时边界上受到给定的表面约束时,该单元体便会处于一个确定的、唯一的应力和应变状态。摆在我们面前的体力分布和表面约束问题,被称为第一类边界值问题。

同样,我们还可以推测出,当物体内部的单元体受到给定的体力分布,同时边界上受到一个给定的位移时,该单元体便会处于一个确定的、唯一的应力和应变状态。摆在我们面前的体力分布和表面位移问题,被称为第二类边界值问题。

在本书中,我们只考虑第一类边界值问题。

我们现在来推导表面约束的公式。为此,我们引入矢量 T 来表示边界上某一点处单位面积上的力。如图 5-3 所示,微小边界表面单元体 ABC 上的表面合力为

$$F = T dA \tag{5-16}$$

我们需要清楚知道的是,由于 T 同时包含单元体表面上的摩擦力和正应力,所以 T 不一定和矢量 dA 共线。接下来我们将得出 T 与接近边界的物体内部应力的关系,其中 T 作为第一类边界值问题中的已知函数。为此,我们对图 5-3 中的单元体应用牛顿定律,现将其放大成如图 5-4 所示,并将坐标平面上的应力表示出来。在 x 方向上对力求和,可得到

$$T_x ABC = \sigma_x COB + \tau_{xy} COA + \tau_{xz} BOA \tag{5-17}$$

用 l,m 和 n 表示平面 ABC 外法线的方向余弦,那么我们可以分别用 $lABC,mABC$ 和 $nABC$ 来替换 COB,COA 和 BOA,再同时除以 ABC,就得到

$$T_x = \sigma_x l + \tau_{xy} m + \tau_{xz} n \tag{5-18}$$

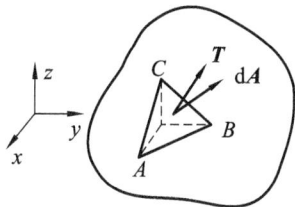

图 5-3 接近边界的一个单元体   图 5-4 放大的单元体

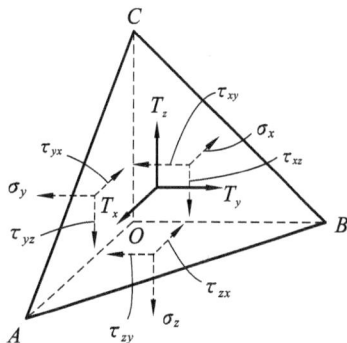

通过类似的方法处理其他坐标轴,并利用剪应力互等定理,我们可以得出以下方程:

$$T_x = \sigma_x l + \tau_{xy} m + \tau_{xz} n \tag{5-19a}$$

$$T_y = \tau_{yx} l + \sigma_y m + \tau_{yz} n \tag{5-19b}$$

$$T_z = \tau_{zx} l + \tau_{zy} m + \sigma_z n \tag{5-19c}$$

上述方程就是所谓的边界条件,对于给定的问题必须满足这些条件。此外,我们须指出的是,当整个物体为自由体时,T 和 B 不能是任意指定的,它们之间的关系需要满足平衡条件(这就回到了刚体静力学)。

5.5 分析方法

我们已经考虑了连续介质力学的基本定律,并且确定对于目前正在考虑的无摩擦、等温的弹性体分析只需要符合牛顿定律即可。此外,我们得到了基于小变形的几何方程,最终我们还得到了线弹性材料的本构定律。对于具有如前列举的所有性能和几何限制的材料,我们提出了可以求解的两类边界值问题。接下来我们会简单探讨如何着手解决这些问题。在现阶段的学习中,可以概括出三种求解方法。

(1)对具有特定边界值问题的相关微分方程直接进行积分。由于所涉及方程复杂,这种"直接"求解法仅适用于相对简单的情况。

(2)选择满足相关微分方程可能的应力或应变分布函数,并且检查一般边界条件,得到可以用函数表达的边界值问题,这就是所谓的逆解法。通过这种方法得到几个简单解,然后进行叠加,形成我们实际关注的问题的解。

(3)将方法(1)与方法(2)结合,这就是所谓的半逆解法。选择应力、应变公式的某部分,而其他部分直接采用积分求得。

有时,对于在指定荷载下的某些几何体,我们可以很容易地获得解析解。根据这些解,我们有时可

以对一般荷载作用下的变形性质做出合理的假设。利用这些近似的方法,结合部分基本定律和本构定律,我们可以建立一些非常有用和有价值的公式。例如,我们对梁、轴、板、壳等结构的处理就是如此。通过这种方式,形成了一个广泛的知识体系,我们称之为材料力学。了解材料力学公式使用范围并进行计算,我们可以对许多工程问题进行快速且精确的计算。

我们将根据弹性理论来分析材料力学的一些基本假设。根据弹性理论求解材料力学问题,从某些角度来说,我们将具有看起来更简单、更方便的公式。也就是说,我们将有办法更真实地了解这些简单公式的应用范围以及预期误差。

5.6　圣维南原理

解决刚体力学的问题时我们发现,当力分布在一个小区域时,采用集中力的概念是有效的。另外,我们处理问题时会采用分布力的刚体合力系统,这样可以得出合理、准确且直接的解。

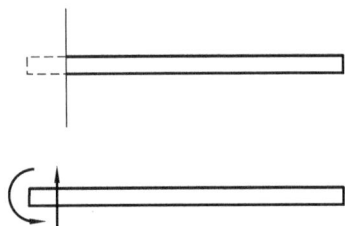

类似于这个过程,当研究物体弹性行为时,我们采用圣维南原理。该原理表明,如果将物体上施加的荷载变换成另一种基于刚体力学角度的等效荷载,那么与施加荷载合理距离外的应力、应变也不会有明显变化。我们可以将第二种荷载称为静力等效荷载,如图 5-5 所示。我们认为对于距离边界表面约束合理远处力学行为的影响仅依赖于该表面荷载的刚体合力。根据这一原理,为了简化支座右侧区域的应力、应变的计算,我们可以用图 5-5 中所示的单个力和力偶来代替墙体对悬臂梁施加的复杂支反力。

图 5-5　悬臂梁及静力等效荷载

本书中的许多问题都可以使用圣维南原理来帮助简化计算,并且不会出现严重的误差。

5.7　唯一性条件

我们现在将研究关于第一类和第二类边界值问题数学解的唯一性这一重要问题。假设对于某时刻任一种边界值问题都有两种可能的解,我们把这些解以如下形式给出

解 1：$\qquad(\tau_{ij})_1,(u_i)_1$

解 2：$\qquad(\tau_{ij})_2,(u_i)_2$ $\qquad$ (5-20)

假定叠加原理成立,我们可以将这些可能的变形叠加,得到一种新的可能变形。相应地,我们可以将新解 τ_{ij} 和 u_i 表示如下

$$\tau_{ij} = (\tau_{ij})_1 - (\tau_{ij})_2 \tag{5-21a}$$

$$u_i = (u_i)_1 - (u_i)_2 \tag{5-21b}$$

显然,这个新解将对应于一个体力分布为零的问题。进一步看,如果解 1 和解 2 对应第一类边界值问题,那么新解对应表面约束为 0 的情况,如果解 1 和解 2 对应第二类边界值问题,那么新解就对应边界位移为 0 的情况。

对于物体而言,如果体力和表面约束消失,或者体力和表面位移消失,将意味着应力和应变同时消失,那么对于新解,我们必然可以推断出

$$\tau_{ij} = 0 \tag{5-22a}$$

$$u_i = 0 \tag{5-22b}$$

相应地，

$$(\tau_{ij})_1 = (\tau_{ij})_2 \tag{5-23a}$$

$$(u_i)_1 = (u_i)_2 \tag{5-23b}$$

在这种情况下，我们可以看到前面提出的两个解确实是相同的，从而证明了唯一性条件。

　　现在，我们更仔细地思考叠加原理，因为它与我们目前的讨论有关。对于大变形情况，我们已经在第3章中明确说明了叠加原理是不成立的。然而，需要指出的是，即使在小变形的情况下叠加原理也可能失效。例如，虽然外部荷载作用下产生的变形很小，但其产生的应力分布明显不同于物体初始形状未发生变化而产生的应力分布。如图5-6中的柱体应力就会受变形δ的影响，尽管这个变形可能非常小。在这种情况下，荷载增量对应力或变形的影响按理应该取决于已经施加的荷载值。而实际上我们并不能像在证明中所做的那样，即根据每个作用在原始几何体上的荷载增量产生的变形来计算最终变形。接下来我们将会发现，对于这些问题，我们会得到一系列的数学解，而并非唯一的解。

　　在我们的证明过程中，有一个条件是被默认的，那就是我们假定有效位移场$(u_i)_1$和$(u_i)_2$是单值函数，它们相减后得到另外一个单值函数u_i。我们在第3章中已经指出，对于空间中一点完全固定的单连体，相容方程成立是确保位移场函数u_i单值性的充分条件。然而对于多连体，为了确保位移场是单值的，就必须满足相容方程和其他一些要求。在这里，我们不加证明地指出，一旦位移场的单值性条件得到确保，唯一性定理就可以使用，且不需要考虑物体的连体形式。

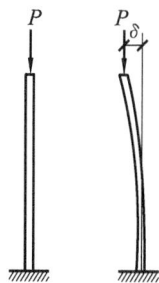

图 5-6　受到变形 δ 影响的柱体应力

5.8　重力作用下悬挂棱柱杆的弹性力学问题

　　我们现在将使用基本方程和圣维南原理来分析一个简单的问题。考虑一个悬挂的均质直棱柱杆，受重力作用，如图5-7所示。体力分布仅来源于重力，可由下式给出

$$\boldsymbol{B} = -\gamma \boldsymbol{k} \tag{5-24}$$

表面约束由暴露在大气中的棱柱杆所有面上受到的均匀大气压力和合力为γaLt的支反力构成。如果在不太接近支承的区域内，根据圣维南原理，我们可以利用该合力进行求解，且不需要确切知道与此相关的应力分布情况。因此，如果我们想计算远离支承区域内的物体内部的应力、应变和位移，这就属于第一类边界值问题。可能的应力分布如下：

$$\sigma_z = \gamma z \quad \sigma_x = \sigma_y = \tau_{xy} = \tau_{xz} = \tau_{yz} = 0 \tag{5-25}$$

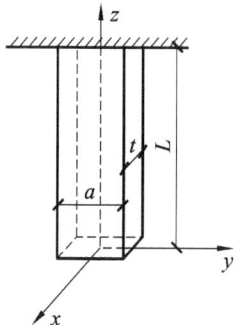

很明显，这样一种应力分布是满足平衡方程(5-3)的。接下来，我们分析该问题的边界条件。如果忽略大气压力的影响，我们可以发现，暴露在大气中的那部分杆的表面约束在所有点上都为零。根据方程(5-19)，我们发现除了应力σ_z外，所有方程右侧的应力均为零。所以，我们在这里仅需考虑方程(5-19c)。那么，我们可以得到

$$T_z = 0 = \sigma_z n \tag{5-26}$$

观察图5-7，我们看到平行于z轴的各杆表面的方向余弦n为零。根据方程(5-25)，我们可以看到，在杆的底面也就是$z=0$

图 5-7　重力作用下的均质直棱柱杆

处,$\sigma_z=0$。因此,对于所有暴露在外的表面都满足边界条件。在上部边界,我们从方程(5-25)中得到唯一的非零应力,$\sigma_z=\gamma L$,那么该上部边界面上均匀应力的合力为

$$\int_A \sigma_z \mathrm{d}A = \int_A \gamma L \mathrm{d}A = \gamma aLt \tag{5-27}$$

这就是该截面上的合力,并且我们发现以上提出的应力分布也满足上部边界条件。

接下来,我们利用应力-应变关系确定应变分布。假设在弹性范围内,可以得到

$$\begin{cases} \varepsilon_{xx}=\dfrac{1}{E}(-\nu\gamma z) \\[2mm] \varepsilon_{yy}=\dfrac{1}{E}(-\nu\gamma z) \quad \gamma_{xy}=\gamma_{xz}=\gamma_{yz}=0 \\[2mm] \varepsilon_{zz}=\dfrac{1}{E}\gamma z \end{cases} \tag{5-28}$$

然后,我们检查该应变分布是否满足相容方程(5-9)。将上述结果代入相容方程,可以很容易地看到是满足相容方程的,从而确保在这种情况下应变来自单值、连续的位移场。

因此可以保证,至少在远离支承的区域内,对于正在研究的问题,我们已经得到了唯一解。

为了完成分析,现在我们为这个问题建立一个位移场。根据方程(5-7)和方程(5-28),我们得到

$$\frac{\partial u_x}{\partial x}=-\frac{\nu\gamma z}{E} \tag{5-29a}$$

$$\frac{\partial u_y}{\partial y}=-\frac{\nu\gamma z}{E} \tag{5-29b}$$

$$\frac{\partial u_z}{\partial z}=\frac{\gamma z}{E} \tag{5-29c}$$

$$\frac{\partial u_x}{\partial y}+\frac{\partial u_y}{\partial x}=0 \tag{5-29d}$$

$$\frac{\partial u_y}{\partial z}+\frac{\partial u_z}{\partial y}=0 \tag{5-29e}$$

$$\frac{\partial u_x}{\partial z}+\frac{\partial u_z}{\partial x}=0 \tag{5-29f}$$

首先对方程(5-29c)进行积分,得到

$$u_z=\frac{\gamma z^2}{2E}+f(x,y) \tag{5-30}$$

其中 $f(x,y)$ 是关于坐标 x 和 y 的任意函数。将上述结果代入方程(5-29e)和方程(5-29f),我们得到

$$\frac{\partial u_y}{\partial z}=-\frac{\partial f}{\partial y} \tag{5-31a}$$

$$\frac{\partial u_x}{\partial z}=-\frac{\partial f}{\partial x} \tag{5-31b}$$

对这些方程进行积分,注意函数 f 只是关于 x 和 y 的函数。因此

$$u_y=-\frac{\partial f}{\partial y}z+g(x,y) \tag{5-32a}$$

$$u_x=-\frac{\partial f}{\partial x}z+h(x,y) \tag{5-32b}$$

其中函数 g 和 h 是关于坐标 x 和 y 的两个任意函数。现在将上述结果代入方程(5-29a)和方程(5-29b),得到

$$-\frac{\partial^2 f}{\partial x^2}z+\frac{\partial h}{\partial x}=-\frac{\nu\gamma z}{E} \tag{5-33a}$$

$$-\frac{\partial^2 f}{\partial y^2}z+\frac{\partial g}{\partial y}=-\frac{\nu\gamma z}{E} \tag{5-33b}$$

由于函数 f,g 和 h 在任何情况下都与 z 无关,故从前面的方程中我们可以推导出以下关系:

$$\frac{\partial h}{\partial x}=0 \tag{5-34a}$$

$$\frac{\partial g}{\partial y}=0 \tag{5-34b}$$

$$\frac{\partial^2 f}{\partial x^2}=\frac{\nu\gamma}{E} \tag{5-34c}$$

$$\frac{\partial^2 f}{\partial y^2}=\frac{\nu\gamma}{E} \tag{5-34d}$$

现在,还没有考虑方程(5-29d)。将方程(5-32)代入方程(5-29d)中,可得到

$$-\frac{\partial^2 f}{\partial y\partial x}z+\frac{\partial h}{\partial y}-\frac{\partial^2 f}{\partial x\partial y}z+\frac{\partial g}{\partial x}=0 \tag{5-35}$$

合并同类项后有

$$-2\frac{\partial^2 f}{\partial y\partial x}z+\frac{\partial h}{\partial y}+\frac{\partial g}{\partial x}=0 \tag{5-36}$$

由于 $\partial h/\partial y$ 和 $\partial g/\partial x$ 仅是关于 x 和 y 的函数,所以我们可以从方程(5-36)得到以下结果:

$$\frac{\partial h}{\partial y}+\frac{\partial g}{\partial x}=0 \tag{5-37a}$$

$$\frac{\partial^2 f}{\partial y\partial x}=0 \tag{5-37b}$$

现在,根据方程(5-34a)和方程(5-34b),我们将函数 g 和 h 表示为

$$\begin{cases}g=C_1\alpha(x)+C_2\\ h=C_3\beta(y)+C_4\end{cases} \tag{5-38}$$

式中,C_1、C_2、C_3 和 C_4 为任意常数;α 和 β 分别是关于 x 和 y 的函数。

方程(5-37a)需要满足

$$C_3\frac{\mathrm{d}\beta(y)}{\mathrm{d}y}+C_1\frac{\mathrm{d}\alpha(x)}{\mathrm{d}x}=0 \tag{5-39}$$

如果 $\beta(y)=y,\alpha(x)=x$,且 $C_3=-C_1$,那么,前面的方程是成立的。因此,函数 g 和 h 表示为

$$g=C_1 x+C_2 \tag{5-40a}$$

$$h=-C_1 y+C_4 \tag{5-40b}$$

根据方程(5-34c)、方程(5-34d)和方程(5-37b),f 可写成以下形式:

$$f=\frac{\nu\gamma}{2E}(x^2+y^2)+C_5 x+C_6 y+C_7 \tag{5-41}$$

可以用 6 个任意常数给出位移场,通过代入来进行验证。将方程(5-40)和方程(5-41)代入方程(5-30)和方程(5-32),可以得到

$$u_x=-\frac{\nu\gamma}{E}xz-C_5 z-C_1 y+C_4 \tag{5-42a}$$

$$u_y=-\frac{\nu\gamma}{E}yz-C_6 z+C_1 x+C_2 \tag{5-42b}$$

$$u_z=\frac{\gamma z^2}{2E}+\frac{\nu\gamma}{2E}(x^2+y^2)+C_5 x+C_6 y+C_7 \tag{5-42c}$$

通过确保该棱柱杆不发生刚体平移或转动的条件,求解这 6 个积分常数。对于位置坐标为 $x=0$,$y=0,z=L$ 的点,有 $u_x=u_y=u_z=0$。那么,所有常数应该满足以下条件

$$\begin{cases}-C_5 L+C_4=0\\ -C_6 L+C_2=0\\ \dfrac{\gamma L^2}{2E}+C_7=0\end{cases} \tag{5-43}$$

此外,在该点处不存在转动,也就是说,对于点$(0,0,L)$有

$$\begin{cases} \Phi_x = \dfrac{1}{2}\left(\dfrac{\partial u_z}{\partial y} - \dfrac{\partial u_y}{\partial z}\right)_{(0,0,L)} = 0 \\[2mm] \Phi_y = \dfrac{1}{2}\left(\dfrac{\partial u_x}{\partial z} - \dfrac{\partial u_z}{\partial x}\right)_{(0,0,L)} = 0 \\[2mm] \Phi_z = \dfrac{1}{2}\left(\dfrac{\partial u_y}{\partial x} - \dfrac{\partial u_x}{\partial y}\right)_{(0,0,L)} = 0 \end{cases} \tag{5-44}$$

因此,对于这些常数我们得到了以下附加条件:

$$\begin{cases} C_5 = -C_6 \\ -C_5 = C_5 \\ C_1 = 0 \end{cases} \tag{5-45}$$

可得 $C_1 = C_5 = C_6 = 0$,回到方程(5-43),得到

$$\begin{cases} C_4 = 0 \\ C_2 = 0 \\ C_7 = -\dfrac{\gamma L^2}{2E} \end{cases} \tag{5-46}$$

使用这些常量值可以确保物体不发生刚体转动,那么位移场变为

$$u_x = -\frac{\nu\gamma}{E}xz \tag{5-47a}$$

$$u_y = -\frac{\nu\gamma}{E}yz \tag{5-47b}$$

$$u_z = \frac{\gamma z^2}{2E} + \frac{\nu\gamma}{2E}(x^2 + y^2) - \frac{\gamma}{2E}L^2 \tag{5-47c}$$

注意,沿 z 轴方向上点只有竖向位移,其表达式为

$$u_z = \frac{\gamma}{2E}(z^2 - L^2) \tag{5-48}$$

其他点由于该构件收缩会产生水平位移。作为练习,你可以证明棱柱杆表面由水平面变为抛物面。

5.9 结　语

在本章中,我们已经阐明了等温的线弹性体的基本定律。我们假定材料的性质不会随温度发生显著变化。此外,我们还假定外部荷载在物体内部产生的应力不会影响温度。

如果我们想要在研究中考虑与温度相关的性能,或荷载导致温度变化这些因素,我们的工作将变得非常困难。除了考虑牛顿定律外,我们还必须认真考虑热力学第一定律。该定律包括热传递、变形能等。此外,还需要考虑材料的状态方程。然而,与流体不同,虽然物理学家与工程师们正在积极地开展这类研究,但我们对固体的状态方程知之甚少。

如果考虑大变形情况,我们就必须对正应变和剪应变重新进行定义,而且我们基本不能使用本章中的胡克定律。这些研究已经远超本书所涉及的范围。

下一章中,我们将运用弹性理论去分析一些关于平面应变与平面应力的问题。

习题

5-1　给定以下应力场

$$\sigma_x = 80x^3 + y \text{ Pa} \qquad \tau_{xy} = 1000 + 100y^2 \text{ Pa}$$

$$\sigma_y = 100x^3 + 1600 \text{ Pa} \quad \tau_{yz} = 0 \text{ Pa}$$

$$\sigma_z = 90y^2 + 100z^3 \text{ Pa} \qquad \tau_{xz} = xz^3 + 100x^2 y \text{ Pa}$$

平衡时所需的体力分布是什么？点$(1,1,5)$处的应力和体力分别是多少？

5-2　在习题 5-1 中，位置$(2,2,5)$的应变是多少（取 $E = 2.1 \times 10^5 \text{ MPa}; \nu = 0.3$）？

5-3　习题 5-1 给出的应力分布满足相容方程吗？

5-4　写出平面应力情况下的平衡方程。

5-5　平面应力的情况下，给定的极坐标下的平面应力分布如下：

$$\sigma_r = \frac{(p_o - p_i)a^2 b^2}{b^2 - a^2}\frac{1}{r^2} + \frac{p_i a^2 - p_o b^2}{b^2 - a^2}$$

$$\sigma_\theta = -\frac{(p_o - p_i)a^2 b^2}{b^2 - a^2}\frac{1}{r^2} + \frac{p_i a^2 - p_o b^2}{b^2 - a^2}$$

$$\tau_{r\theta} = 0$$

证明其满足平衡方程。现在考虑边界条件，证明这种分布对应于在内外表面受到均匀压力的厚壁长圆筒中的应力分布，如图 5-8 所示。

5-6　在习题 5-5 中，取

$$p_o = 3.5 \text{ MPa} \qquad a = 600 \text{ mm}$$

$$p_i = 0.35 \text{ MPa} \quad b = 1200 \text{ mm}$$

按弹性理论预测的最大应力是多少？假设整个圆筒的切向应力是均匀分布的，将求得的最大应力与薄壁圆筒理论计算的应力进行比较。

5-7　写出适用于平面应力情况的胡克定律。首先用应变表示关于应力的函数，接着得到基于应变来表示应力的适用于平面应力情况的胡克定律。

5-8　给定以下应力分布

$$\sigma_x = C_1$$

$$\sigma_y = C_2$$

$$\tau_{xy} = C_3$$

$$\sigma_z = \tau_{xz} = \tau_{yz} = 0$$

其中 C_1、C_2、C_3 为常数。这是一个什么样边界值问题的解？

5-9　给定以下应力分布

$$\sigma_x = C_1 x + C_2 y$$

$$\sigma_y = C_3 x + C_4 y$$

$$\tau_{xy} = C_5 x + C_6 y$$

$$\sigma_z = \tau_{xz} = \tau_{yz} = 0$$

在什么条件下，这种应力分布满足弹性力学基本方程？

5-10　在习题 5-9 中，选取除 C_1 外所有常数等于 0，对应于这种应力状态的边界值问题是什么？

5-11　考虑图 5-9 所示的端部加载的悬臂梁。假设平面应力如下：

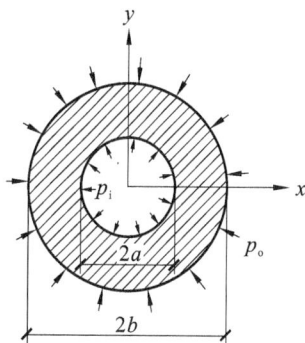

图 5-8　习题 5-5 图

$$\sigma_x = C_1 xy$$
$$\sigma_y = 0$$
$$\tau_{xy} = C_2 + C_3 y^2$$

请使用圣维南原理确定满足边界条件与牛顿定律的常数 C_1,C_2 和 C_3,并确定该问题的应变。证明计算所得的应变分布不满足其中一个相容方程(因此,上述解不是精确解。然而,我们将在下一章中可以看到,这是一个很好的近似解)。

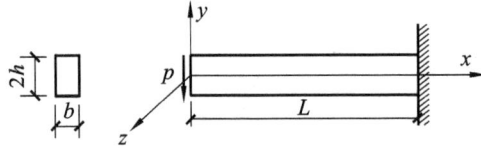

图 5-9　习题 5-11 图

6 弹性理论中平面应变和平面应力问题

6.1 引　　言

在第 5 章中,我们介绍了弹性力学的基本方程,并求解了一个三维线弹性问题。现在,我们想要用弹性理论来解决另外几类问题。首先,我们将讨论平面应变问题,可以看到,利用基本方程可以顺利地求解这类问题。接着,我们将讨论平面应力问题。对于此问题,严格的分析过程是非常复杂的,超出了本书的范围。但是,我们可以通过弹性理论建立近似的求解过程。

6.2　艾　里　函　数

在第 3 章中,我们将平面应变视为一类应变状态,对于某轴方向,通常选择 z 轴,相关区域内应变 γ_{xx},γ_{yz} 和 ε_{zz} 为零。在下列条件下,远离端部约束的棱柱体部分可视为处于平面应变状态。

(1)棱柱体的轴与 z 轴方向一致,表面约束和体力都垂直于该轴。

(2)表面约束和体力仅是关于 x 和 y 的函数,且合力为零,以免产生弯曲变形。

因此,在远离刚性约束的区域中,第 3 章中所示的大坝可以被视为处于平面应变状态。此外,对于两端受约束,并受到内压或外压的厚壁圆筒(图 6-1),在远离端部的区域内可以被视为平面应变问题(我们将在后面研究该问题)。现在,我们考虑在上述限制条件下平面应变的基本方程。

在这个情况下,我们先考虑胡克定律,有

$$\varepsilon_{xx} = \frac{1}{E}\left[\sigma_x - \nu(\sigma_y + \sigma_z)\right] \qquad (6\text{-}1a)$$

$$\varepsilon_{yy} = \frac{1}{E}\left[\sigma_y - \nu(\sigma_x + \sigma_z)\right] \qquad (6\text{-}1b)$$

$$0 = \frac{1}{E}\left[\sigma_z - \nu(\sigma_x + \sigma_y)\right] \qquad (6\text{-}1c)$$

$$\gamma_{xy} = \frac{1}{G}\tau_{xy} \qquad \tau_{xz} = \tau_{yz} = 0 \qquad (6\text{-}1d)$$

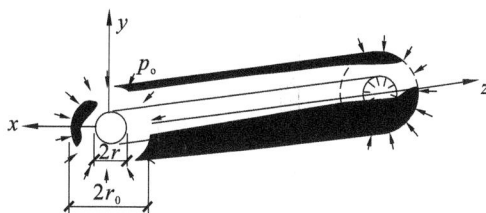

图 6-1　两端受约束的厚壁圆筒

根据方程(6-1c)得到

$$\sigma_z = \nu(\sigma_x + \sigma_y) \qquad (6\text{-}2)$$

再考虑牛顿定律,有

$$\frac{\partial \sigma_x}{\partial x} + \frac{\partial \tau_{xy}}{\partial y} + B_x = 0 \qquad (6\text{-}3a)$$

$$\frac{\partial \tau_{yx}}{\partial x}+\frac{\partial \sigma_y}{\partial y}+B_y=0 \qquad (6\text{-}3\text{b})$$

$$\frac{\partial \sigma_z}{\partial z}=0 \qquad (6\text{-}3\text{c})$$

注意,由于上述条件(1)中对体力的设定,体力分量 B_z 为零。方程(6-3c)表明,σ_z 不是关于 z 的函数。根据对该问题施加的条件,我们可以进一步推断出 σ_x,σ_y 和 τ_{xy} 也不是大于 z 的函数。现在,我们将进一步阐明体力分布是保守力系的条件,即

$$\boldsymbol{B}=-\mathrm{grad}\ \boldsymbol{V} \qquad (6\text{-}4)$$

其中 $\boldsymbol{V}$ 是所谓的势能函数。那么,这种情况下牛顿定律方程变成

$$\frac{\partial \sigma_x}{\partial x}+\frac{\partial \tau_{xy}}{\partial y}-\frac{\partial V}{\partial x}=0 \qquad (6\text{-}5\text{a})$$

$$\frac{\partial \tau_{yx}}{\partial x}+\frac{\partial \sigma_y}{\partial y}-\frac{\partial V}{\partial y}=0 \qquad (6\text{-}5\text{b})$$

$$\sigma_z=f(x,y) \qquad (6\text{-}5\text{c})$$

我们已确定我们所分析的所有应力 τ_{ij},或是常数,或是关于坐标 x 和 y 的函数。那么根据胡克定律,应变项 ε_{ij} 也应具有相同的条件。当考虑相容方程(5-9)时,除了其中一个方程不成立外,其他所有方程都是恒成立的。这个方程是

$$\frac{\partial^2 \varepsilon_{xx}}{\partial y^2}+\frac{\partial^2 \varepsilon_{yy}}{\partial x^2}=\frac{\partial^2 \gamma_{xy}}{\partial x \partial y} \qquad (6\text{-}6)$$

为了得到关于应力的方程,我们将方程(6-1)中的应变代入方程(6-6),可得到

$$\frac{1}{E}\frac{\partial^2 \sigma_x}{\partial y^2}-\frac{\nu}{E}\frac{\partial^2 \sigma_y}{\partial y^2}-\frac{\nu}{E}\frac{\partial^2 \sigma_z}{\partial y^2}+\frac{1}{E}\frac{\partial^2 \sigma_y}{\partial x^2}-\frac{\nu}{E}\frac{\partial^2 \sigma_x}{\partial x^2}-\frac{\nu}{E}\frac{\partial^2 \sigma_z}{\partial x^2}=\frac{1}{G}\frac{\partial^2 \tau_{xy}}{\partial x \partial y} \qquad (6\text{-}7)$$

在上述方程的右侧,我们将 G 用第 4 章中得到的关系替换,有

$$G=\frac{E}{2(1+\nu)} \qquad (6\text{-}8)$$

此外,我们还可以用以下方式替换方程(6-7)中的 $\partial^2 \tau_{xy}/\partial x \partial y$。先将方程(6-5a)对 x 求偏导,再将方程(6-5b)对 y 求偏导。最后,将两个方程相加并求解 $\partial^2 \tau_{xy}/\partial x \partial y$,即

$$\frac{\partial^2 \tau_{xy}}{\partial x \partial y}=\frac{1}{2}\left(\frac{\partial^2 V}{\partial x^2}+\frac{\partial^2 V}{\partial y^2}-\frac{\partial^2 \sigma_x}{\partial x^2}-\frac{\partial^2 \sigma_y}{\partial y^2}\right) \qquad (6\text{-}9)$$

现在将方程(6-8)和方程(6-9)代入方程(6-7)中,消去 $1/E$ 后再重新排列,得到

$$\boldsymbol{\nabla}^2(\sigma_x+\sigma_y)=\frac{1}{1-\nu}\boldsymbol{\nabla}^2 V \qquad (6\text{-}10)$$

我们引入一个函数 Φ,称之为艾里函数,用如下稍间接的方式来定义它。

$$\sigma_x=V+\frac{\partial^2 \Phi}{\partial y^2} \qquad (6\text{-}11\text{a})$$

$$\sigma_y=V+\frac{\partial^2 \Phi}{\partial x^2} \qquad (6\text{-}11\text{b})$$

$$\tau_{xy}=-\frac{\partial^2 \Phi}{\partial x \partial y} \qquad (6\text{-}11\text{c})$$

将这些应力代入方程(6-5a)和方程(6-5b)中,我们发现这些方程是恒等的。因此,当把应力写成上述形式时,平衡方程会自动满足。现在,将这些刚给出的应力公式代入方程(6-10)中,得到

$$\boldsymbol{\nabla}^2\left(\frac{\partial^2 \Phi}{\partial x^2}+\frac{\partial^2 \Phi}{\partial y^2}\right)=-\frac{1-2\nu}{1-\nu}\boldsymbol{\nabla}^2 V \qquad (6\text{-}12)$$

方程(6-12)可以写成以下形式

$$\boldsymbol{\nabla}^4 \Phi=-\frac{1-2\nu}{1-\nu}\boldsymbol{\nabla}^2 V \qquad (6\text{-}13)$$

式中，$\mathbf{\nabla}^4$ 称为拉普拉斯算子，在二维坐标空间中定义为

$$\mathbf{\nabla}^4 = \mathbf{\nabla}^2(\mathbf{\nabla}^2) = \frac{\partial^4}{\partial x^4} + 2\frac{\partial^4}{\partial x^2 \partial y^2} + \frac{\partial^4}{\partial y^4} \tag{6-14}$$

如果不考虑体力，方程(6-13)变为

$$\mathbf{\nabla}^4 \Phi = 0 \tag{6-15}$$

我们称之为双调和方程。

引入艾里函数得到了什么？如果不考虑体力，我们现在仅需要通过方程(6-15)求解艾里函数，而不需要求解三个未知数 $\sigma_x, \sigma_y, \tau_{xy}$，这样就简化了我们的工作。在知道 Φ 的情况下，我们就可以通过方程(6-11)确定应力分布。

对于棱柱体横截面上平面应变问题的边界条件，满足下面给定的表面约束分量关系：

$$T_x = \sigma_x l + \tau_{xy} m \tag{6-16a}$$

$$T_y = \tau_{yx} l + \sigma_y m \tag{6-16b}$$

$$T_z = 0 = \sigma_z n \tag{6-16c}$$

由于 $n=0$，最后一个条件是恒成立的。将 $\sigma_x, \sigma_y, \tau_{xy}$ 用方程(6-11)中的各项代替，我们得到

$$\begin{cases} T_x = \left(V + \dfrac{\partial^2 \Phi}{\partial y^2}\right) l - \dfrac{\partial^2 \Phi}{\partial x \partial y} m \\[3mm] T_y = -\dfrac{\partial^2 \Phi}{\partial x \partial y} l + \left(V + \dfrac{\partial^2 \Phi}{\partial x^2}\right) m \end{cases} \tag{6-17}$$

不考虑体力，上述边界条件变成

$$\begin{cases} T_x = \dfrac{\partial^2 \Phi}{\partial y^2} l - \dfrac{\partial^2 \Phi}{\partial x \partial y} m \\[3mm] T_y = -\dfrac{\partial^2 \Phi}{\partial x \partial y} l + \dfrac{\partial^2 \Phi}{\partial x^2} m \end{cases} \tag{6-18}$$

6.3 柱坐标问题

平面应变中存在一些值得关注的问题，它们最好使用柱坐标而不是直角坐标进行研究。因此，我们将以柱坐标形式建立 6.2 节中所提出的基本方程。为了方便，我们不考虑体力。我们所需要的转换公式如下：

$$\begin{cases} x = r\cos\theta & r = (x^2 + y^2)^{\frac{1}{2}} \\[2mm] y = r\sin\theta & \theta = \arctan \dfrac{y}{x} \end{cases} \tag{6-19}$$

第一步，我们将通过应力 $\sigma_x, \sigma_y, \tau_{xy}$ 来表示 σ_r。因此我们有

$$\sigma_r = l^2 \sigma_x + m^2 \sigma_y + n^2 \sigma_z + 2(lm\tau_{yx} + mn\tau_{yz} + ln\tau_{zx}) \tag{6-20}$$

注意到 $n=0, l=\cos\theta$ 以及 $m=\sin\theta$，代入方程(6-20)得到

$$\sigma_r = \sigma_x \cos^2\theta + \sigma_y \sin^2\theta + 2\tau_{xy}\sin\theta\cos\theta \tag{6-21}$$

在不考虑体力的情况下，我们将方程(6-21)右边的应力用方程(6-11)代替。

$$\sigma_r = \frac{\partial^2 \Phi}{\partial y^2}\cos^2\theta + \frac{\partial^2 \Phi}{\partial x^2}\sin^2\theta - 2\frac{\partial^2 \Phi}{\partial x \partial y}\sin\theta\cos\theta \tag{6-22}$$

要使上述方程中的 σ_r 完全用柱坐标表示，就要计算函数 Φ 对柱坐标求偏导。因此，考虑到函数 Φ 与 z 无关，根据方程(6-19)，我们得到

$$\frac{\partial \Phi}{\partial x} = \frac{\partial \Phi}{\partial r}\frac{\partial r}{\partial x} + \frac{\partial \Phi}{\partial \theta}\frac{\partial \theta}{\partial x} = \frac{\partial \Phi}{\partial r}\frac{x}{r} + \frac{\partial \Phi}{\partial \theta}\left(-\frac{y}{r^2}\right) = \frac{\partial \Phi}{\partial r}\cos\theta - \frac{\partial \Phi}{\partial \theta}\frac{\sin\theta}{r}$$

$$\begin{aligned}\frac{\partial^2 \Phi}{\partial x^2} &= \left[\frac{\partial}{\partial r}\left(\frac{\partial \Phi}{\partial x}\right)\right]\frac{\partial r}{\partial x} + \left[\frac{\partial}{\partial \theta}\left(\frac{\partial \Phi}{\partial x}\right)\right]\frac{\partial \theta}{\partial x} \\ &= \left[\frac{\partial}{\partial r}\left(\frac{\partial \Phi}{\partial x}\right)\right]\cos\theta - \left[\frac{\partial}{\partial \theta}\left(\frac{\partial \Phi}{\partial x}\right)\right]\frac{\sin\theta}{r} \\ &= \frac{\partial^2 \Phi}{\partial r^2}\cos^2\theta - 2\frac{\partial^2 \Phi}{\partial r \partial \theta}\frac{\sin\theta\cos\theta}{r} + 2\frac{\partial \Phi}{\partial \theta}\frac{\sin\theta\cos\theta}{r^2} + \frac{\partial \Phi}{\partial r}\frac{\sin^2\theta}{r} + \frac{\partial^2 \Phi}{\partial \theta^2}\frac{\sin^2\theta}{r^2}\end{aligned} \tag{6-23}$$

通过类似的步骤,也可以得到

$$\frac{\partial^2 \Phi}{\partial y^2} = \frac{\partial^2 \Phi}{\partial r^2}\sin^2\theta + 2\frac{\partial^2 \Phi}{\partial r \partial \theta}\frac{\sin\theta\cos\theta}{r} - 2\frac{\partial \Phi}{\partial \theta}\frac{\sin\theta\cos\theta}{r^2} + \frac{\partial \Phi}{\partial r}\frac{\cos^2\theta}{r} + \frac{\partial^2 \Phi}{\partial \theta^2}\frac{\cos^2\theta}{r^2} \tag{6-24}$$

$$\frac{\partial^2 \Phi}{\partial x \partial y} = \frac{\partial^2 \Phi}{\partial r^2}\sin\theta\cos\theta + \frac{\partial^2 \Phi}{\partial r \partial \theta}\frac{1-2\sin^2\theta}{r} - \frac{\partial^2 \Phi}{\partial \theta^2}\frac{\sin\theta\cos\theta}{r^2} - \frac{\partial \Phi}{\partial r}\frac{\sin\theta\cos\theta}{r} - \frac{\partial \Phi}{\partial \theta}\frac{1-2\sin^2\theta}{r^2} \tag{6-25}$$

将方程(6-23)至方程(6-25)代入方程(6-22),并对各项进行重新排列和消除后,可以得到以下结果

$$\sigma_r = \frac{1}{r}\frac{\partial \Phi}{\partial r} + \frac{1}{r^2}\frac{\partial^2 \Phi}{\partial \theta^2} \tag{6-26}$$

通过类似步骤,也可以得到关于 σ_θ 和 $\tau_{r\theta}$ 的相应关系,现在我们给出这组关系。

$$\sigma_r = \frac{1}{r}\frac{\partial \Phi}{\partial r} + \frac{1}{r^2}\frac{\partial^2 \Phi}{\partial \theta^2} \tag{6-27a}$$

$$\sigma_\theta = \frac{\partial^2 \Phi}{\partial r^2} \tag{6-27b}$$

$$\tau_{r\theta} = \frac{1}{r^2}\frac{\partial \Phi}{\partial \theta} - \frac{1}{r}\frac{\partial^2 \Phi}{\partial r \partial \theta} \tag{6-27c}$$

通过在柱坐标系中引入拉普拉斯算子,我们可以得到在柱坐标系中的相容方程,表达式如下

$$\mathbf{\nabla}^4 \Phi = \mathbf{\nabla}^2(\mathbf{\nabla}^2 \Phi) = \left(\frac{\partial^2}{\partial r^2} + \frac{1}{r}\frac{\partial}{\partial r} + \frac{1}{r^2}\frac{\partial^2}{\partial \theta^2}\right)\left(\frac{\partial^2 \Phi}{\partial r^2} + \frac{1}{r}\frac{\partial \Phi}{\partial r} + \frac{1}{r^2}\frac{\partial^2 \Phi}{\partial \theta^2}\right) = 0 \tag{6-28}$$

现在,我们已经得到了柱坐标系中不考虑体力时的平面应变问题中的应力的基本方程。

6.4 轴对称应力分布

在本节中,我们将仅限于讨论关于 z 轴对称的问题。这就意味着对于所有关于参数 θ 的偏导数都为零。在不考虑体力的情况下,基本方程简化为以下形式

$$\sigma_r = \frac{1}{r}\frac{\partial \Phi}{\partial r} \tag{6-29a}$$

$$\sigma_\theta = \frac{\partial^2 \Phi}{\partial r^2} \tag{6-29b}$$

$$\tau_{r\theta} = 0 \tag{6-29c}$$

$$\mathbf{\nabla}^4 \Phi = \left(\frac{\partial^2}{\partial r^2} + \frac{1}{r}\frac{\partial}{\partial r}\right)\left(\frac{\partial^2 \Phi}{\partial r^2} + \frac{1}{r}\frac{\partial \Phi}{\partial r}\right) = \frac{\mathrm{d}^4 \Phi}{\mathrm{d}r^4} + \frac{2}{r}\frac{\mathrm{d}^3 \Phi}{\mathrm{d}r^3} - \frac{1}{r^2}\frac{\mathrm{d}^2 \Phi}{\mathrm{d}r^2} + \frac{1}{r^3}\frac{\mathrm{d}\Phi}{\mathrm{d}r} = 0 \tag{6-29d}$$

现在让我们考虑函数 Φ 的双调和微分方程[方程(6-29d)]。在微分方程课程中,这是著名的欧拉-柯西方程。通过对自变量进行以下变换,可以将该方程变换为常系数方程

$$r = \mathrm{e}^t \tag{6-30}$$

然后得到

$$\frac{\mathrm{d}\Phi}{\mathrm{d}r}=\frac{\mathrm{d}\Phi}{\mathrm{d}t}\frac{\mathrm{d}t}{\mathrm{d}r}=\mathrm{e}^{-t}\frac{\mathrm{d}\Phi}{\mathrm{d}t}$$

$$\frac{\mathrm{d}^2\Phi}{\mathrm{d}r^2}=\mathrm{e}^{-t}\frac{\mathrm{d}}{\mathrm{d}t}\Big(\mathrm{e}^{-t}\frac{\mathrm{d}\Phi}{\mathrm{d}t}\Big)=\mathrm{e}^{-t}\Big(\mathrm{e}^{-t}\frac{\mathrm{d}^2\Phi}{\mathrm{d}t^2}-\mathrm{e}^{-t}\frac{\mathrm{d}\Phi}{\mathrm{d}t}\Big)=\mathrm{e}^{-2t}\Big(\frac{\mathrm{d}^2\Phi}{\mathrm{d}t^2}-\frac{\mathrm{d}\Phi}{\mathrm{d}t}\Big)$$

继续推导，得到

$$\frac{\mathrm{d}^3\Phi}{\mathrm{d}r^3}=\mathrm{e}^{-3t}\Big(\frac{\mathrm{d}^3\Phi}{\mathrm{d}t^3}-3\frac{\mathrm{d}^2\Phi}{\mathrm{d}t^2}+2\frac{\mathrm{d}\Phi}{\mathrm{d}t}\Big)$$

$$\frac{\mathrm{d}^4\Phi}{\mathrm{d}r^4}=\mathrm{e}^{-4t}\Big(\frac{\mathrm{d}^4\Phi}{\mathrm{d}t^4}-6\frac{\mathrm{d}^3\Phi}{\mathrm{d}t^3}+11\frac{\mathrm{d}^2\Phi}{\mathrm{d}t^2}-6\frac{\mathrm{d}\Phi}{\mathrm{d}t}\Big)$$

现在把上述结果代入方程(6-29)，得到

$$\mathrm{e}^{-4t}\Big(\frac{\mathrm{d}^4\Phi}{\mathrm{d}t^4}-6\frac{\mathrm{d}^3\Phi}{\mathrm{d}t^3}+11\frac{\mathrm{d}^2\Phi}{\mathrm{d}t^2}-6\frac{\mathrm{d}\Phi}{\mathrm{d}t}\Big)+2\mathrm{e}^{-4t}\Big(\frac{\mathrm{d}^3\Phi}{\mathrm{d}t^3}-3\frac{\mathrm{d}^2\Phi}{\mathrm{d}t^2}+2\frac{\mathrm{d}\Phi}{\mathrm{d}t}\Big)-$$

$$\mathrm{e}^{-4t}\Big(\frac{\mathrm{d}^2\Phi}{\mathrm{d}t^2}-\frac{\mathrm{d}\Phi}{\mathrm{d}t}\Big)+\mathrm{e}^{-4t}\Big(\frac{\mathrm{d}\Phi}{\mathrm{d}t}\Big)=0$$

消去 e^{-4t} 并合并各项，得到下述常系数微分方程：

$$\frac{\mathrm{d}^4\Phi}{\mathrm{d}t^4}-4\frac{\mathrm{d}^3\Phi}{\mathrm{d}t^3}+4\frac{\mathrm{d}^2\Phi}{\mathrm{d}t^2}=0 \tag{6-31}$$

方程(6-31)的附属方程是

$$p^4-4p^3+4p^2=0 \tag{6-32}$$

我们可以将前面的方程因式分解为如下形式：

$$p^2(p-2)^2=0 \tag{6-33}$$

那么微分方程(6-31)的通解为

$$\Phi=C_1+C_2t+C_3\mathrm{e}^{2t}+C_4t\mathrm{e}^{2t} \tag{6-34}$$

其中有 4 个任意的积分常数。用方程(6-30)代替 t，那么就有了关于 z 轴对称情况下函数 Φ 的通解，即

$$\Phi=C_1+C_2\ln r+C_3r^2+C_4r^2\ln r \tag{6-35}$$

利用方程(6-29)，上述函数对应的应力分布为

$$\begin{cases}\sigma_r=\dfrac{C_2}{r^2}+2C_3+C_4(1+2\ln r)\\[2mm]\sigma_\theta=-\dfrac{C_2}{r^2}+2C_3+C_4(3+2\ln r)\\[2mm]\tau_{r\theta}=0\end{cases} \tag{6-36}$$

　　现在将研究厚壁圆筒(图 6-1)分别在内压 p_i 和外压 p_o 作用下的特殊情况的问题。内半径和外半径分别表示为 r_i 和 r_o。对于这个问题，边界条件[方程(6-17)]可以简化为下列形式。

$$-p_i=(\sigma_r)_{r_i} \tag{6-37a}$$

$$-p_o=(\sigma_r)_{r_o} \tag{6-37b}$$

将这些条件施加于方程(6-36)所给出的应力分布中，我们得到下列方程

$$-p_i=\frac{C_2}{r_i^2}+2C_3+C_4(1+2\ln r_i) \tag{6-38a}$$

$$-p_o=\frac{C_2}{r_o^2}+2C_3+C_4(1+2\ln r_o) \tag{6-38b}$$

到目前为止，我们得到的结论表明，用两个方程来计算三个常数将有无数组解。为了得到唯一解，我们考虑位移条件，那么常数 C_4 必须为零，然后我们可以解出常数 C_2 和 C_3。

$$C_2=\frac{r_i^2r_o^2(p_o-p_i)}{r_o^2-r_i^2} \tag{6-39a}$$

$$2C_3 = \frac{p_i r_i^2 - p_o r_o^2}{r_o^2 - r_i^2} \tag{6-39b}$$

因此,我们可以得到应力分布

$$\sigma_r = \frac{r_i^2 r_o^2 (p_o - p_i)}{r_o^2 - r_i^2} \frac{1}{r^2} + \frac{p_i r_i^2 - p_o r_o^2}{r_o^2 - r_i^2} \tag{6-40a}$$

$$\sigma_\theta = -\frac{r_i^2 r_o^2 (p_o - p_i)}{r_o^2 - r_i^2} \frac{1}{r^2} + \frac{p_i r_i^2 - p_o r_o^2}{r_o^2 - r_i^2} \tag{6-40b}$$

如果把上面的一对方程相加,我们会发现 σ_r 和 σ_θ 的和是一个常数,即

$$\sigma_r + \sigma_\theta = 2 \frac{p_i r_i^2 - p_o r_o^2}{r_o^2 - r_i^2} \tag{6-41}$$

但是根据胡克定律我们得出

$$\varepsilon_{zz} = 0 = \frac{1}{E} \left[\sigma_z - \nu(\sigma_r + \sigma_\theta) \right] \tag{6-42}$$

利用方程(6-41),我们可以知道对于这种情况,σ_z 一定是一个常数,其值如下

$$\sigma_z = \nu(\sigma_r + \sigma_\theta) = \frac{2\nu(p_i r_i^2 - p_o r_o^2)}{r_o^2 - r_i^2} \tag{6-43}$$

这就完成了对有约束的厚壁圆筒在内压和外压作用下的应力分布情况的讨论。在下一节中,我们将发现该解对具有内压和外压作用下的无约束的厚壁圆筒也是适用的。为了证明这一点,下一步我们将研究平面应力理论。

6.5　基本方程的讨论

你可能记得在第 2 章中我们定义平面应力分布为

$$\tau_{zx} = \tau_{yz} = \sigma_z = 0 \tag{6-44}$$

我们曾经指出过,与上述应力公式接近的最简单的物理问题是对称面上施加荷载的板,如图 6-2 所示。

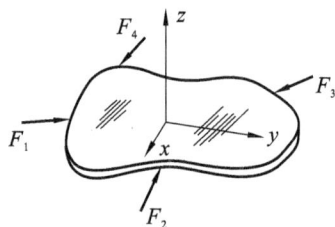

图 6-2　对称面上施加荷载的板

为了处理这些问题,我们将进一步假设,非零应力不随垂直于板的 z 坐标变化而变化。因此,应力被认为仅仅是关于 x 和 y 的函数。现在我们写出这个应力状态下的牛顿定律方程。

$$\frac{\partial \sigma_x}{\partial x} + \frac{\partial \tau_{xy}}{\partial y} + B_x = 0 \tag{6-45a}$$

$$\frac{\partial \tau_{yx}}{\partial x} + \frac{\partial \sigma_y}{\partial y} + B_y = 0 \tag{6-45b}$$

$$B_z = 0 \tag{6-45c}$$

如果我们依然像前文一样,要求体力场是一个保守场,则方程(6-45)变为

$$\frac{\partial \sigma_x}{\partial x} + \frac{\partial \tau_{xy}}{\partial y} - \frac{\partial V}{\partial x} = 0 \tag{6-46a}$$

$$\frac{\partial \tau_{yx}}{\partial x} + \frac{\partial \sigma_y}{\partial y} - \frac{\partial V}{\partial y} = 0 \tag{6-46b}$$

$$\frac{\partial V}{\partial z} = 0 \tag{6-46c}$$

从方程(6-46c)中我们可以明显看出,函数 V 一定是一个仅关于 x 和 y 的函数。同时,这些方程与平面应变问题中对应的方程是相同的,因此我们可以用艾里函数表示的应力来满足这些方程。那么,像之前一样,我们得到

$$\begin{cases} \sigma_x = \dfrac{\partial^2 \Phi}{\partial y^2} + V \\[2mm] \sigma_y = \dfrac{\partial^2 \Phi}{\partial x^2} + V \\[2mm] \tau_{xy} = -\dfrac{\partial^2 \Phi}{\partial x \partial y} \end{cases} \tag{6-47}$$

对于应变,利用广义胡克定律,有

$$\begin{cases} \varepsilon_{xx} = \dfrac{1}{E}(\sigma_x - \nu \sigma_y) \\[2mm] \varepsilon_{yy} = \dfrac{1}{E}(\sigma_y - \nu \sigma_x) \\[2mm] \varepsilon_{zz} = -\dfrac{\nu}{E}(\sigma_x + \sigma_y) \\[2mm] \gamma_{xy} = \dfrac{1}{G}\tau_{xy} \quad \gamma_{xz} = 0 \quad \gamma_{yz} = 0 \end{cases} \tag{6-48}$$

我们现在可以得出结论,应变也与 z 坐标无关。将方程(6-47)代入方程(6-48),可以得到用函数 Φ 和 V 表示的应变,即

$$\begin{cases} \varepsilon_{xx} = \dfrac{1}{E}\left[\left(\dfrac{\partial^2 \Phi}{\partial y^2} - \nu \dfrac{\partial^2 \Phi}{\partial x^2}\right) + (1-\nu)V\right] \\[3mm] \varepsilon_{yy} = \dfrac{1}{E}\left[\left(\dfrac{\partial^2 \Phi}{\partial x^2} - \nu \dfrac{\partial^2 \Phi}{\partial y^2}\right) + (1-\nu)V\right] \\[3mm] \varepsilon_{zz} = -\dfrac{\nu}{E}\left[\left(\dfrac{\partial^2 \Phi}{\partial x^2} + \dfrac{\partial^2 \Phi}{\partial y^2}\right) + 2V\right] \\[3mm] \gamma_{xy} = -\dfrac{1}{G}\dfrac{\partial^2 \Phi}{\partial x \partial y} \\[3mm] \gamma_{xz} = \gamma_{yz} = 0 \end{cases} \tag{6-49}$$

现在让我们回到相容方程(3-97)。首先检查相容方程的第一个式子,将上述方程代入,有

$$\frac{1}{E}\left[\frac{\partial^4 \Phi}{\partial y^4} - \nu \frac{\partial^4 \Phi}{\partial y^2 \partial x^2} + (1-\nu)\frac{\partial^2 V}{\partial y^2} + \frac{\partial^4 \Phi}{\partial x^4} - \nu \frac{\partial^4 \Phi}{\partial x^2 \partial y^2} + (1-\nu)\frac{\partial^2 V}{\partial x^2}\right] = -\frac{1}{G}\frac{\partial^4 \Phi}{\partial y^2 \partial x^2}$$

方程两边同乘 E,并根据方程(6-8),用 $2(1+\nu)$ 替换 E/G,合并同类项后,得到

$$\frac{\partial^4 \Phi}{\partial x^4} + 2\frac{\partial^4 \Phi}{\partial y^2 \partial x^2} + \frac{\partial^4 \Phi}{\partial y^4} = -(1-\nu)\boldsymbol{\nabla}^2 V \tag{6-50}$$

然而,并非所有剩余的相容方程都能像在平面应变问题中一样恒满足。因此,相容方程(3-97)中(b)、(c)和(f)需要满足

$$\frac{\partial^4 \Phi}{\partial y^4} + \frac{\partial^4 \Phi}{\partial y^2 \partial x^2} + 2\frac{\partial^2 V}{\partial y^2} = 0 \tag{6-51a}$$

$$\frac{\partial^4 \Phi}{\partial x^4} + \frac{\partial^4 \Phi}{\partial x^2 \partial y^2} + 2\frac{\partial^2 V}{\partial x^2} = 0 \tag{6-51b}$$

$$\frac{\partial^4 \Phi}{\partial y \partial x^3} + \frac{\partial^4 \Phi}{\partial x \partial y^3} + 2\frac{\partial^2 V}{\partial x \partial y} = 0 \tag{6-51c}$$

可以得出结论,平面应力问题比平面应变问题更难。然而,结果表明,如果忽略方程(6-51)给出的关于函数 Φ 的相容要求,仅考虑方程(6-50)给出的相容要求,我们就可以得到薄板应力分布较接近的近似解。即得到以下方程:

$$\sigma_x = \frac{\partial^2 \Phi}{\partial y^2} + V \tag{6-52a}$$

$$\sigma_y = \frac{\partial^2 \Phi}{\partial x^2} + V \tag{6-52b}$$

$$\tau_{xy} = -\frac{\partial^2 \Phi}{\partial x \partial y} \tag{6-52c}$$

$$\nabla^4 \Phi = -(1-\nu)\nabla^2 V \tag{6-52d}$$

注意,在不考虑体力的情况下,平面应变和平面应力问题的方程完全一样,这两种情况下函数 Φ 都需满足双调和方程。因此,在不考虑体力的情况下,根据这些方程,我们可以得到关于平面应变问题的一个精确解,以及关于平面应力问题的一个近似解。后面我们将利用这个结论。最后,在上述柱坐标中得到的与平面应变有关的结论也适用于不考虑体力的平面应力问题。

6.6 带孔平板的平面应力求解

在平面应变状态下,对于上述方程,我们已经给出了一组用柱坐标表示的解。方程(6-40)所研究的应力分布情况与沿轴线被完全约束的厚壁圆筒相对应。对于厚壁圆筒来说,平面应变要求 $\varepsilon_{zz} = 0$,由此计算出 σ_z,其值为一个常数。

从我们之前的解释来看,利用方程(6-40)得到的解是一个特定平面应力问题较接近的近似解。毫无疑问,在充分考虑边界条件后,对应的平面应力问题就是一个带有小孔的薄圆盘,其在内部和外部的表面上分别受到压力 p_i 和压力 p_o。简言之,可以发现我们所研究的平面应力问题就是一个厚壁圆筒上的切片,其切片的垂直方向完全不存在约束,只在边界上承受压力。对于两端受约束的厚壁圆筒,$\varepsilon_{zz} = 0$,且 σ_z 是常数。对于带有孔的薄圆盘,有 $\sigma_z = 0$,且根据胡克定律,ε_{zz} 是一个常数,即

$$\varepsilon_{zz} = \frac{1}{E}\left[-\nu(\sigma_r + \sigma_\theta)\right] \tag{6-53}$$

将方程(6-40)中的 σ_r 和 σ_θ 代入,得

$$\varepsilon_{zz} = \frac{2\nu}{E}\frac{p_o r_o^2 - p_i r_i^2}{r_o^2 - r_i^2} \tag{6-54}$$

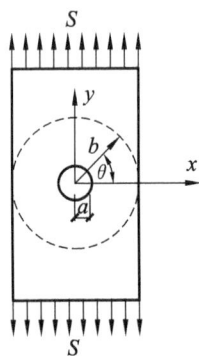

现在,我们研究第二个相关问题。考虑一个受拉伸荷载的薄板,如图 6-3 所示,在这个板的中心有一个小孔,要求计算这种几何物体的应力分布。简单起见,我们假设作用在板端部的拉伸荷载为均匀应力,记作 S,如图 6-3 所示。

如果板中没有孔,我们将得到一个均匀的应力场,有 $\sigma_y = S$,$\sigma_x = \tau_{yx} = 0$。小孔的出现将导致小孔附近的应力分布不均匀,但根据圣维南原理,在远离小孔的区域内的应力接近预设的均匀应力值。针对这一点,我们考虑一个半径为 b 用虚线表示的假想大圆内的区域,如图 6-3 所示,并采用极坐标。为了表示在 $r=b$ 处的 σ_r 和 $\tau_{r\theta}$,我们可以采用方程(2-26)给出的平面应力的变换公式。因此,考虑 r 方向与 x 的对应关系,θ 方向与 y 的对应关系,可得到

图 6-3 承受拉伸荷载的薄板

$$(\sigma_r)_{r=b} = \frac{S}{2}(1-\cos 2\theta) \tag{6-55a}$$

$$(\tau_{r\theta})_{r=b} = \frac{S}{2}\sin 2\theta \tag{6-55b}$$

以上边界条件可以很方便地看作由两部分组成。第一部分是均匀的径向应力,其值为 $S/2$,对于这一部分,我们把同心圆之间的区域看作无约束厚壁圆筒的切片,其内压 $p_i = 0$,外压 $p_o = -S/2$。因此,可以使用前面得到的厚壁圆筒的解来处理这个边界条件。剩余的边界应力变成了薄的带孔圆盘,且该圆盘在外

边缘存在可变的径向正应力$-(S/2)\cos2\theta$和可变剪应力$(S/2)\sin2\theta$,而其内边缘不存在径向应力和剪应力。这两个问题如图 6-4 所示。第一个问题的解可以从我们之前的结果中直接得到。因此,根据方程(6-40)可得

$$\sigma'_r = -\frac{a^2 b^2 \frac{S}{2}}{b^2-a^2}\frac{1}{r^2} + \frac{\frac{S}{2}b^2}{b^2-a^2} \tag{6-56a}$$

$$\sigma'_\theta = \frac{a^2 b^2 \frac{S}{2}}{b^2-a^2}\frac{1}{r^2} + \frac{\frac{S}{2}b^2}{b^2-a^2} \tag{6-56b}$$

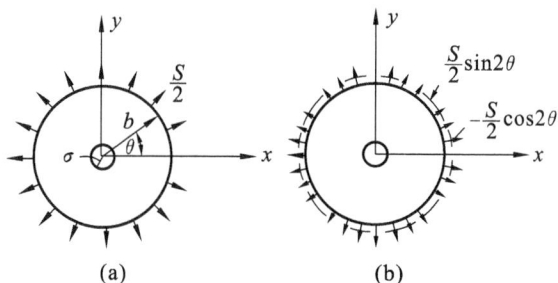

图 6-4 半径为 b 的假想大圆内的区域的扩大

现在我们讨论第二个问题。首先,假定对于这个问题,艾里函数可采用如下形式:

$$\Phi = [f(r)]\cos2\theta \tag{6-57}$$

式中,$f(r)$是关于 r 的待定函数。这种情况下我们可以使用相容方程(6-28)。将前面的函数 Φ 代入相容方程,得到

$$\left(\frac{\partial^2}{\partial r^2} + \frac{1}{r}\frac{\partial}{\partial r} + \frac{1}{r^2}\frac{\partial^2}{\partial\theta^2}\right)\cos2\theta\left(\frac{\partial^2 f}{\partial r^2} + \frac{1}{r}\frac{\partial f}{\partial r} - \frac{4}{r^2}f\right) = 0 \tag{6-58}$$

对方程(6-58)进一步微分,并合并同类项,得到

$$\frac{\mathrm{d}^4 f}{\mathrm{d}r^4} + \frac{2}{r}\frac{\mathrm{d}^3 f}{\mathrm{d}r^3} - \frac{9}{r^2}\frac{\mathrm{d}^2 f}{\mathrm{d}r^2} + \frac{9}{r^3}\frac{\mathrm{d}f}{\mathrm{d}r} = 0 \tag{6-59}$$

于是我们再次得到了欧拉-柯西方程。下面的步骤与 6.4 节中开头的步骤一样。因此,我们将其留作习题,让你来证明上述方程的通解为

$$f(r) = C_1 + C_2 r^2 + C_3 r^4 + \frac{C_4}{r^2} \tag{6-60}$$

那么,应力函数就变为

$$\Phi = \left(C_1 + C_2 r^2 + C_3 r^4 + \frac{C_4}{r^2}\right)\cos2\theta \tag{6-61}$$

现在将上述函数 Φ 代入方程(6-27)以确定相应的应力。合并同类项后,有

$$\begin{cases} \sigma_r = -\left(2C_2 + \frac{4C_1}{r^2} + \frac{6C_4}{r^4}\right)\cos2\theta \\ \sigma_\theta = \left(2C_2 + 12C_3 r^2 + \frac{6C_4}{r^4}\right)\cos2\theta \\ \tau_{r\theta} = \left(2C_2 + 6C_3 r^2 - \frac{2C_1}{r^2} - \frac{6C_4}{r^4}\right)\sin2\theta \end{cases} \tag{6-62}$$

让这个应力分布来满足这个问题的边界条件。因此

当 $r=a$ 时

$$\sigma_r = \tau_{r\theta} = 0$$

当 $r=b$ 时

$$
\begin{cases}
\sigma_r = -\dfrac{S}{2}\cos2\theta \\[2mm]
\tau_{r\theta} = \dfrac{S}{2}\sin2\theta
\end{cases}
\tag{6-63}
$$

那么,我们得到

$$
\frac{4}{a^2}C_1 + 2C_2 + 0 + \frac{6}{a^4}C_4 = 0
\tag{6-64a}
$$

$$
\frac{4}{b^2}C_1 + 2C_2 + 0 + \frac{6}{b^4}C_4 = \frac{S}{2}
\tag{6-64b}
$$

$$
-\frac{2}{a^2}C_1 + 2C_2 + 6a^2 C_3 - \frac{6}{a^4}C_4 = 0
\tag{6-64c}
$$

$$
-\frac{2}{b^2}C_1 + 2C_2 + 6b^2 C_3 - \frac{6}{b^4}C_4 = \frac{S}{2}
\tag{6-64d}
$$

用纯代数方法求解这 4 个积分常数,结果如下:

$$
\begin{cases}
C_1 = \dfrac{72S\left(\dfrac{a^2}{b^2}-\dfrac{b^2}{a^4}\right)}{-\dfrac{576}{b^4}+\dfrac{864}{a^2 b^2}+144\dfrac{a^2}{b^6}+144\dfrac{b^2}{a^6}-\dfrac{576}{a^4}} \\[8mm]
C_2 = \dfrac{36S\left(\dfrac{4}{b^4}+\dfrac{3}{a^2 b^2}+\dfrac{b^2}{a^6}\right)}{-\dfrac{576}{b^4}+\dfrac{864}{a^2 b^2}+144\dfrac{a^2}{b^6}+144\dfrac{b^2}{a^6}-\dfrac{576}{a^4}} \\[8mm]
C_3 = \dfrac{24S\left(\dfrac{1}{a^6}-\dfrac{1}{b^2 a^4}\right)}{-\dfrac{576}{b^4}+\dfrac{864}{a^2 b^2}+144\dfrac{a^2}{b^6}+144\dfrac{b^2}{a^6}-\dfrac{576}{a^4}} \\[8mm]
C_4 = \dfrac{36S\left(\dfrac{b^2}{a^2}-\dfrac{a^2}{b^2}\right)}{-\dfrac{576}{b^4}+\dfrac{864}{a^2 b^2}+144\dfrac{a^2}{b^6}+144\dfrac{b^2}{a^6}-\dfrac{576}{a^4}}
\end{cases}
\tag{6-65}
$$

接下来,考虑 $a/b=0$ 的情况,也就是半径趋向无穷大的情况,于是,我们就得到了这 4 个积分常数:

$$
C_1 = -\frac{a^2}{2}S \quad C_2 = \frac{S}{4} \quad C_3 = 0 \quad C_4 = \frac{a^4}{4}S
\tag{6-66}
$$

那么,第二个相关问题的解是

$$
\begin{cases}
\sigma_r'' = \dfrac{S}{2}\left(-1+4a^2\dfrac{1}{r^2}-3a^4\dfrac{1}{r^4}\right)\cos2\theta \\[2mm]
\sigma_\theta'' = \dfrac{S}{2}\left(1+3a^4\dfrac{1}{r^4}\right)\cos2\theta \\[2mm]
\tau_{r\theta}'' = \dfrac{S}{2}\left(1+2a^2\dfrac{1}{r^2}-3a^4\dfrac{1}{r^4}\right)\sin2\theta
\end{cases}
\tag{6-67}
$$

现在我们可以通过组合两个相关问题的解来解决目前的问题,首先,调整方程(6-56),使其公式中半径 $b\rightarrow\infty$。再将方程两边同时除以 b^2,并令 $a/b=0$,我们得到

$$
\begin{cases}
\sigma_r' = \dfrac{S}{2}\left(1-\dfrac{a^2}{r^2}\right) \\[2mm]
\sigma_\theta' = \dfrac{S}{2}\left(1+\dfrac{a^2}{r^2}\right) \\[2mm]
\tau_{r\theta}' = 0
\end{cases}
\tag{6-68}
$$

因此,总解就是

$$\begin{cases} \sigma_r = \sigma_r' + \sigma_r'' = \dfrac{S}{2}\left[\left(1 - \dfrac{a^2}{r^2}\right) + \left(-1 + 4\dfrac{a^2}{r^2} - 3\dfrac{a^4}{r^4}\right)\cos 2\theta\right] \\[2mm] \sigma_\theta = \sigma_\theta' + \sigma_\theta'' = \dfrac{S}{2}\left[\left(1 + \dfrac{a^2}{r^2}\right) + \left(1 + 3\dfrac{a^4}{r^4}\right)\cos 2\theta\right] \\[2mm] \tau_{r\theta} = \tau_{r\theta}' + \tau_{r\theta}'' = \dfrac{S}{2}\left(1 + 2\dfrac{a^2}{r^2} - 3\dfrac{a^4}{r^4}\right)\sin 2\theta \end{cases} \tag{6-69}$$

我们再来研究上述远离小孔的区域和小孔本身的应力分布情况。注意,当远离小孔时,我们可以在前面的方程中去掉分母为 r 的项。因此,当 r 变大时,应力分布接近以下应力状态:

$$\begin{cases} \sigma_r = \dfrac{S}{2}(1 - \cos 2\theta) \\[2mm] \sigma_\theta = \dfrac{S}{2}(1 + \cos 2\theta) \\[2mm] \tau_{r\theta} = \dfrac{S}{2}\sin 2\theta \end{cases} \tag{6-70}$$

该应力状态与均匀应力场,即 $\sigma_x = 0, \sigma_y = S, \tau_{xy} = 0$ 相对应。接下来,考虑小孔处的应力状态。因此,在方程(6-69)中令 $r = a$,有

$$\begin{cases} (\sigma_r)_{r=a} = 0 \\[2mm] (\sigma_\theta)_{r=a} = S + 2S\cos 2\theta \\[2mm] (\tau_{r\theta})_{r=a} = 0 \end{cases} \tag{6-71}$$

径向和切向应力显然是整个小孔圆周上的主应力。最大正应力出现在 $\theta = 0$ 处。我们发现这时应力 $\sigma_x = 3S$。这是我们讨论疲劳时出现应力集中危险的一个生动例子。在外行人看来,一个小孔可能不存在危害,但在此处产生的应力是无孔状态下在此处产生最大应力的 3 倍。正如在先前讨论中我们所指出的那样,对于这种情况,我们必须十分小心、仔细地处理。

对于带孔平板问题,我们已经能够计算出应力集中系数 $K = 3$。因此,最大应力 $\sigma_{\max}$ 可以表示为

$$\sigma_{\max} = K\sigma = 3S \tag{6-72}$$

式中,σ 为无孔时的最大应力。

你可以在手册中找到其他常见情况下的应力集中系数表。这些值有些是根据理论计算的,但大多数是通过实验应力分析得到的。

6.7 曲梁弹性力学求解案例

在本章后面的习题中,你将有机会使用已经学习的方法来求解矩形截面直梁问题。在本节中,我们将对曲梁进行研究。该曲梁上下边缘由同心圆弧构成,且在梁两端受到纯力偶荷载作用,如图 6-5 所示。该梁横截面为矩形,与梁的其他部分尺寸相比,其厚度 t 较小。

通过考虑沿径向截面分离的曲梁各部分的自由体,如图 6-6 所示,根据平衡条件,我们可以得到外露内表面上的合力为纯力偶,其大小为 M。因此,在内截面上的合力不是关于 θ 的函数。而且,我们假设,这些外露面上的应力本身不会随 θ 变化。这是一个关于 O 的轴对称的平面应力问题。因为不考虑体力,我们可以利用在 6.4 节中提出的轴对称情况下双调和方程的一般解。因此,根据方程(6-36)所给出满足该问题边界条件的应力分布为

$$r = a \quad r = b \quad \sigma_r = 0 \tag{6-73a}$$

$$r = a \quad r = b \quad \tau_{r\theta} = 0 \tag{6-73b}$$

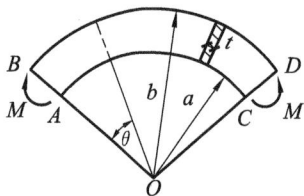

图 6-5　两端作用纯力偶的同心圆弧曲梁

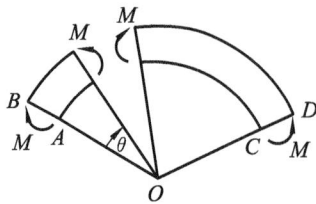

图 6-6　曲梁各部分的自由体

在端面

$$\int_A \sigma_\theta dA = 0 \tag{6-73c}$$

$$\int_A r\sigma_\theta dA = -M \tag{6-73d}$$

根据方程(6-73a)给出的条件,有

$$\frac{C_2}{a^2} + 2C_3 + C_4(1 + 2\ln a) = 0 \tag{6-74a}$$

$$\frac{C_2}{b^2} + 2C_3 + C_4(1 + 2\ln b) = 0 \tag{6-74b}$$

由于各点处 $\tau_{r\theta}=0$,方程(6-73b)给出的条件也是恒等的。接下来,我们检查由方程(6-73c)所给出的条件。我们可以代入方程(6-27b)表达如下:

$$\int_A \sigma_\theta dA = t\int_a^b \frac{d^2\Phi}{dr^2}dr = 0$$

对于上面的等式,消去 t 后得到

$$\int_a^b \frac{d^2\Phi}{dr^2}dr = \frac{d\Phi}{dr}\Big|_a^b = b\left[\frac{C_2}{b^2} + 2C_3 + C_4(1+2\ln b)\right] - a\left[\frac{C_2}{a^2} + 2C_3 + C_4(1+2\ln a)\right] = 0 \tag{6-75}$$

但这个方程没有增加任何新的未知数,因为只要方程(6-74)成立,这个方程就会成立。现在我们考虑方程(6-73d)给出的最后一个边界条件。再次使用方程(6-27b),我们得到

$$\int_A r\sigma_\theta dA = t\int_a^b r\frac{d^2\Phi}{dr^2}dr = -M \tag{6-76}$$

方程两边同时除以 t,然后分部积分,得到

$$\int_a^b r\frac{d^2\Phi}{dr^2}dr = \int_a^b r d\left(\frac{d\Phi}{dr}\right) = r\frac{d\Phi}{dr}\Big|_a^b - \int_a^b \frac{d\Phi}{dr}dr = -\frac{M}{t} \tag{6-77}$$

根据方程(6-74)和方程(6-75)可知,$d\Phi/dr$ 在梁上、下极限处为零,因此方程(6-77)变为

$$\int_a^b \frac{d\Phi}{dr}dr = \Phi\Big|_a^b = \frac{M}{t} \tag{6-78}$$

我们接着使用方程(6-35),得到

$$C_2\ln\left(\frac{b}{a}\right) + C_3(b^2-a^2) + C_4(b^2\ln b - a^2\ln a) = \frac{M}{t} \tag{6-79}$$

现在,我们利用方程(6-74a)、方程(6-74b)和方程(6-79)求解常数 C_2、C_3 和 C_4。得到以下结果:

$$C_2 = -\frac{4M}{St}a^2b^2\ln\frac{b}{a} \tag{6-80a}$$

$$C_3 = \frac{M}{St}[b^2-a^2 + 2(b^2\ln b - a^2\ln a)] \tag{6-80b}$$

$$C_4 = -\frac{2M}{St}(b^2-a^2) \tag{6-80c}$$

其中

$$S=(b^2-a^2)^2-4a^2b^2\left(\ln\frac{b}{a}\right)^2 \tag{6-80d}$$

将上述积分常数的值代入方程中,就可以得到如下某点应力状态的公式:

$$\begin{cases} \sigma_r=-\dfrac{4M}{St}\left(\dfrac{a^2b^2}{r^2}\ln\dfrac{b}{a}+b^2\ln\dfrac{r}{b}+a^2\ln\dfrac{a}{r}\right) \\ \sigma_\theta=-\dfrac{4M}{St}\left(-\dfrac{a^2b^2}{r^2}\ln\dfrac{b}{a}+b^2\ln\dfrac{r}{b}+a^2\ln\dfrac{a}{r}+b^2-a^2\right) \\ \tau_{r\theta}=0 \end{cases} \tag{6-81}$$

如果施加的力矩与该解相对应的应力分布相同,那么可以认为上述应力分布对于整个梁都是正确的。如果施加的力矩与该解相对应的应力分布不一致(通常是这种情况),根据圣维南原理,我们得到的结果在远离梁端的区域内是有效的。

在图 6-7 中,我们已经绘制了横截面上的应力情况。请注意,最大应力出现在较下面的纤维处。我们也画出了由材料力学近似方法计算得到的应力。这些应力将在后面的章节中讨论。应注意,更精确的分析得到的应力比应用材料力学简单公式计算得到的应力更大。如果$(b-a)$的值与a相比较并不小,可以预料到采用材料力学公式得到的应力可能会存在相当大的误差。这就是一个通过更精确的弹性理论去阐述更简单、更方便的材料力学公式的使用范围和精度的例子。

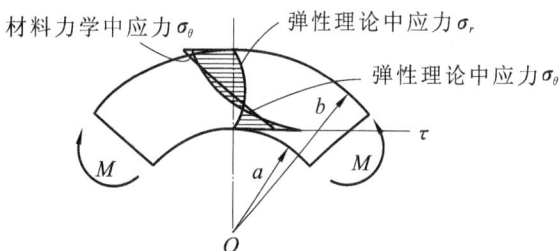

图 6-7 曲梁的应力

6.8 结　语

在本章中,我们主要关注的是弹性理论。在第 5 章中,我们使用该理论解决了一个三维问题,而在本章中,我们将该理论应用于平面应变与平面应力问题。许多其他相关的问题都可以借助该理论得以解决。

对弹性理论问题求解的简洁介绍表明,这个过程无论如何都不简单。由于我们研究的许多结构都是相对简单的物体形状,所以有更多可利用的方法,我们统称为材料力学。

通过回顾材料力学中的细节,我们可以注意到弹性理论与材料力学的关系。以下相关的评论可能有助于我们正确看待问题:

(1)在材料力学中,我们对各类物体的变形做了一些简单假设。通过运用弹性理论中的一些定律,来对这些假设进行检验,可以得到一些有用的计算公式。弹性理论使我们了解到材料力学中近似公式的有效范围和一般精度。

(2)材料力学中的公式有明确的应用范围。当问题超出应用范围时,我们必须求助于更加全面的理论,经常需要使用数值方法及计算机。因此,对被应用在过于简化的材料力学计算的集中系数 K,有时是通过更一般的理论得到的,就如本章对带小孔受拉板说明的那样。

(3)有大量的实验方法用于计算复杂几何形状中的应力和应变。这些方法包括光弹性法、应变测量方法、脆性漆法和各种有用的类似技术。为了能够有效地将这些复杂的方法应用于其他的琐碎问题,这要求我们需要具备弹性理论的知识。

习题

6-1 假定函数 Φ 为函数 f 和 g 的乘积,即

$$\Phi = fg$$

证明

$$\nabla^2\Phi = (\nabla^2 f)g + 2\left(\frac{\partial f}{\partial x}\frac{\partial g}{\partial x} + \frac{\partial f}{\partial y}\frac{\partial g}{\partial y}\right) + (\nabla^2 g)f$$

假设 g 是调和函数,且 $f = x$。证明 $\nabla^4\Phi = 0$,即 x 与一个调和函数相乘会得到双调和函数。

6-2 在习题 6-1 中,取函数 $f = r^2 = x^2 + y^2 + z^2$。

(1)证明

$$\nabla^2\Phi = 6g + 4\left(x\frac{\partial g}{\partial x} + y\frac{\partial g}{\partial y} + z\frac{\partial g}{\partial z}\right)$$

(2)证明 $\nabla^4\Phi = 0$。

即一个调和函数 g 与 r^2 的乘积是一个双调和函数。

6-3 考虑一个内径为 300 mm、外径为 450 mm 的厚壁圆筒。圆筒内保持 140 Pa 的压力。如果该圆筒体沿其轴线完全受约束,最大正应力是多少?最大剪应力是多少?

6-4 考虑一个外径为 1500 mm 的圆筒,设内径记为 d_i。如果圆筒内保持 350 Pa 的压力,若按薄壁圆筒理论计算得到的该圆筒最大应力的值与按弹性理论公式计算值相比,其误差在 10% 内,则内径 d_i 是多少?

6-5 证明方程(6-60)给出了微分方程(6-59)的一个一般解。

6-6 证明当 b 趋于无穷大时,方程(6-65)中的常数 C_1、C_2、C_3 和 C_4 可以简化为方程(6-66)中给出的常数。

6-7 有一块带有小孔的板,其受力状况如图 6-8 所示,证明该板的应力集中系数为 4。

6-8 考虑一个曲梁,如图 6-9 所示,其中

$$a = 1200 \text{ mm}$$
$$b = 1500 \text{ mm}$$
$$M = 1.40 \text{ kN} \cdot \text{m}$$
$$t = 90 \text{ mm}$$

确定由力偶作用产生的最大应力。将该结果与材料力学中的挠曲公式中的最大应力进行比较

$$\sigma_\theta = \frac{My}{I}$$

其中 y 是任意截面与中性轴的距离,I 是截面绕该轴的二次惯性矩。

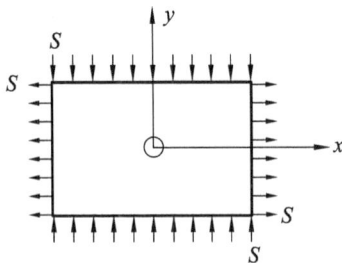

图 6-8 习题 6-7 图

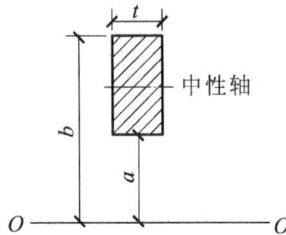

图 6-9 习题 6-8 图

6-9 给定多项式

$$\Phi = C_1 x^2 + C_2 xy + C_3 y^2$$

作为艾里函数,什么边界条件下,可以视它为平面应力和平面应变的解?

6-10 使用下列多项式作为艾里函数

$$\Phi = C_1 x^3 + C_2 x^2 y + C_3 x y^2 + C_4 y^3$$

试计算如图 6-10 所示的梁在纯弯矩作用下的常数 C_1, C_2, C_3, C_4。

6-11 使用习题 6-10 中给出的多项式,找到其他有解的边界值问题。

6-12 给定以下多项式

$$\Phi = C_1 x^4 + C_2 x y + C_3 x y^3 + C_4 y^4$$

通常对系数进行调整可以使之成为一个双调和函数。现在调整系数,以便能得到如图 6-11 所示的悬臂梁问题的解,其中剪应力均匀地分布在梁的上端和下端,在梁端施加集中荷载 P。

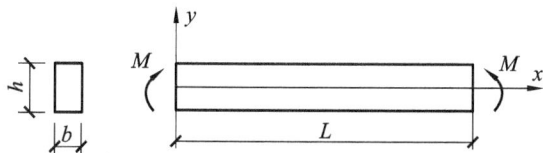

图 6-10 习题 6-10 图

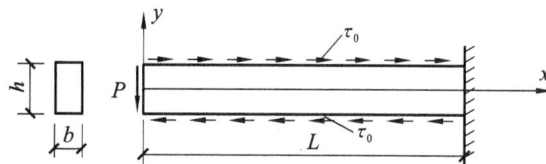

图 6-11 习题 6-12 图

7 塑性理论

7.1 引　言

弹性理论的发展具有两个基本特性:一个特性表明加载过程具有完全可逆性,那么,当导致物体产生变形的力被移除时,该物体会立即恢复到最初的未变形状态。另一个特性表明物体在荷载作用下的变形或应变仅取决于最终应力,而不取决于应力加载过程或应变路径。既然任何产生的应变都可由初始应力、最终应力以及特定的比例常数确定,那么弹性性能可被视为点函数。当物体出现塑性或永久变形时,以上两个特性则变得不再明显。

为了产生塑性变形或塑性流动,应力必须超过被称为屈服应力的应力水平。对于许多固体(如延性金属)材料,如果加载过程中产生的应力大大超过屈服应力,其变形或形状变化能够继续变化到很大程度。此外,当达到最终变形时,特定的应变单元体可能在达到其最终状态前经历过不同的加载过程。因此,它不仅在卸载时未发现像弹性一样的完全可逆性,而且最终应变取决于加载过程,而不只是初始应力和最终应力。这一发现意味着塑性性能是一个路径函数,确定总应变时需要将应变路径上的应变增量进行累加。

在塑性研究中,至少有三种不同的求解方法,分别是:

(1)数学求解法。采用材料性能的理想模型,并且主要关注满足规定边界条件的应力和应变分布。这可以适当地被称为宏观塑性理论,它与长期存在的弹性理论最为相似。

(2)金属物理学中使用的求解方法。真实固体中单晶体的变形方式构成研究的基础,其目的是揭示各单晶体基本性能的关系,并将单晶体基本性能推广到工程师们通常使用的组成固体的多晶体聚合物中,这被称为微观塑性理论。

(3)技术求解方法。通过引用某些现象学准则,将宏观尺度上通过实验观察到的真实固体性能与数学表达式结合起来。这可以使我们在一般的设计领域中做出一些有用的预测,它可以适当地被称为宏观工程塑性。本章我们将主要研究这种求解方法。

7.2　弹性与塑性比较

为方便起见,上述许多结论在课本中将以表格形式呈现。通过这种方式,可以表明这两种性能的主要差异。

因为屈服开始以及随后可能发生的行为是我们主要关注的问题,下面将使用不同的模型来说明所涉及的物理过程。对于后面提出的任何一种模型,都需要做出一些假设:

（1）固体是各向同性且均匀的。

（2）拉伸和压缩屈服点相同。这意味着不产生包申格效应。

（3）体积变化可以忽略不计。因此,体积膨胀为零,并且泊松比为 0.5。虽然该比值是一个弹性常数,但将其引入塑性变形中时,不会引起混淆。

（4）应力状态下的平均正应力或静水应力分量的大小不影响屈服。

（5）应变率的影响可以忽略不计。

（6）不考虑温度效应。

注意:假设(3)和假设(4)在许多延性金属的实验中得到合理的证实,但对许多固体聚合物并不适用。

7.3 塑性变形模型

7.3.1 理想刚塑性固体模型

理想刚塑性性能已被广泛应用于许多分析研究中。这意味着在达到一定的应力水平(E 为无穷大)前不会发生变形,那么只要必要的流动应力被施加,变形就会无限地进行。注意:这并不意味有关固体的潜在断裂现象。对应的理想模型如图 7-1 所示。

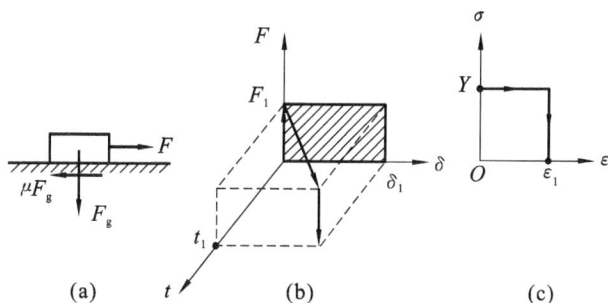

图 7-1 理想模型

(a)对理想刚塑性固体模型的描述;(b)力-位移-时间三维图;(c)应力-应变图

注意以下几点:

（1）随着施加荷载 F 的增加,在达到某个临界力 F_1 之前,物体不会发生位移,一旦荷载达到临界力 F_1,变形将随着时间不断进行。力 F_1 与屈服应力或流动应力 Y 直接相关。

（2）移除荷载 F_1 后,塑性功没有恢复(如 $F\text{-}\delta$ 平面中的阴影区域所示),而永久变形 δ_1 存在。

（3）在变形过程中,固体屈服应力不会提高。这意味着没有发生应变硬化效应。

7.3.2 具有线性硬化的刚塑性固体模型

具有线性硬化的刚塑性固体模型比前面的模型更符合实际,因为它结合了在许多固体中观察到的应变硬化的影响,尤其是延性金属,同样,在塑性变形开始之前必须达到一定的临界应力水平,但要持续变形需要继续施加应力,如图 7-2 所示。注意以下几点影响:

（1）仅当所施加荷载 F 达到临界力 F_0 时,变形才会开始,并产生初始流动应力 Y_0。

（2）变形仅在施加荷载 Y 不断增加的情况下才持续发生,其中 $Y=Y_0+f(\varepsilon)$,$f(\varepsilon)$ 与直线的斜率有关,注意这与模量 E 具有相似性。在该模型中,应变硬化现象出现,并且意味着塑性变形将会导致进一步变形所需的应力增加。

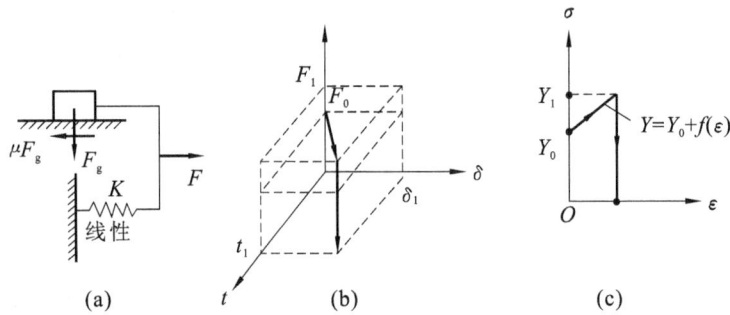

图 7-2　图 7-1 中所述的线性硬化刚塑性固体模型的描述

7.3.3　具有非线性硬化的刚塑性固体模型

具有幂律形式应变硬化的刚塑性固体模型,可以为许多固体材料提供更好的描述。图 7-3 中描述了这种模型。

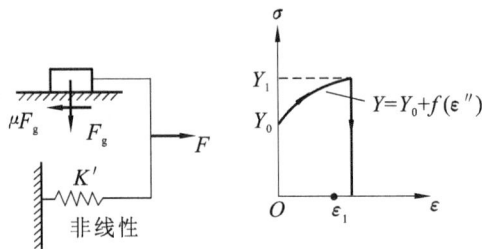

图 7-3　遵循幂律硬化行为的刚体应变硬化模型描述

这里需要注意的是,除了应变硬化以非线性速率发生外,该塑性性能与前面的模型性能相同,指数 n 大于 0 且小于 1。

最后,通过在上述三个模型的变形初始阶段增加直线段,就可以包含弹性效应,其中直线的斜率表示一定数值的弹性模量(小于无穷大)。因为许多情况下涉及的塑性应变值的量级比弹性应变值的量级要大,所以为了方便,忽略弹性应变。进行这样的处理,我们应该认识三个事实:

(1)体积变化只能通过包括 ν 小于 0.5 的弹性效应确定。忽略这些影响,我们可以引入体积不变的概念。

(2)加载后变形恢复只存在于弹性阶段。因此,如果直接关注这点,上述模型将不描述弹性行为。还要注意,在涉及弹性恢复效应的情况下,不断增加的弹性应变会伴随着不断增加的塑性流动。

(3)如果弹性应变和塑性应变的值是相同量级,除非上述模型包括弹性部分,否则上述模型将不再适用。

7.4　屈服轨迹与屈服面

假设材料是各向同性的,不存在包申格效应,塑性流动期间具有不可压缩性且屈服不受静水效应的影响,那么这些固有的条件必包含在预测屈服开始的任何准则中。

引入二维应力空间图说明上述假设的一些结果。在这里可以看到,将单个应力视为总应力的分量,为此可将其作为向量来处理。我们要明白,这与坐标轴的变换是无关的。在所有情况下,这种讨论仅限于有一个主应力为 0 的主应力空间,我们将使用 σ_1-σ_2 平面来绘图。假设一个方向上被施加拉伸应力,且

$0<\sigma_1<Y$,仅描述弹性行为,由于拉伸和压缩具有等效性,弹性范围可以扩大到$-Y<\sigma_1<Y$,并由于各向同性,所以$-Y<\sigma_2<Y$。因此,在σ_1-σ_2应力平面中存在四个点,表示屈服的开始,但为了得到一个可接受的屈服理论,就必须包括更多的复杂应力状态。这需要概括弹性范围和屈服点的含义,并使用某些应力极限。我们将在图7-4中展示这是如何开始的。

在二维应力空间中,这四个点位于$\pm Y$处且处于屈服轨迹上。现在假设材料应力达到如图7-4所示的点A,然后保持应力不变且同时增加应力σ_2。在某点处,如点B,弹性行为结束,我们将点B作为该应力空间中的屈服点。注意:沿着一个或两个方向上同时加载也可能沿着线OB进行,以至于在点B处再次发生屈服。因此,为了到达点B,我们可能有许多加载路径,在达到屈服点前,所有行为都是弹性的。利用多条加载路径,由产生的屈服点描述的轨迹被分成两部分,即弹性阶段(轨迹内部)和屈服开始阶段(即轨迹本身)。考虑含有应变硬化的模型,将意味着这种效应倾向于增加随后的屈服强度或新的流动应力。在本书中,我们假定任何这类趋势将以同一方式增大初始屈服轨迹,这被称为各向同性硬化。

在这里引入三维应力空间的概念是合适的。在图7-5中,假设在1、2和3坐标方向上作用的应力a、b和c的组合正好导致屈服,定义总的应力状态为σ,其中起点在原点,终点为屈服点。如果进行足够的实验,就可以得到所有这些点都将位于屈服表面上这一结论。由单个矢量如σ,其所描述的任何应力状态位于屈服表面内仅产生弹性效应,当矢量终点到达屈服表面时开始发生屈服。注意:屈服轨迹是通过将某主应力为常数的平面剖开屈服表面所得到的曲线(如前面研究$\sigma_3=0$时的情况)。

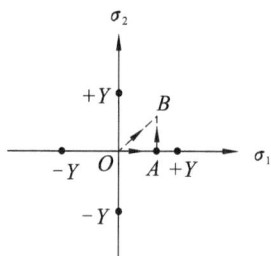

图 7-4 二维应力空间中的屈服点

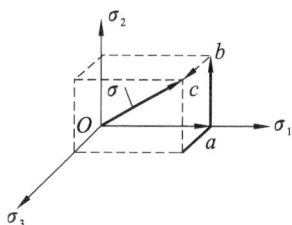

图 7-5 三维应力空间中的应力合成

考虑平均正应力σ_m的大小不影响屈服,屈服表面的概念可以得到更充分的解释。参考图7-6阐明σ_m的含义,单元左侧表明施加的应力状态。如图7-6所示,σ_m等于三个正应力代数和的三分之一,压应力在代数和中为负值。如果每个施加的应力减去平均应力,则产生应力偏量。从字面来说,这些值偏离了平均值,如果屈服不受σ_m的影响,那么这些应力偏量在某种程度上必然会导致屈服。结果表明,它们只是剪应力的函数。

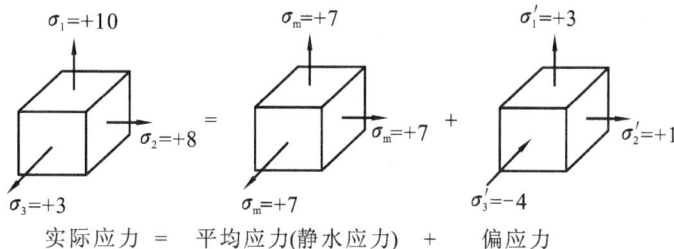

实际应力 = 平均应力(静水应力) + 偏应力

图 7-6 应力状态下的平均应力和偏应力分量

因此,如果应力组合$(\sigma_1,\sigma_2,\sigma_3)$刚好导致屈服,那么$(\sigma_1+\sigma_0,\sigma_2+\sigma_0,\sigma_3+\sigma_0)$也必然导致屈服,并且$(\sigma_1,\sigma_2,\sigma_3)$对于不同$\sigma_0$的应力组合将在屈服表面形成一条直线,该条直线与$\sigma_1=\sigma_2=\sigma_3$的直线平行。这条直线由关于1,2,3坐标系的等方向余弦来定义。现在,由于假设各向同性且不存在包申格效应,这条直线绕空间对角线($\sigma_1=\sigma_2=\sigma_3$)旋转必然产生一个棱柱,它的表面就是屈服面。为了充分说明该棱柱,必须明确其横截面形状和尺寸。

所有垂直于空间对角线的平面通过方程 $\sigma_1+\sigma_2+\sigma_3=C$(常数)来定义,对于任何一组正应力,该常数为 $3\sigma_m$。如果常数为零,则该平面通过原点且与棱柱轴线垂直,通常被称为 π 平面,与屈服表面相交形成的曲线称为 C 曲线。这表明,只要选取合适的 σ_0,屈服面上的任何点可以降至 C 曲线上的对应点,σ_0 值使得初始点沿着屈服表面上下移动。最后,考虑一组应力状态如下:

$$(\sigma_1,\sigma_2,\sigma_3)=(6,-2,1) \quad \text{并且} \quad \sigma_m=\frac{5}{3}$$

如果该应力状态减去 σ_m,则

$$(\sigma_1',\sigma_2',\sigma_3')=\left(\frac{13}{3},-\frac{11}{3},-\frac{2}{3}\right) \quad \text{并且} \quad \sum\sigma_i'=0$$

因此,三维应力矢量由位于 π 平面的应力偏量和垂直于 π 平面的平均应力或静水应力分量组成。既然我们已经研究了屈服轨迹和屈服表面的物理意义,那么我们将考虑一些已经提出的可能屈服表面。

7.5 屈 服 准 则

如第 2 章所述,对于任何三维应力状态都存在一个三次方程,其三个根表示主应力。这个方程表达式为

$$\sigma_p^3-I_1\sigma_p^2-I_2\sigma_p-I_3=0 \tag{7-1}$$

式中,不变量 I_1,I_2 和 I_3 是用主应力表示的函数,如下所示

$$\begin{cases} I_1=\sigma_1+\sigma_2+\sigma_3 \\ I_2=-(\sigma_1\sigma_2+\sigma_2\sigma_3+\sigma_3\sigma_1) \\ I_3=\sigma_1\sigma_2\sigma_3 \end{cases} \tag{7-2}$$

可以立即注意到,$I_1=3\sigma_m$,因此,第一个不变量是静水应力分量或平均正应力分量的函数,不会影响屈服。对那些已经发现屈服行为与 σ_m 无关的固体,任何可接受的屈服准则都不应涉及 I_1。

假设提出如下屈服准则:当 $\sigma_1-\sigma_2-\sigma_3$ 的值为常数 10 时,将发生屈服。如果这是一个可接受的准则,那么 $\sigma_1=+5$,$\sigma_2=-2$,$\sigma_3=-3$ 提供了一个产生屈服的应力状态。现在叠加一个应力 $\sigma_0=+10$,这意味着新的应力状态是 $(15,8,7)$,根据提出的准则,由于 $\sigma_1-\sigma_2-\sigma_3$ 的值为 0 而不是 10,所以该应力状态将不会导致屈服。然而,平均应力发生了变化。这个准则与实验观察的结果并不一致,因此不能认为它具有所需的一般性。两个最广泛使用的准则都满足与 I_1 无关的条件,并且当对延性金属进行实验时,该准则和实验结果具有极好的一致性。

7.6 特雷斯卡屈服准则

特雷斯卡屈服准则提出,当最大剪应力对应的某个函数达到临界值时,将发生屈服。只要可能,我们将约定 $\sigma_1>\sigma_2>\sigma_3$,但在某些情况下,这种相对的比较关系事先是未知的。此外,当考虑在二维或三维应力空间中绘图时,将无法严格保持此约定。

回想一下,绘制的三个莫尔圆,最大圆的半径就是最大剪应力的值。考虑代数符号,该准则写为如下形式。

如果 $\sigma_1>\sigma_2>\sigma_3$,那么

$$|\sigma_{\max}-\sigma_{\min}|=C=|\sigma_1-\sigma_3| \tag{7-3}$$

无论施加的应力状态如何,如果该准则都能被普遍接受,那么其常数就可以通过简单的标准实验确定。

(1)对于单轴拉伸应力状态,当 σ_1 达到单轴屈服应力 Y 时就发生屈服。因此

$$\sigma_1 = Y, \quad \sigma_2 = \sigma_3 = 0 \quad \tau_{\max} = \frac{Y}{2}$$

根据方程(7-3)有

$$|\sigma_1 - 0| = C = Y$$

(2)对于纯剪切应力状态,即 $\sigma_1 = -\sigma_3 = \tau_{\max}$,$\sigma_2 = 0$。为方便起见,将最大容许剪应力设定为剪切屈服应力 k。根据方程(7-3),我们得到

$$|\sigma_1 - (-\sigma_1)| = C = 2\sigma_1 = 2k$$

因此,特雷斯卡屈服准则可以表示为

如果 $\sigma_1 > \sigma_2 > \sigma_3$,那么

$$|\sigma_{\max} - \sigma_{\min}| = Y = 2k = |\sigma_1 - \sigma_3| \tag{7-4}$$

如果一个固体完全满足这一准则,那么拉伸和剪切屈服应力的比值为 2∶1。这并不意味这个比值是观察得到的,相反,它是通过该准则预测得到的。

【例 7-1】 某材料的拉伸屈服强度 Y 为 50 MPa,承受 30 MPa 的轴向压应力(图 7-7)。根据特雷斯卡屈服准则预测产生屈服时,确定在初始压应力垂直方向上应施加拉伸应力的大小,并绘制出该状况对应的莫尔圆。

【解】 $|\sigma_1 - \sigma_3| = Y$,其中 $\sigma_1 > \sigma_2 > \sigma_3$。

这里 σ_2 为 0,σ_3 为负数,σ_1 为未知应力但符号为正。

$|\sigma_1 - (-30)| = 50$,所以 $\sigma_1 = 20$ MPa。

注意该圆的直径为 50 MPa。

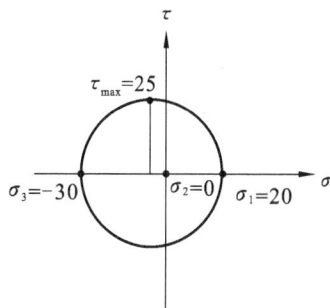

图 7-7 例 7-1 图(单位:MPa)

7.7 米泽斯屈服准则

可能由于特雷斯卡屈服准则产生的屈服表面在应力空间中具有棱角,并忽略了中间应力 σ_2,米泽斯提出了一个涉及光滑函数的准则。尽管数学式简化为 $6J_2 = C$(J_2 为第二应力偏量不变量,C 为常数),但这几乎没有给出具体的解释。

$J_2 = -(\sigma'_x \sigma'_y + \sigma'_y \sigma'_z + \sigma'_x \sigma'_z - \tau^2_{xy} - \tau^2_{yz} - \tau^2_{zx})$ 或

$J_2 = -[(\sigma_x - \sigma_m)(\sigma_y - \sigma_m) + (\sigma_y - \sigma_m)(\sigma_z - \sigma_m) + (\sigma_x - \sigma_m)(\sigma_z - \sigma_m) - (\tau^2_{xy} + \tau^2_{yz} + \tau^2_{zx})]$,则

$6J_2 = (\sigma_x - \sigma_y)^2 + (\sigma_y - \sigma_z)^2 + (\sigma_x - \sigma_z)^2 + 6(\tau^2_{xy} + \tau^2_{yz} + \tau^2_{zx})$,或 $6J_2 = (\sigma_1 - \sigma_2)^2 + (\sigma_2 - \sigma_3)^2 + (\sigma_3 - \sigma_1)^2$

对此,我们提出了几种物理解释,即扭曲能和等八面体剪应力理论。需要强调的是,它们是按照提出的数学假设来进行解释,但就关心该准则使用而言,不需要沉迷于历史上先后顺序。米泽斯屈服准则最广泛使用的形式是依据主应力,按照方程(7-5)预测屈服发生。

$$(\sigma_1 - \sigma_2)^2 + (\sigma_2 - \sigma_3)^2 + (\sigma_3 - \sigma_1)^2 = C \tag{7-5}$$

一般形式为

$$(\sigma_x - \sigma_y)^2 + (\sigma_y - \sigma_z)^2 + (\sigma_z - \sigma_x)^2 + 6(\tau^2_{xy} + \tau^2_{yz} + \tau^2_{zx}) = C \tag{7-6}$$

这些公式的等价性证明作为习题由感兴趣的读者完成。

对于方程(7-5),由于所有的主应力的权重相等,所以我们最初不需要知道主应力的相对代数关系。还需注意,每组应力差都对应三个莫尔圆中的一个剪应力,因此就像应用特雷斯卡屈服准则一样,剪应力与屈服的开始是相关联的。为了确定常数,利用与 7.6 节中相同的过程。

(1)对于单轴拉伸状态,当 $\sigma_1 = Y$,$\sigma_2 = \sigma_3 = 0$ 时屈服发生,应用方程(7-5)有

$$2\sigma_1^2 = C = 2Y^2$$

(2)对于纯剪切状态,当 $\sigma_1 = -\sigma_3 = k$,$\sigma_2 = 0$ 时屈服发生,应用方程(7-5)有

$$\sigma_1^2 + \sigma_1^2 + 4\sigma_1^2 = C = 6\sigma_1^2 = 6k^2$$

因此,米泽斯屈服准则可以写成

$$(\sigma_1 - \sigma_2)^2 + (\sigma_2 - \sigma_3)^2 + (\sigma_3 - \sigma_1)^2 = 2Y^2 = 6k^2 \tag{7-7}$$

根据该准则,拉伸屈服和剪切屈服应力的关系如下:

$$Y = \sqrt{3}k$$

这里第一次暗示两个准则可能会导致不同的预测结果。

将每个准则看作关于等效应力 $\bar{\sigma}$ 的函数会变得方便,其中 $\bar{\sigma}$ 是所施加应力的函数。当其大小达到单轴拉伸屈服强度时,施加的应力状态将导致屈服发生(即它已达到等效应力水平)。因此

米泽斯屈服准则:

$$\bar{\sigma} = \frac{1}{\sqrt{2}} \left[(\sigma_1 - \sigma_2)^2 + (\sigma_2 - \sigma_3)^2 + (\sigma_3 - \sigma_1)^2 \right]^{\frac{1}{2}} \tag{7-8}$$

特雷斯卡屈服准则:

$$\bar{\sigma} = |\sigma_{max} - \sigma_{min}| \tag{7-9}$$

受 σ_1,σ_2 和 σ_3 的影响,当 $\bar{\sigma}$ 达到 Y 值时,每一准则都可以预测发生屈服。然而,根据米泽斯屈服准则,当 $\bar{\sigma} = \sqrt{3}k$ 时,预测发生屈服,而根据特雷斯卡屈服准则;当 $\bar{\sigma} = 2k$ 时,预测发生屈服。

这两条准则有多种表达方式,但重要的是要认识到它们本身不是自然规律,它们的初始假设是由于数学推理而非物理推理。令人惊讶的是,这两个准则与物理观测结果如此接近。

【例 7-2】 使用米泽斯屈服准则对例 7-1 重新进行计算(图 7-8)。

【解】 根据方程(7-7)有

$$(\sigma_1 - 0)^2 + [0 - (-30)]^2 + (-30 - \sigma_1)^2 = 2 \times 50^2$$

$$\sigma_1^2 + 900 + 900 + 60\sigma_1 + \sigma_1^2 = 5000$$

$$\sigma_1^2 + 30\sigma_1 - 1600 = 0$$

$$\sigma_1 = \frac{-30 \pm (900 + 6400)^{\frac{1}{2}}}{2} = \frac{-30 \pm 85.44}{2}$$

所以

$$\sigma_1 = -57.72 \text{ MPa } \text{或 } 27.72 \text{ MPa}.$$

根据该问题的要求,拉应力为 27.72 MPa 是正确答案。负值 -57.72 MPa 是导致屈服时所需的压应力。注意:此处 τ_{max} 为 $Y/\sqrt{3}$,而在例 7-1 中,τ_{max} 为 $Y/2$。

参考图 7-9,两种准则均不受 σ_m 的影响,然而应力偏量(作为剪应力函数)至关重要。假设在原有应力状态基础上,叠加一个 $+9$ 的应力分量,则 $\sigma_1 = 48$ MPa,$\sigma_2 = 17$ MPa,$\sigma_3 = 13$ MPa 以及新的 $\sigma_m = 26$ MPa。最大圆的圆心位于 $\sigma = 30.5$ MPa,$\tau_{max} = 17.5$ MPa。注意,$\sigma_1' = 22$ MPa,$\sigma_2' = -9$ MPa,$\sigma_3' = -13$ MPa,与原有应力状态应力偏量相同。因此,σ_m 的变化会移动圆的位置,但既不会改变圆的大小,也不会影响应力偏量的值。注意

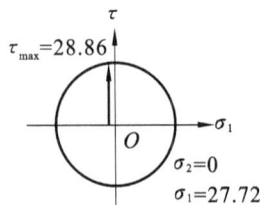

图 7-8 例 7-2 图(单位:MPa)

$$\sigma_1' = \sigma_1 - \sigma_m = \frac{(\sigma_1 - \sigma_2) + (\sigma_1 - \sigma_3)}{3}$$

类比可得到 σ_2' 和 σ_3'，因此应力偏量是剪应力的函数。

许多实际问题可以简化为双轴(平面)应力情况进行近似处理，假设 $\sigma_2 = 0$，若在 σ_1-σ_3 应力空间中绘图，则特雷斯卡屈服准则可以用图 7-10 表示，米泽斯屈服准则可以用图 7-11 表示。如果将两个图合并，结果如图 7-12 所示。

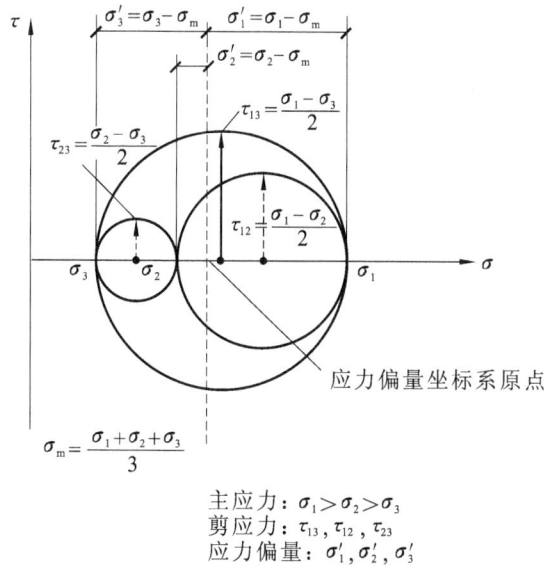

主应力：$\sigma_1 > \sigma_2 > \sigma_3$
剪应力：τ_{13}，τ_{12}，τ_{23}
应力偏量：σ_1'，σ_2'，σ_3'

假设 $\sigma_1 = 39$，$\sigma_2 = 8$，以及 $\sigma_3 = 4$(单位均为MPa)，那么 $\sigma_m = 17$，最大圆的圆心在 $\sigma = 21.5$，$\tau_{max} = 17.5$处。注意 $\sigma_1' = 22$，$\sigma_2' = -9$，$\sigma_3' = -13$。

图 7-9　表明主应力、平均正应力和应力偏量分量的三维应力状态莫尔圆

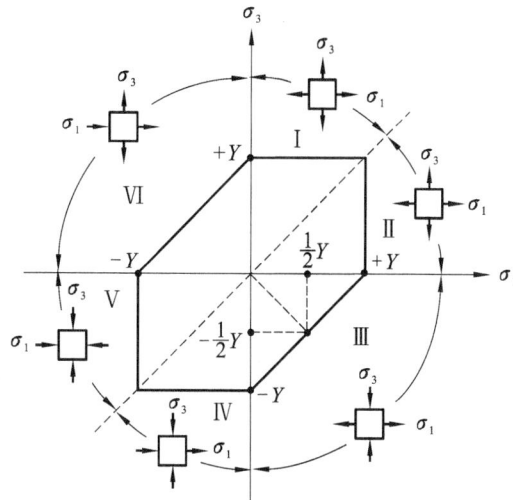

区	条件	边界	
Ⅰ	$\sigma_3 > \sigma_1 > 0$	$\sigma_3 = +Y$	位于上述粗线(边界)内的任何应力状态都不会导致屈服。
Ⅱ	$\sigma_1 > \sigma_3 > 0$	$\sigma_1 = +Y$	
Ⅲ	$\sigma_1 = -\sigma_3 = k = Y/2$	如图45°斜线	
Ⅳ	$0 > \sigma_1 > \sigma_3$	$\sigma_3 = -Y$	
Ⅴ	$0 > \sigma_3 > \sigma_1$	$\sigma_1 = -Y$	
Ⅵ	看第Ⅲ象限以上		

图 7-10　特雷斯卡屈服准则下屈服轨迹

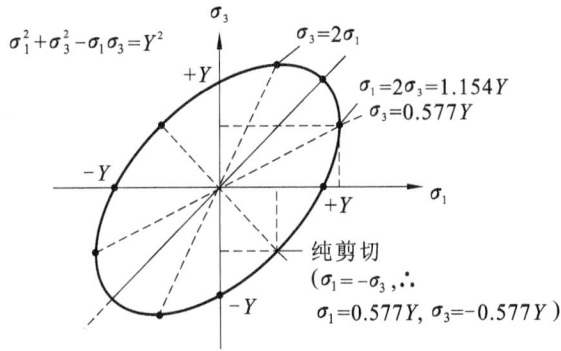

图 7-11　米泽斯屈服准则下屈服轨迹

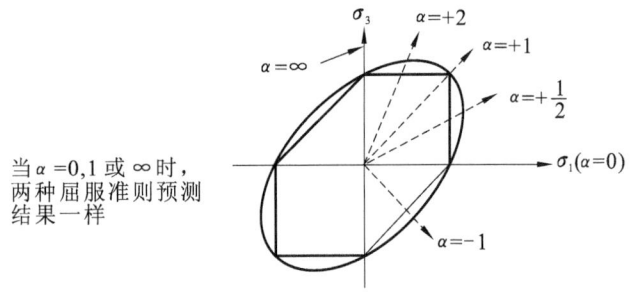

图 7-12　表明特定加载路径特雷斯卡和
米泽斯屈服准则下的屈服轨迹

图 7-12 中不同的恒应力比例加载路径表明，当 $\alpha=-1$（纯剪切）或 $\alpha=1/2$ 或 2 时，这两种准则预测结果出现了最大差异。我们经常会忽略，两个准则在六边形外接椭圆的某些特定点处会发生重合，这是因为当方程（7-8）和方程（7-9）中的常数只有通过单轴拉伸或压缩来确定时，每个准则才能预测屈服。

7.8　扭　曲　能

米泽斯屈服准则的一种解释是，当引起扭曲的弹性能达到临界值时，就会发生屈服。该应变能一般通过从总弹性应变能中减去体积膨胀应变能求出，因此，单位体积的总应变能为

$$W_v = \frac{1}{2}(\sigma_x \varepsilon_{xx} + \sigma_y \varepsilon_{yy} + \sigma_z \varepsilon_{zz} + \tau_{xy}\gamma_{xy} + \tau_{yz}\gamma_{yz} + \tau_{zx}\gamma_{zx}) \tag{7-10}$$

或者在主应力情况下表示为

$$W_v = \frac{1}{2}(\sigma_1 \varepsilon_{11} + \sigma_2 \varepsilon_{22} + \sigma_3 \varepsilon_{33}) \tag{7-11}$$

为了将方程（7-11）表示为应力的函数，利用方程（4-23）中给出的广义胡克定律，我们得到

$$W_v = \frac{1}{2E}(\sigma_1^2 + \sigma_2^2 + \sigma_3^2) - \frac{\nu}{E}(\sigma_1\sigma_2 + \sigma_2\sigma_3 + \sigma_3\sigma_1) \tag{7-12}$$

由于只有正应力会引起体积变化，因此膨胀应变是

$$\Delta = \varepsilon_{11} + \varepsilon_{22} + \varepsilon_{33} = \frac{1-2\nu}{E}(\sigma_1 + \sigma_2 + \sigma_3) = \frac{3}{E}(1-2\nu)\sigma_m \tag{7-13}$$

现在，根据正应变与 σ_m 的等式关系，以及 $\Delta=3\varepsilon_m$，有

$$\varepsilon_m = \frac{1-2\nu}{E}\sigma_m \tag{7-14}$$

观察由于膨胀而产生的功，$W_d = 3\sigma_m\varepsilon_m/2$，那么 $W_d = 3(1-2\nu)\sigma_m^2/(2E)$，最后得到

$$W_d = \frac{1-2\nu}{6E}(\sigma_1 + \sigma_2 + \sigma_3)^2 \tag{7-15}$$

将方程（7-12）减去方程（7-15）就可以得到剪切应变能 W_s，经过多次运算后的结果是

$$W_s = \frac{1}{12G}\big[(\sigma_1 - \sigma_2)^2 + (\sigma_2 - \sigma_3)^2 + (\sigma_3 - \sigma_1)^2\big] \tag{7-16}$$

那么在单轴拉伸过程中，即 $\sigma_2 = \sigma_3 = 0$ 时产生的剪切应变能为

$$W_s = \frac{\sigma_1^2}{6G} \tag{7-17}$$

当 $\sigma_1 = Y$ 时，得到导致屈服的临界值 W_{sc}。如果方程(7-16)和该临界值相等，则有

$$\frac{1}{12G}\big[(\sigma_1-\sigma_2)^2+(\sigma_2-\sigma_3)^2+(\sigma_3-\sigma_1)^2\big]=\frac{Y^2}{6G} \tag{7-18}$$

这与方程(7-7)相同。这就解释了为什么米泽斯屈服准则通常被称为扭曲能理论，它的物理意义就是当引起扭曲的弹性能达到临界值时，发生屈服。

7.9　等八面体剪应力

我们提出米泽斯屈服准则的第二种物理解释。为了简单起见，考虑一个由主应力方向定义的坐标系和一条方向余弦为 $l=m=n$ 且经过原点的直线。与该直线和空间其他区域中等角直线垂直的平面称为等八面体平面，而八个等效平面相交形成等八面体。对于这种情况，前面第 2 章中就已证明：

$$\sigma_n=\sigma_1 l^2+\sigma_2 m^2+\sigma_3 n^2 \tag{7-19}$$

由于 $l=m=n=\cos 54°44'=1/\sqrt{3}$，$\sigma_n=1/3(\sigma_1+\sigma_2+\sigma_3)$，因此垂直于等八面体平面的应力为 σ_m，并且由于 σ_m 对屈服没有影响，因此作用于该平面的剪应力(τ_0)必须达到临界值时，才会发生屈服。此处不做详细推导，该应力可以表示为

$$\tau_0=\frac{1}{3}\big[(\sigma_1-\sigma_2)^2+(\sigma_2-\sigma_3)^2+(\sigma_3-\sigma_1)^2\big]^{\frac{1}{2}} \tag{7-20}$$

与方程(7-8)比较得到

$$\tau_0=\frac{\sqrt{2}}{3}\bar{\sigma} \tag{7-21}$$

米泽斯屈服准则的另一个表示形式可以通过乘适当的系数得到。而在本书的其余部分中，只要用到该准则，我们都用方程(7-8)来表示。

7.10　流动法则或塑性应力-应变关系

正如弹性区域可以用广义胡克定律表示一样，塑性区域也需要有类似的关系表达：

$$\varepsilon_{11}^e=\frac{1}{E}\big[\sigma_1-\nu(\sigma_2+\sigma_3)\big]$$

回顾一下在塑性变形期间使用应变增量表示加载路径或加载历史，这些流动法则可以用如下简单的方法制定。

考虑单轴拉伸下的塑性流动，如图 7-13 所示。

$$\sigma_m=\frac{\sigma_1+\sigma_2+\sigma_3}{3}=\frac{1}{3}\sigma_1$$

现在，在 1 方向上的应力偏量为 $\sigma_1'=\sigma_1-\sigma_m$，并且在特定时刻(图 7-13)。

$$\sigma_1'=\frac{2}{3}\sigma_1 \quad 且 \quad \sigma_2'=\sigma_3'=0-\frac{1}{3}\sigma_1=-\frac{1}{3}\sigma_1$$

所以

$$\sigma_1'=-2\sigma_2'=-2\sigma_3'$$

由于体积恒定，塑性应变增量之和一定为零，因此

$$d\varepsilon_{11}^p+d\varepsilon_{22}^p+d\varepsilon_{33}^p=0$$

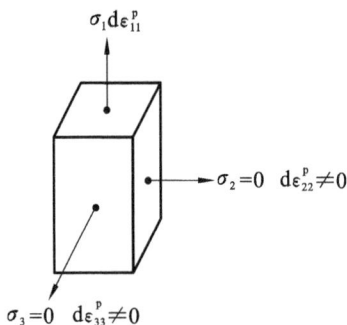

图 7-13　单轴拉伸状态下的应力和塑性应变增量

并且在这种情况下由于对称性有 $\mathrm{d}\varepsilon_{22}^{p}=\mathrm{d}\varepsilon_{33}^{p}$，因此 $\mathrm{d}\varepsilon_{11}^{p}=-2\mathrm{d}\varepsilon_{22}^{p}=-2\mathrm{d}\varepsilon_{33}^{p}$，据此可以得到 $\mathrm{d}\varepsilon_{11}^{p}/\mathrm{d}\varepsilon_{22}^{p}=-2=\sigma_{1}'/\sigma_{2}'$，以此类推，可以写成下述形式：

$$\frac{\mathrm{d}\varepsilon_{11}^{p}}{\sigma_{1}'}=\frac{\mathrm{d}\varepsilon_{22}^{p}}{\sigma_{2}'}=\frac{\mathrm{d}\varepsilon_{33}^{p}}{\sigma_{3}'}=C=\mathrm{d}\lambda \quad \text{（通常情况下）} \tag{7-22}$$

式中，常数并不总为 -2，但比值总是为某个常数。

这意味着，当前塑性应变增量与当前应力偏量的比值是常数，但不知道两者的值。有以下几点值得注意：

(1)前面我们用简单的方法得到了流动法则。实际上，这可能被视为一个必要但不充分条件，因为没有提供真正的证据说明其他应力状态也会导致相同的结果。

(2)方程(7-22)表示省略了弹性应变增量的普朗特流动法则。

(3)方程(7-22)与莱维-米泽斯方程(比普朗特-罗伊斯提出早)相同，其中假设总应变增量等于塑性应变增量。因此，莱维-米泽斯方程可以被视为一般表达式的特殊形式。

为方便起见，流动法则除方程(7-22)形式以外还可以用其他各种形式表示，即

$$\frac{\mathrm{d}\varepsilon_{11}^{p}-\mathrm{d}\varepsilon_{22}^{p}}{\sigma_{1}-\sigma_{2}}=\mathrm{d}\lambda \tag{7-23}$$

$$\mathrm{d}\varepsilon_{11}^{p}=\frac{2}{3}\mathrm{d}\lambda\left[\sigma_{1}-\frac{1}{2}(\sigma_{2}+\sigma_{3})\right] \tag{7-24}$$

$$\mathrm{d}\varepsilon_{11}^{p}=\frac{\mathrm{d}\overline{\varepsilon^{p}}}{\overline{\sigma}}\left[\sigma_{1}-\frac{1}{2}(\sigma_{2}+\sigma_{3})\right] \tag{7-25}$$

注意方程(7-25)和广义胡克定律之间极其相似，其中 $1/E$ 被 $\mathrm{d}\overline{\varepsilon^{p}}/\overline{\sigma}$ 代替且由于不可压缩性 ν 用 $1/2$ 表示。某种意义上 E 是一个常数，系数 $\mathrm{d}\overline{\varepsilon^{p}}/\overline{\sigma}$ 是一个可变比例系数。定义等效应变增量 $\mathrm{d}\overline{\varepsilon^{p}}$，其形式为

$$\mathrm{d}\overline{\varepsilon^{p}}=\frac{\sqrt{2}}{3}\left[(\mathrm{d}\varepsilon_{11}^{p}-\mathrm{d}\varepsilon_{22}^{p})^{2}+(\mathrm{d}\varepsilon_{22}^{p}-\mathrm{d}\varepsilon_{33}^{p})^{2}+(\mathrm{d}\varepsilon_{33}^{p}-\mathrm{d}\varepsilon_{11}^{p})^{2}\right]^{\frac{1}{2}} \tag{7-26}$$

注意式(7-26)与方程(7-8)给出的等效应力函数非常相似。像方程(7-8)选择系数 $1/\sqrt{2}$ 使在单轴拉伸条件下 $\overline{\sigma}$ 等于 σ_{1} 一样，此处选择系数 $\sqrt{2}/3$，以便在单轴拉伸条件下，$\mathrm{d}\overline{\varepsilon^{p}}$ 的值等于 $\mathrm{d}\varepsilon_{11}^{p}$。关键要认识到，方程(7-8)和方程(7-26)中，$\overline{\sigma}$ 和 $\mathrm{d}\overline{\varepsilon^{p}}$ 始终为正，因此它们在方程(7-25)中的比值也一定为正。

对于任何屈服流动法则都可以用塑性势能的概念推导出来。该方法提出由应力 σ_{ij} 产生的应变增量由下式得到：

$$\mathrm{d}\varepsilon_{ij}^{p}=\frac{\partial f}{\partial\sigma_{ij}}\mathrm{d}\lambda' \tag{7-27}$$

式中，f 为选取的屈服函数。如果使用米泽斯屈服准则

$$f(\sigma_{ij})=(\sigma_{1}-\sigma_{2})^{2}+(\sigma_{2}-\sigma_{3})^{2}+(\sigma_{3}-\sigma_{1})^{2}=C$$

然后

$$\frac{\partial f}{\partial \sigma_1} = 2(\sigma_1 - \sigma_2) - 2(\sigma_3 - \sigma_1) = 4\sigma_1 - 2(\sigma_2 + \sigma_3)$$

$$3\sigma_m = \sigma_1 + \sigma_2 + \sigma_3$$

所以

$$\sigma_2 + \sigma_3 = 3\sigma_m - \sigma_1$$

现在

$$\frac{\partial f}{\partial \sigma_1} = 6(\sigma_1 - \sigma_m) = 6\sigma_1'$$

最后

$$d\varepsilon_{11}^p = 6\sigma_1' d\lambda' \quad 或 \quad \frac{d\varepsilon_{11}^p}{\sigma_1'} = d\lambda$$

这与方程(7-22)一样。与特雷斯卡屈服准则相关联的流动法则几乎没有用途。

特定选用的塑性势能对应的塑性流动法则可以很容易得到。

$$d\varepsilon_{ii}^p = \sigma_i' d\lambda, \quad d\gamma_{ij}^p = \tau_{ij} d\lambda$$

或

$$\frac{d\varepsilon_{xx}^p}{\sigma_x'} = \frac{d\varepsilon_{yy}^p}{\sigma_y'} = \frac{d\varepsilon_{zz}^p}{\sigma_z'} = \frac{d\gamma_{xy}^p}{\tau_{xy}} = \frac{d\gamma_{yz}^p}{\tau_{yz}} = \frac{d\gamma_{zx}^p}{\tau_{zx}} = d\lambda \tag{7-28}$$

这种形式的应力-应变关系是由莱维和米泽斯独立提出的,他们使用塑性应变增量而不是用总应变增量。普朗特提出针对平面应变状态,罗伊斯提出针对任意应变状态的考虑弹性应变增量的修正方程(7-28)。根据方程(7-28),塑性剪应变增量随对应的剪应力消失,因此主应力轴和塑性应变增量一致。方程(7-28)还表明,应力莫尔圆可用于塑性应变增量,只要原点沿 σ 轴方向移动的量与静水应力相等。

对于弹塑性材料,完整的普朗特-罗伊斯方程可表示为

$$d\varepsilon = d\varepsilon^e + d\varepsilon^p$$

$$d\varepsilon_{ii} = \frac{1}{E}\left[(1+\nu)d\sigma_i - \nu\delta_{ii}d\sigma_k\right] + \left(\sigma_i - \frac{1}{3}\sigma_{ii}\delta_k\right)d\lambda \tag{7-29}$$

$$d\varepsilon_{ij} = \frac{1}{E}\left[(1+\nu)d\tau_{ij}\right] + \frac{\tau_{ij}}{2}d\lambda$$

其中方程(7-23)和方程(7-24)中的 $d\lambda = 3d\overline{\varepsilon^p}/(2\overline{\sigma})$。对于无应变硬化材料,$d\lambda$ 可被视为基本未知数。方程(7-29)由两种类型的方程组构成,每种类型的方程组由三个方程组成。

$$d\varepsilon_{xx} = \frac{1}{E}\left[d\sigma_x - \nu(d\sigma_y + d\sigma_z)\right] + \frac{2}{3}d\lambda\left[\sigma_x - \frac{1}{2}(\sigma_y + \sigma_z)\right]$$

$$d\varepsilon_{yy} = \frac{1}{E}\left[d\sigma_y - \nu(d\sigma_x + d\sigma_z)\right] + \frac{2}{3}d\lambda\left[\sigma_y - \frac{1}{2}(\sigma_x + \sigma_z)\right]$$

$$d\varepsilon_{zz} = \frac{1}{E}\left[d\sigma_z - \nu(d\sigma_x + d\sigma_y)\right] + \frac{2}{3}d\lambda\left[\sigma_z - \frac{1}{2}(\sigma_x + \sigma_y)\right]$$

$$d\varepsilon_{xy} = \frac{d\tau_{xy}}{2G} + \frac{\tau_{xy}}{2}d\lambda$$

$$d\varepsilon_{yz} = \frac{d\tau_{yz}}{2G} + \frac{\tau_{yz}}{2}d\lambda$$

$$d\varepsilon_{zx} = \frac{d\tau_{zx}}{2G} + \frac{\tau_{zx}}{2}d\lambda$$

在许多实际问题中,加载过程中弹性应变比塑性应变小。那么普朗特-罗伊斯方程将可由更易于处理的莱维-米泽斯方程代替,其对应方程(7-28)省略了上标。这相当于假定为刚塑性材料。

注意,塑性应变与总应力有关,而弹性应变与应力增量变化有关。图 7-14 可以证明这一点。在

图 7-14(a)中,由于存在线性关系,应力增量变化导致应变增量变化,且增量变化相同。然而,在图 7-14(b)中,一个相等的应力增量变化并不导致相同的应变增量变化,相反,它们与斜率的变化有关,因此特定的增量大小取决于给定时刻的总应力。

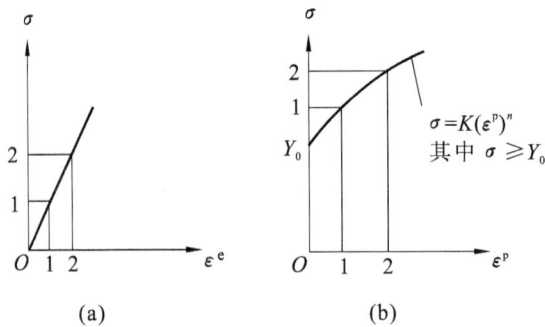

图 7-14 弹性和塑性变形的应力-应变曲线

【**例 7-3**】 考虑一个两端密闭的薄壁圆管,其内部受压为 7 MPa;假设这将导致塑性变形发生,并且忽略端部效应。在这种荷载作用下,满足的应力关系如下:

$$\sigma_\theta = \sigma_1 = \frac{pr}{t}, \quad \sigma_l = \sigma_2 = \frac{pr}{2t}, \quad \sigma_r = \sigma_3 = 0 \tag{7-30}$$

式中,p 为内压;r 为薄壁圆管的半径。令 r/t 值为 20,并注意到 $\sigma_2 = \sigma_1/2 = \sigma_m$。求圆管环向的总应变。

【**解**】 应用方程(7-25)有

$$d\varepsilon_{11}^p = -d\varepsilon_{33}^p, \quad d\varepsilon_{22}^p = 0 \tag{7-31}$$

采用增量形式,"1"方向上的总应变为

$$d\varepsilon_{11} = d\varepsilon_{11}^p + d\varepsilon_{11}^e \quad (即塑性应变加弹性应变)$$

根据方程(7-25)和胡克定律有

$$d\varepsilon_{11} = \frac{d\,\overline{\varepsilon}^p}{\overline{\sigma}}\left[\sigma_1 - \frac{1}{2}(\sigma_2 + \sigma_3)\right] + \frac{1}{E}[d\sigma_1 - \nu(d\sigma_2 + d\sigma_3)]$$

使用方程(7-30)中的关系,简化后有

$$d\varepsilon_{11} = \frac{d\,\overline{\varepsilon}^p}{\overline{\sigma}}\left(\frac{3}{4}\sigma_1\right) + \frac{d\sigma_1}{E}\left(1 - \frac{\nu}{2}\right)$$

将方程(7-30)的应力关系代入方程(7-8)中,并将方程(7-31)的应变关系代入方程(7-26),得出

$$\overline{\sigma} = \frac{\sqrt{3}}{2}\sigma_1 \quad 以及 \quad d\,\overline{\varepsilon}^p = \frac{2}{\sqrt{3}}d\varepsilon_{11}^p \tag{7-32}$$

根据该例题中给出的值有 $\sigma_1 = 140$ MPa,因此 $\overline{\sigma} = 70\sqrt{3}$ MPa。现在知道等效应力和等效应变的关系是十分重要的。假设 $\overline{\sigma} = K(\overline{\varepsilon}^p)^n$ 适用于塑性变形部分,其中 $K = 175$ MPa 且 $n = 0.25$。那么该关系式为

$$70\sqrt{3} = 175\,(\overline{\varepsilon}^p)^{0.25}$$

结果为

$$\overline{\varepsilon}^p = 0.693^4 = 0.23$$

其中利用方程(7-26)进行积分 $\int_0^{\overline{\varepsilon}^p} d\,\overline{\varepsilon}^p$,因此

$$\int_0^{\varepsilon_{11}^p} d\varepsilon_{11}^p = \int_0^{\overline{\varepsilon}^p} \frac{\sqrt{3}}{2} d\,\overline{\varepsilon}^p$$

得

$$\varepsilon_{11}^p = 0.199$$

为了计算弹性应变部分,注意 $d\sigma_1$ 等于 $20dp$,其中 dp 为 7 MPa。对于铝(其 K 值和 n 值采用上述数字表

示是非常合理的),取 $E=7\times10^4$ MPa 且 $\nu=\dfrac{1}{3}$。因为 $\mathrm{d}\varepsilon_{11}^{\mathrm{p}}=\mathrm{d}\sigma_1(1-\nu/2)/E$,所以 $\varepsilon_{11}^{\mathrm{p}}\approx0.002$,因此 $\varepsilon_{11}=$ 0.199+0.002=0.201。

有以下几点相关的要点:

(1)当遇到较大的塑性应变时,忽略弹性应变几乎不会带来误差,这通常会大大简化分析过程。

(2)在许多实际问题中必须使用近似法,因为实际产生的变形并不遵循像例题中一样的简单加载路径。

(3)如果需要确定的数值答案,则必须得到某种 $\bar{\sigma}\bar{\varepsilon}$ 的关系。

(4)上述积分的下限值被视为零。从物理角度讲,这意味着压力在开始时为零,并且在施加荷载之前,材料中没有产生弹性或塑性应变。

7.11　塑性力学变形理论(全量理论)

到目前为止,本章给出的应力-应变关系是增量形式,被称为塑性力学的增量理论。这些增量理论的基础是塑性变形取决于加载路径,因此为了得到最终的变形状态,我们必须对增量应变-应力关系沿加载路径进行积分。

然而还有提供应力和应变之间全量关系的一些塑性理论,这些理论被称为塑性力学变形理论或塑性力学全量理论。亨奇在 1924 年首先提出塑性力学全量理论,伊留辛分别在 1943 年及 1947 年对该塑性力学全量理论做进一步完善,其内容简单描述如下。

亨奇-伊留辛塑性力学全量理论指出,塑性应变分量与应力偏量分量成正比:

$$\frac{\varepsilon_{xx}^{\mathrm{p}}}{\sigma_x'}=\frac{\varepsilon_{yy}^{\mathrm{p}}}{\sigma_y'}=\frac{\varepsilon_{zz}^{\mathrm{p}}}{\sigma_z'}=\frac{\gamma_{xy}^{\mathrm{p}}}{\tau_{xy}}=\frac{\gamma_{yz}^{\mathrm{p}}}{\tau_{yz}}=\frac{\gamma_{xz}^{\mathrm{p}}}{\tau_{xz}}=\lambda \tag{7-33}$$

$$\begin{cases}\varepsilon_{xx}^{\mathrm{p}}=\lambda\sigma_x' \\ \varepsilon_{yy}^{\mathrm{p}}=\lambda\sigma_y' \\ \varepsilon_{zz}^{\mathrm{p}}=\lambda\sigma_z'\end{cases}\quad\begin{cases}\gamma_{xy}^{\mathrm{p}}=\lambda\tau_{xy} \\ \gamma_{yz}^{\mathrm{p}}=\lambda\tau_{yz} \\ \gamma_{xz}^{\mathrm{p}}=\lambda\tau_{xz}\end{cases} \tag{7-34}$$

除了使用塑性应变分量 $\varepsilon_{ij}^{\mathrm{p}}$ 代替塑性应变增量 $\mathrm{d}\varepsilon_{ij}^{\mathrm{p}}$ 外,这个方程类似普朗特-罗伊斯方程,这意味着塑性应变分量与应力偏量分量同轴。注意该方程中给出的 σ_x'、σ_y'、σ_z' 是应力偏量分量,而 τ_{xy}、τ_{yz}、τ_{xz} 是应力分量。如果等效应力及等效塑性应变为

$$\bar{\sigma}=\frac{1}{\sqrt{2}}\big[(\sigma_x-\sigma_y)^2+(\sigma_y-\sigma_z)^2+(\sigma_z-\sigma_x)^2+6(\tau_{xy}^2+\tau_{yz}^2+\tau_{xz}^2)\big]^{\frac{1}{2}}$$

$$=\frac{1}{\sqrt{2}}\big[(\sigma_1-\sigma_2)^2+(\sigma_2-\sigma_3)^2+(\sigma_3-\sigma_1)^2\big]^{\frac{1}{2}} \tag{7-35}$$

$$\overline{\varepsilon^{\mathrm{p}}}=\frac{\sqrt{2}}{3}\big[(\varepsilon_{xx}^{\mathrm{p}}-\varepsilon_{yy}^{\mathrm{p}})^2+(\varepsilon_{yy}^{\mathrm{p}}-\varepsilon_{zz}^{\mathrm{p}})^2+(\varepsilon_{zz}^{\mathrm{p}}-\varepsilon_{xx}^{\mathrm{p}})^2+6(\gamma_{xy}^2+\gamma_{yz}^2+\gamma_{xz}^2)\big]^{\frac{1}{2}}$$

$$=\frac{\sqrt{2}}{3}\big[(\varepsilon_{11}^{\mathrm{p}}-\varepsilon_{22}^{\mathrm{p}})^2+(\varepsilon_{22}^{\mathrm{p}}-\varepsilon_{33}^{\mathrm{p}})^2+(\varepsilon_{33}^{\mathrm{p}}-\varepsilon_{11}^{\mathrm{p}})^2\big]^{\frac{1}{2}} \tag{7-36}$$

比例函数 λ 可以根据 $\bar{\sigma}$ 和 $\overline{\varepsilon^{\mathrm{p}}}$ 进行求解:

$$\lambda=\frac{3\,\overline{\varepsilon^{\mathrm{p}}}}{2\bar{\sigma}} \tag{7-37}$$

因此方程(7-34)变为

$$\begin{cases} \varepsilon_{xx}^{p} = \dfrac{3}{2}\dfrac{\overline{\varepsilon^{p}}}{\overline{\sigma}}\sigma_{x}' & \gamma_{xy}^{p} = \dfrac{3}{2}\dfrac{\overline{\varepsilon^{p}}}{\overline{\sigma}}\tau_{xy} \\[2mm] \varepsilon_{yy}^{p} = \dfrac{3}{2}\dfrac{\overline{\varepsilon^{p}}}{\overline{\sigma}}\sigma_{y}' & \gamma_{yz}^{p} = \dfrac{3}{2}\dfrac{\overline{\varepsilon^{p}}}{\overline{\sigma}}\tau_{yz} \\[2mm] \varepsilon_{zz}^{p} = \dfrac{3}{2}\dfrac{\overline{\varepsilon^{p}}}{\overline{\sigma}}\sigma_{z}' & \gamma_{zx}^{p} = \dfrac{3}{2}\dfrac{\overline{\varepsilon^{p}}}{\overline{\sigma}}\tau_{zx} \end{cases} \tag{7-38}$$

作为该理论的一部分,假设在等效应力和等效塑性应变之间存在一个广义函数

$$\overline{\sigma} = \overline{\sigma}(\overline{\varepsilon^{p}}) \tag{7-39}$$

该函数与加载路径无关,因此可以通过简单的单轴拉伸或压缩试验得到。

弹性应变通过胡克定律控制,由于

$$\frac{\varepsilon_{ii}^{e'}}{\sigma_{i}'} = \frac{1}{2G} \tag{7-40}$$

式中,$\varepsilon_{ii}^{e'} = \varepsilon_{ii}^{e} - \varepsilon_{m}^{e}$,$\varepsilon_{m}^{e}$ 为平均弹性应变。

因此,总的应变可以认为是弹性应变和塑性应变之和:

$$\varepsilon_{ii} = \frac{\sigma_{i}'}{2G} + \frac{\sigma_{m}}{3K} + \frac{3}{2}\frac{\overline{\varepsilon^{p}}}{\overline{\sigma}}\sigma_{i}' \tag{7-41}$$

$$\gamma_{ij} = \frac{\tau_{ij}}{G} + \frac{3}{2}\frac{\overline{\varepsilon^{p}}}{\overline{\sigma}}\tau_{ij} \tag{7-42}$$

式中,$K = \dfrac{E}{3(1-2\nu)}$ 是体积弹性模量。如果弹性应变与塑性应变相比较小,弹性应变可以忽略,则塑性力学全量理论的基本方程为

$$\varepsilon_{ii} = \frac{3}{2}\frac{\overline{\varepsilon^{p}}}{\overline{\sigma}}\sigma_{i}' \tag{7-43}$$

$$\gamma_{ij} = \frac{3}{2}\frac{\overline{\varepsilon^{p}}}{\overline{\sigma}}\tau_{ij} \tag{7-44}$$

并且

$$\overline{\sigma} = \overline{\sigma}(\overline{\varepsilon^{p}}) \tag{7-45}$$

由于塑性变形一般取决于加载路径,塑性力学全量理论的应用是非常有限的。正如前面所讨论的,塑性变形有两个特征:加载的非线性及塑性流动的不可逆性。不可逆性是对加载路径依赖的必要响应,因此产生了对增量基本方程的需求。然而,对于比例加载路径,基本满足应力和应变之间非线性关系的塑性力学全量理论可以很好地描述塑性变形。事实上,当所有应力分量以相同的比例加载时,塑性力学增量理论被简化为全量理论。设按相同速率增加加载应力

$$\sigma_{ij} = k\sigma_{ij}^{0} \tag{7-46}$$

式中,k 是单调增加比例因子;σ_{ij}^{0} 是任意应力状态。

因此,我们可以得到

$$\overline{\sigma} = k\overline{\sigma^{0}} \tag{7-47}$$

增量理论的普朗特-罗伊斯方程可以写为如下形式:

$$d\varepsilon_{ii}^{p} = \frac{3d\overline{\varepsilon^{p}}}{2\overline{\sigma}}\sigma_{i}' = \frac{3d\overline{\varepsilon^{p}}}{2\overline{\sigma^{0}}}\sigma_{i}^{0'} \tag{7-48a}$$

$$d\gamma_{ij}^{p} = \frac{3d\overline{\varepsilon^{p}}}{2\overline{\sigma^{0}}}\tau_{ij} = \frac{3d\overline{\varepsilon^{p}}}{2\overline{\sigma^{0}}}\tau_{ij}^{0} \tag{7-48b}$$

上式通过积分很容易变成

$$\varepsilon_{ii}^{p} = \frac{3}{2}\frac{\overline{\varepsilon^{p}}}{\overline{\sigma^{0}}}\sigma_{i}^{0'} = \frac{3}{2}\frac{\overline{\varepsilon^{p}}}{\overline{\sigma}}\sigma_{i}' \tag{7-49a}$$

$$\gamma_{ij}^{\mathrm{p}} = \frac{3}{2} \frac{\overline{\epsilon^{\mathrm{p}}}}{\overline{\sigma^0}} \tau_{ij}^0 = \frac{3}{2} \frac{\overline{\epsilon^{\mathrm{p}}}}{\overline{\sigma^0}} \tau_{ij} \qquad (7\text{-}49\mathrm{b})$$

这就是亨奇-伊留辛塑性力学全量理论基本模型,该理论表明塑性应变仅是最终应力状态的函数并且不取决于加载路径。塑性力学全量理论的应用并不限于按比例加载,但事实上全量理论应用范围仍然有一些问题。需要强调的是,由于塑性变形具有不可逆性,塑性力学全量理论必须慎重使用。当真实加载路径与比例加载路径差别很大时,将会导致严重的误差。

7.12 平衡微分方程和屈服准则联立求解

当求解平面轴对称等简单塑性问题时,可以将平衡微分方程和屈服准则进行联立求解,以求出物体塑性变形时的应力分布,其积分常数根据自由表面和接触面上的边界条件确定。

受内压塑性圆筒应力计算过程如下。

圆筒的内壁受均匀压力 p 作用,圆筒的尺寸如图 7-15 所示。由于圆筒很长,如压力容器、管道等,可以按平面应变问题简化处理。显然,该问题属于轴对称平面问题,因此 τ_{rz}、$\tau_{z\theta}$ 为零,σ_θ、σ_r 都是主应力,它们随 r 变化,因此以偏微分形式表示的方程(5-12)变为常微分方程,并且忽略体力的作用,此时柱坐标平衡微分方程为

$$\frac{\mathrm{d}\sigma_r}{\mathrm{d}r} + \frac{\sigma_r - \sigma_\theta}{r} = 0 \qquad (7\text{-}50)$$

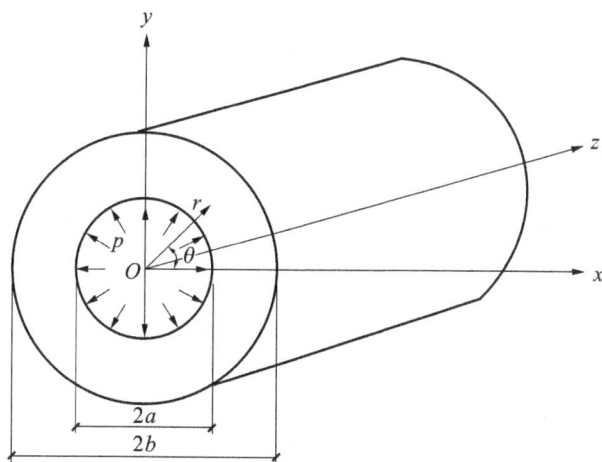

图 7-15 受内压作用的塑性圆筒

根据米泽斯屈服准则,有

$$\bar{\sigma} = \frac{1}{\sqrt{2}} \sqrt{(\sigma_\theta - \sigma_r)^2 + (\sigma_\theta - \sigma_z)^2 + (\sigma_r - \sigma_z)^2} = Y \qquad (7\text{-}51)$$

由于该问题是平面应变问题,有 $\sigma_z = \frac{1}{2}(\sigma_r + \sigma_\theta)$,代入方程(7-51),得

$$\sigma_\theta - \sigma_r = \frac{2}{\sqrt{3}} Y \qquad (7\text{-}52)$$

将方程(7-52)代入方程(7-50)得

$$\mathrm{d}\sigma_r = \frac{2}{\sqrt{3}} Y \frac{\mathrm{d}r}{r} \qquad (7\text{-}53)$$

积分得 $\sigma_r = \frac{2}{\sqrt{3}} Y \ln Cr$，利用边界条件确定积分常数 C，当 $r=b$，$\sigma_r=0$ 时，$C=\frac{1}{b}$。联立方程(7-52)得到塑性圆筒的应力解

$$\begin{cases} \sigma_r = \dfrac{2}{\sqrt{3}} Y \ln \dfrac{r}{b} \\[3mm] \sigma_\theta = \dfrac{2}{\sqrt{3}} Y \left(1 + \ln \dfrac{r}{b}\right) \end{cases} \tag{7-54}$$

分析式(7-54)可知，截面的 σ_θ 总为拉应力，σ_r 总为压应力。当 $r=a$ 时，有最大的压力 p，即

$$p = \frac{2}{\sqrt{3}} Y \ln \frac{b}{a} \tag{7-55}$$

7.13 圆柱体镦粗变形力计算的主应力法

主应力法实际上仍然是平衡微分方程与屈服准则的联立求解。但在工程应用中，为了简化计算，需采用一些基本假设。

主应力法基本步骤如下：

(1)将问题近似按轴对称问题或平面问题来处理，并选用相应的坐标系。

(2)根据某瞬时变形体的变形趋向，截取包括接触平面在内的典型单元，在接触面上有正应力和剪应力(摩擦力)，且假设在其他截面(非接触面)上仅有均匀分布的正应力(主应力)，从而使平衡方程的数量缩减至一个，而且将偏微分方程变为常微分方程。

(3)确定单元塑性条件时，忽略了摩擦切向应力的影响，假设其上的正应力为主应力，从而得到简化的屈服准则。

(4)联立求解上述平衡微分方程与屈服准则，可求得接触面上的应力分布。

平行模板间圆柱体镦粗受力分析示意图如图 7-16 所示，由于镦粗物体是轴对称的，所受载荷也是轴对称的，因此属于轴对称问题。

考虑金属沿径向流动，选取图 7-16 所示的单元，瞬时高度为 h，厚度为 dr 的扇形坯料，其应力分别为 σ_z，σ_r，σ_θ 及 τ_{zr}。

由沿 r 方向平衡微分方程可得

$$(\sigma_r + d\sigma_r)(r + dr)d\theta h - \sigma_r d\theta h - 2\sigma_\theta \sin\frac{d\theta}{2} dr h + 2\tau_{zr} r d\theta dr = 0 \tag{7-56}$$

重新排列并略去高阶项，方程(7-56)可写为

$$\frac{d\sigma_r}{dr} + \frac{2\tau_{zr}}{h} + \frac{\sigma_r - \sigma_\theta}{r} = 0 \tag{7-57}$$

对实心圆柱镦粗，径向应变 $\varepsilon_r = dr/r$，而切向应变为

$$\varepsilon_\theta = \frac{2\pi(r + dr) - 2\pi r}{2\pi r} = \frac{dr}{r} \tag{7-58}$$

很显然，两者相等，根据应力-应变关系有

$$\sigma_r = \sigma_\theta \tag{7-59}$$

将方程(7-59)代入方程(7-57)，可得

$$\frac{d\sigma_r}{dr} + \frac{2\tau_{zr}}{h} = 0 \tag{7-60}$$

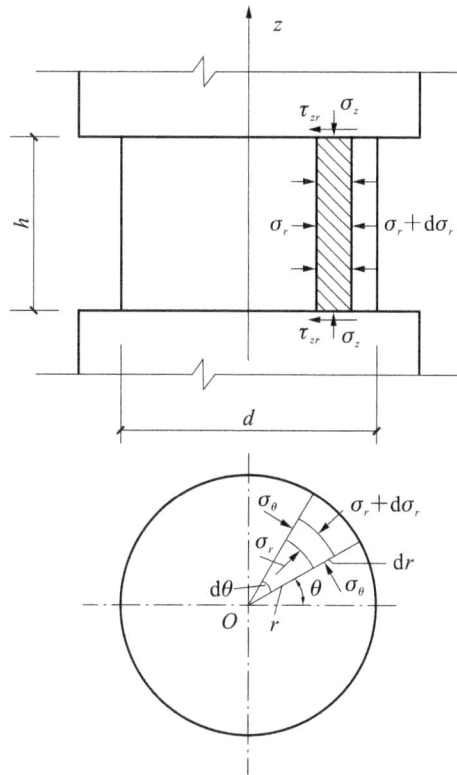

图 7-16 圆柱体镦粗单元受力分析

假设摩擦条件为 $\tau_{zr} = \begin{cases} \mu Y \\ 0 \\ \mu\sigma_z \end{cases}$ 。

若 τ_{zr} 取 $Y/2$，则方程(7-60)变为

$$\frac{\mathrm{d}\sigma_r}{\mathrm{d}r} = -\frac{Y}{h} \tag{7-61}$$

由于 $\varepsilon_r = \varepsilon_\theta > 0$，$\varepsilon_z < 0$，考虑应力的符号，显然 $(-\sigma_r) = (-\sigma_\theta) > (-\sigma_z)$，忽略摩擦力 τ_{zr}，特雷斯卡屈服准则 $\sigma_{\max} - \sigma_{\min} = Y$ 变为

$$(-\sigma_r) - (-\sigma_z) = Y \tag{7-62}$$

即

$$\sigma_z - \sigma_r = Y \tag{7-63}$$

可得

$$\frac{\mathrm{d}\sigma_z}{\mathrm{d}r} = \frac{\mathrm{d}\sigma_r}{\mathrm{d}r} \tag{7-64}$$

联立求解方程(7-61)与方程(7-64)，得

$$\mathrm{d}\sigma_z = -\frac{Y}{h}\mathrm{d}r \tag{7-65}$$

对方程(7-65)积分得

$$\sigma_z = -\frac{Y}{h}r + C \tag{7-66}$$

如果整个圆柱镦粗体全部发生塑性变形，并且根据边界条件，当 $r = d/2$ 时，$\sigma_r = 0$，则方程(7-63)变为

$$\sigma_z = Y \tag{7-67}$$

那么

$$C = Y + \frac{Y}{h} \frac{d}{2} \tag{7-68}$$

将 C 代入方程(7-66)得圆柱体镦粗接触面应力分布为

$$\sigma_z = Y\left[1 + \frac{1}{h}\left(\frac{d}{2} - r\right)\right] \tag{7-69}$$

若取边界摩擦 $\tau_{zr} = \mu\sigma_z$，可求得圆柱体镦粗接触面上应力分布

$$\sigma_z = Y\exp\left[\frac{2\mu(0.5d - r)}{h}\right] \tag{7-70}$$

若取边界摩擦条件为 $\tau_{zr} = \mu Y$，可求得圆柱体镦粗接触面上应力分布为

$$\sigma_z = Y\left[1 + \frac{2\mu}{h}(0.5d - r)\right] \tag{7-71}$$

7.14 正交性和屈服表面

流动法则的一个物理解释是主应力轴和应变轴重合。为了说明这一点，我们将这些分量在三维空间(屈服面)或在二维空间(屈服轨迹)中用矢量表示是十分有用的。再次强调，这种方法与张量变换无关;相反，每个应力或应变分量的方向和大小定义了所绘制图中对应的相关矢量。

对于正交性问题，我们已经提出了几种研究结果，并得出了总应变矢量必须垂直于屈服面的结论。德鲁克基于塑性功为正的概念进行了最严谨的证明。因此，任何可接受的屈服面一定是绕其原点凸出的，或者穿过该表面的直线与该平面的交点不会多于两个。

这里有两个简单的物理示例。首先考虑图 7-17，假设 σ_1 是主应力，材料是各向同性的。图 7-17 中显示了两种可能的形状变化。仅凭直觉，图 7-17(a)中所示变化远比图 7-17(b)中所示变化发生的可能性要大。因此，应力和应变方向一致。

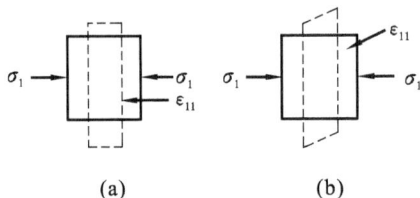

图 7-17 说明主应力和主应变方向的可能情况

现在考虑位于图 7-18 所示表面上的块体，其中图 7-18(a)所示摩擦存在于整个接触面上，俯视图 7-18(b)表示力 F 以一定角度 θ 作用于块体。由于 F 必须克服摩擦力才能使块体运动，且在 θ 减少到 0 时产生最大的功，因此，当力作用方向与块体运动方向相同时，预计块体将沿 x 方向移动且做功最大。这是最大功原理的一个简单例子，做功最大将导致最大的能量耗散，从而降低系统的总能量(在热力学中，外部系统中力做的功通常被认为是正的)。可以推断正在变形的材料提供了最大的抗塑性变形能力。为了实现这一点，应变向量必须与屈服面垂直。

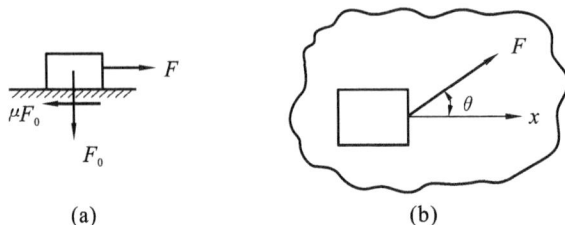

图 7-18 最大功原理说明示意图

现在将合理详细地指出在三维应力空间中屈服表面的真实发展过程。坐标系和主应力方向有关,如图 7-19 所示。点 P 处的应力状态由主应力引起,其中 $\sigma_1=OP_1$,$\sigma_2=OP_2$,$\sigma_3=OP_3$,因此 OP 为该应力空间中总的应力。

在这个坐标系中,考虑直线 OH,具有相同的方向余弦(即 $l=m=n=1/\sqrt{3}$),即 $\alpha=\beta=\gamma=54°44'$。

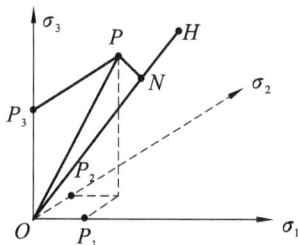

图 7-19 表示总应力 OP、静水应力 ON 及偏应力 NP 的三维应力空间

直线 OP 在 OH 上的投影点为点 N,故 $\angle ONP=90°$,那么

$$ON=\frac{1}{\sqrt{3}}(\sigma_1+\sigma_2+\sigma_3)$$

$$(NP)^2=(OP)^2-(ON)^2=(\sigma_1^2+\sigma_2^2+\sigma_3^2)-\left(\frac{\sigma_1+\sigma_2+\sigma_3}{\sqrt{3}}\right)^2$$

$$=\sigma_1^2+\sigma_2^2+\sigma_3^2-\frac{1}{3}[\sigma_1^2+\sigma_2^2+\sigma_3^2+2(\sigma_1\sigma_2+\sigma_2\sigma_3+\sigma_3\sigma_1)]$$

$$=\frac{2}{3}(\sigma_1^2+\sigma_2^2+\sigma_3^2)-\frac{2}{3}(\sigma_1\sigma_2+\sigma_2\sigma_3+\sigma_3\sigma_1)$$

$$=\frac{1}{3}[(\sigma_1-\sigma_2)^2+(\sigma_2-\sigma_3)^2+(\sigma_3-\sigma_1)^2]$$

根据米泽斯屈服准则,当 $\sum(\sigma_1-\sigma_2)^2=2Y^2$ 时屈服发生,所以

$$3(NP)^2=2Y^2 \quad \text{或} \quad NP=\frac{\sqrt{2}Y}{\sqrt{3}}$$

因此,屈服轨迹可以表示为一个半径为 $NP=\sqrt{2}Y/\sqrt{3}$ 的圆,在主应力空间中,这是一个半径为 $\sqrt{2}Y/\sqrt{3}$ 的圆柱,该圆柱的轴线是一条经过原点并与三个坐标轴方向成等倾角的直线。

图 7-20 展示了三维的米泽斯屈服准则中的"圆柱"和特雷斯卡屈服准则中的"六边形"。注意,$\sigma_2=0$ 平面与该圆柱表面相切将得到更熟悉的两个准则中的屈服轨迹。在同一坐标系上将主应变叠加,应变增量的矢量和表示为 $d\varepsilon^p$,该矢量与屈服表面垂直,并且总是由分量 $d\varepsilon_{11}^p$,$d\varepsilon_{22}^p$ 和 $d\varepsilon_{33}^p$ 组成,且它们的方向与 σ_1,σ_2 和 σ_3 所在轴的方向一致。因此,总应变矢量 $d\varepsilon^p$ 的方向仅取决于屈服表面的形状。只要点 P 位于屈服表面内,就不会发生屈服,但当点 P 接近屈服表面时,马上就会出现塑性流动。注意,ON 表示静水应力分量并会沿 OH 移动(即增加或减少 σ_m),它不影响屈服,NP 表示控制屈服的应力偏量。

如果将该应力投影到通过点 O 且垂直于 OH 的平面上(即从圆柱顶部沿平行于 OH 的方向向下看),结果如图 7-21 所示。如前文所述,这被称为 π 平面,其方程如下:

$$\sigma_1+\sigma_2+\sigma_3=0 \quad (\text{即} \ \sigma_m=0)$$

并且原坐标轴在投影后各轴间角度为 $120°$。由于屈服表面没有膨胀的趋势,任何静水应力分力都垂直于 π 平面,所以不会产生塑性功。然而,垂直于屈服表面的应力偏量分量,如果导致屈服表面膨胀,则会产生塑性功。回到图 7-18 中,当总应变矢量(图 7-20 中的 $d\varepsilon^p$)垂直于屈服表面且与总应力的偏量分量方向一致时,做功最大。

现在根据图 7-22 所示的屈服轨迹考虑上述解释的意义。请注意,虽然 $\sigma_2=0$,但在大多数情况下 $d\varepsilon_{22}^p$

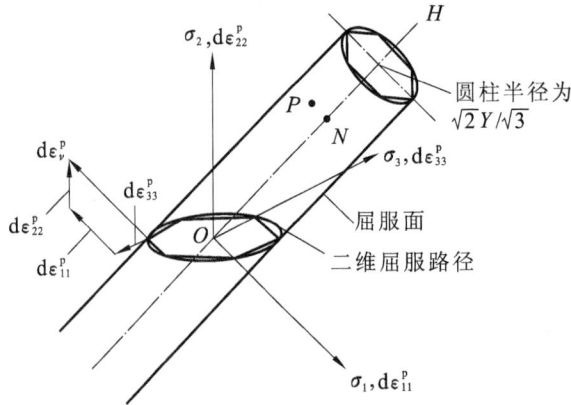

图 7-20　特雷斯卡和米泽斯屈服准则下的屈服表面

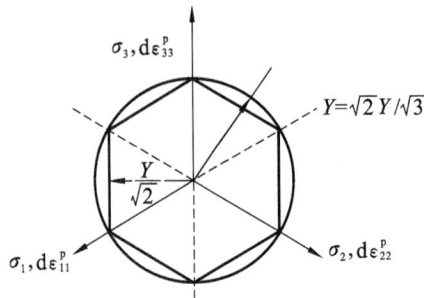

图 7-21　特雷斯卡和米泽斯屈服准则下的屈服表面在 π 平面上的投影

不为零。$d\varepsilon_\nu^p$ 如图 7-22 所示。同时要注意,$\boldsymbol{\sigma}$ 表示 σ_1 和 σ_3 的矢量和,并且可以将它再次分解为另一对表示 σ_m 和应力偏量的分量。

正交性的价值之一是,基于实验结果构建屈服表面和屈服轨迹。贝可芬在许多情况下都证实了这一点。

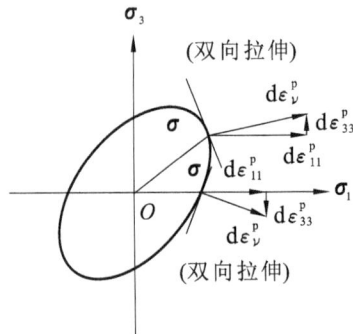

图 7-22　与屈服轨迹相关的正交性原理说明

7.15　塑　性　功

考虑一个长度为 l_0 的棒状体,且在一个面积($w_0 t_0$)上受到拉力 F 作用,并产生塑性伸长 $\mathrm{d}l$。所做的功为 $F\mathrm{d}l$,而单位体积上所做的功为

$$dW_p = \frac{Fdl}{w_0 t_0 l_0} = \frac{F}{w_0 t_0} \cdot \frac{dl}{l_0} = \sigma d\varepsilon^p \tag{7-72}$$

如果剪力引起变形,进行类似的推导可以证明 $\tau d\gamma$ 表示在该剪力作用下每单位体积所做的功。这些单独贡献可以采用方程(7-10)的方法进行叠加得到,注意方程(7-72)中不会出现系数1/2。采用主应力有

$$dW_p = \sigma_1 d\varepsilon_{11}^p + \sigma_2 d\varepsilon_{22}^p + \sigma_3 d\varepsilon_{33}^p \tag{7-73}$$

通过方程(7-8)和方程(7-26)来定义等效应力和等效应变增量函数时,没有涉及塑性功,但现在我们可以期望在此处可以使用此类定义。事实确实如此,表达式是

$$dW_p = \bar{\sigma} d\overline{\varepsilon^p} \tag{7-74}$$

通过两个例子来证明这个等式。首先考虑轴向拉伸状态,其中 $\sigma_1 \neq 0$, $\sigma_2 = \sigma_3 = 0$ 且 $d\varepsilon_{22}^p = d\varepsilon_{33}^p = -\frac{1}{2}d\varepsilon_{11}^p$(因为 $d\varepsilon_{11}^p + d\varepsilon_{22}^p + d\varepsilon_{33}^p = 0$)。由于其他项可以消去,根据方程(7-73)得到 $dW_v = \sigma_1 d\varepsilon_{11}$。等效应力和等效应变增量函数表示为

$$\bar{\sigma} = \sigma_1$$
$$d\overline{\varepsilon^p} = d\varepsilon_{11}^p$$

因此等式证明成立。

接下来考虑纯剪切状态,其中 $\sigma_1 = -\sigma_3$, $\sigma_2 = 0$, $d\varepsilon_{11}^p = -d\varepsilon_{33}^p$ 且 $d\varepsilon_{22}^p = 0$。根据方程(7-73)有

$$dW_p = \sigma_1 d\varepsilon_{11}^p + (-\sigma_1)(-d\varepsilon_{11}^p) = 2\sigma_1 d\varepsilon_{11}^p$$

根据等效应力和等效应变增量的关系,有

$$\bar{\sigma} = \sqrt{3}\sigma_1 \qquad d\overline{\varepsilon^p} = \frac{2}{\sqrt{3}}d\varepsilon_{11}^p$$

因此

$$dW_p = (\sqrt{3}\sigma_1)\left(\frac{2}{\sqrt{3}}d\varepsilon_{11}^p\right) = 2\sigma_1 d\varepsilon_{11}^p$$

与前面结果一样。

因此,除了在预测屈服和流动法则方面发挥作用外,等效应力和等效应变增量的概念为计算塑性变形引起的功提供了一种简便的方法。

7.16 应力和塑性应变增量莫尔圆的比较

由于总应力的静水应力分量不会引起塑性流动,这意味着应力圆图上的 σ_m 值应与应变增量圆的原点一致(即 $d\varepsilon^p = 0$)。在单轴拉伸状态下,$\sigma_m = \sigma_1/3$, $\sigma_2 = \sigma_3 = 0$,而 $d\varepsilon_{22}^p = d\varepsilon_{33}^p = -d\varepsilon_{11}^p/2$。图 7-23(a)表明了应力圆与应变相对比例关系的物理意义,注意应变增量间是 $-2:1$ 的关系。图 7-23(b)表示更加一般情况下 $\sum d\varepsilon_{ii}^p$ 一定等于 0。这对研究平面应变也具有指导意义。

由于塑性流动状态下 $\nu = \frac{1}{2}$,故 $\sigma_2 = \frac{1}{2}(\sigma_1 + \sigma_3)$。

在这种情况下 $\sigma_m = \sigma_2$,所以应变增量圆的圆心与 $\sigma_2 = 0$ 或 $d\varepsilon_{22}^p = 0$ 一致。可以通过方程(7-25)很容易检验该结果。

我们应该认识到,通过叠加相同的附加均匀正应力改变 σ_m,将使应力圆发生平移,但不会对任何一组圆的大小产生影响。因此,应变比并不能唯一地明确目前的应力状态。

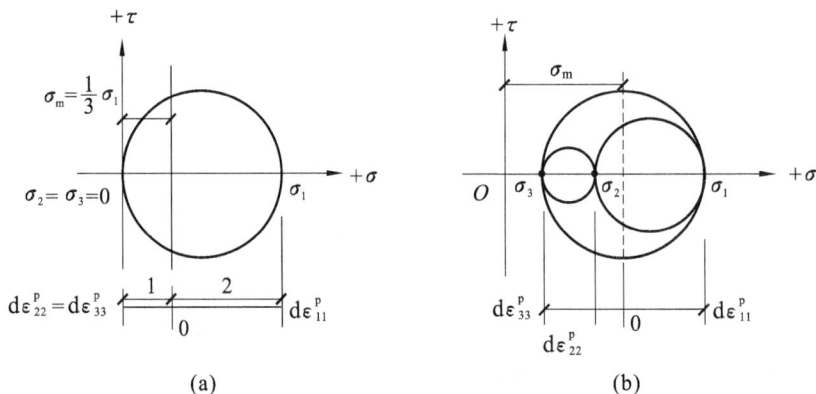

图 7-23 莫尔应力圆与塑性应变增量的关系

(a)单轴拉伸状态下；(b)一般三轴状态下

7.17 几何方面的考虑

应力偏量矢量 **OQ** 可视为矢量 **OL** 分量沿参考坐标轴在偏量平面上投影的和向量,如图 7-24 所示。由于在原来的应力空间中每个应力轴与偏量平面的夹角为 $\arcsin 1/\sqrt{3}$,因此沿每一轴的投影长度是实际长度的 $\sqrt{2/3}$。因此,应力偏量平面中的各矢量分量的长度为 $OL=\sqrt{2/3}\,\sigma_1$,$LM=\sqrt{2/3}\,\sigma_2$,$MQ=\sqrt{2/3}\,\sigma_3$。

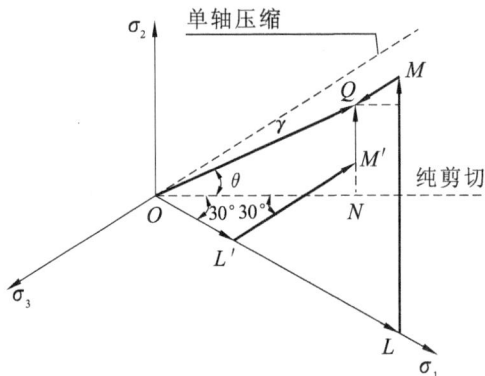

图 7-24 沿投影轴的偏应力矢量及其分量

从几何角度看,**OQ** 相对于水平方向和竖直方向分解得到的互相垂直的两个分量为

$$\begin{cases} ON=\dfrac{\sigma_1-\sigma_3}{\sqrt{2}}=r\cos\theta \\[2mm] QN=\dfrac{2\sigma_2-\sigma_3-\sigma_1}{\sqrt{6}}=r\sin\theta \end{cases} \tag{7-75}$$

式中,(r,θ) 是点 Q 的极坐标。对于屈服准则的实验验证,引入洛德应力参数 μ 会更加方便,其定义如下:

$$\mu=\frac{2\sigma_2-\sigma_3-\sigma_1}{\sigma_3-\sigma_1}=-\sqrt{3}\tan\theta \quad (\sigma_1>\sigma_2>\sigma_3) \tag{7-76}$$

为了获得屈服轨迹,只需施加使 θ 变化范围为 0 至 $\pm\pi/6$,μ 变化范围为 0 至 ∓1 的应力。当 $\mu=0$ 时,$\sigma_2=\frac{1}{2}(\sigma_3+\sigma_1)$,并且此时处于由 $\frac{1}{2}(\sigma_3-\sigma_1,\sigma_1-\sigma_3,0)$ 及静水应力 $\frac{1}{2}(\sigma_3+\sigma_1)$ 表达的纯剪切应力状态。当 $\mu=$

—1 时，$\sigma_1 = \sigma_2$，并且处于单轴压缩状态，其应力状态可表达为 $(\sigma_3 - \sigma_1, 0, 0)$ 及静水应力 σ_1。我们将纯剪切和单轴拉伸（或单轴压缩）应力状态对应的屈服强度分别用 k 和 Y 表示。然后根据方程(7-75)可以得出当 $\theta = 0$ 时，$r = \sqrt{2}k$ 以及当 $\theta = \pi/6$ 时，$r = \sqrt{2/3}Y$。对于大多数金属，k 处于 $Y/2$ 和 $Y/\sqrt{3}$ 之间。方程(7-75)也可以写成以下不同形式：

$$\sigma_1' - \sigma_3' = \sqrt{2}r\cos\theta; \quad \sigma_1' + \sigma_3' = -\sigma_2' = -\sqrt{2/3}r\sin\theta$$

因此，主应力偏量可以表示为

$$\begin{cases} \sigma_1' = \sqrt{\dfrac{2}{3}}r\cos\left(\dfrac{\pi}{6} + \theta\right) \\[2mm] \sigma_2' = \sqrt{\dfrac{2}{3}}r\sin\theta \\[2mm] \sigma_3' = -\sqrt{\dfrac{2}{3}}r\cos\left(\dfrac{\pi}{6} - \theta\right) \end{cases} \tag{7-77}$$

当已知屈服轨迹时，r 是一个关于 θ 的已知函数，方程(7-77)被看作用参数 θ 定义的屈服准则。

当给出矢量 $\boldsymbol{OQ}$ 时，根据 $\sigma_2' = \sqrt{2/3}QN$，我们就可以通过几何图形获得应力偏量。因此，如果点 M' 位于 QN 上，且 $QM'/NM' = 2$，那么 $QM' = \sqrt{2/3}\sigma_2'$。绘制与 $O\sigma_3$ 平行的 $L'M'$。根据几何图形既然 $\boldsymbol{L'M'} = \boldsymbol{OL'} + \boldsymbol{QM'}$，它符合 $OL' = \sqrt{2/3}\sigma_1'$ 和 $L'M' = -\sqrt{2/3}\sigma_3'$。因此，点 L' 和点 M' 定义了应力偏量。静水应力的大小为 $\sqrt{2/3}LL'$，但它并不是由矢量 $\boldsymbol{OQ}$ 定义的。基于方程(7-77)应力偏量的比值为

$$\sigma_1' : \sigma_2' : \sigma_3' = (\sqrt{3} - \tan\theta) : 2\tan\theta : -(\sqrt{3} + \tan\theta) \tag{7-78}$$

将方程(7-77)中的三个方程相乘，并且 $\sigma_1'\sigma_2'\sigma_3' = J_3$，$r = \sqrt{2J_2}$，我们得到

$$J_3^2 = \frac{4}{27}J_2^3 \sin^2 3\theta \tag{7-79}$$

如果给出屈服轨迹的极坐标方程，则 J_2 就是关于 θ 的已知函数。

7.18 各向同性的应变硬化

我们已经看到，当应力偏量矢量的大小增大到一定值，例如应力达到屈服轨迹时，材料发生屈服。如果屈服轨迹不是圆（如米泽斯屈服准则），则屈服应力矢量的值取决于其在应力偏量平面中的最终方向。如果材料未发生应变硬化现象，塑性应力状态总是以应力点始终位于恒定屈服轨迹上的方式发生变化。对于应变硬化材料，屈服轨迹的大小和形状取决于自前次退火后的塑性变形的完整过程。假设材料在退火状态下是各向同性的，并且可以忽略在冷加工过程中产生的各向异性和包申格效应。上述对屈服准则的讨论适用于任何给定的材料的应变硬化状态。

假设应力状态如图 7-25 所示沿着应力空间中的某一路径 P_0P 发生变化，屈服面均匀扩展而不改变形状，则可得到一个简便的应变硬化的数学公式，该应变硬化量由最终塑性状态确定。由于屈服轨迹只是尺寸增加，任何给定的应变硬化状态都可以由当前单轴拉伸的屈服应力来定义。因此，揭示当前屈服应力与初始屈服状态后的塑性变形量之间的关系是十分有必要的。为此，我们将屈服准则中的 Y 用被称为等效应力的 $\bar{\sigma}$ 代替。参考米泽斯屈服准则，我们可以写出

$$\begin{aligned} \bar{\sigma} &= \sqrt{\frac{3}{2}}(\sigma_x'^2 + \sigma_y'^2 + \sigma_z'^2 + 2\tau_{xy}^2 + 2\tau_{yz}^2 + 2\tau_{zx}^2)^{\frac{1}{2}} \\ &= \sqrt{\frac{1}{2}}[(\sigma_x - \sigma_y)^2 + (\sigma_y - \sigma_z)^2 + (\sigma_z - \sigma_x)^2 + 6\tau_{xy}^2 + 6\tau_{yz}^2 + 6\tau_{zx}^2]^{\frac{1}{2}} \end{aligned} \tag{7-80}$$

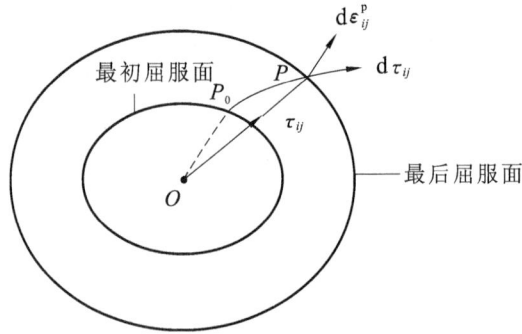

图 7-25　各向同性硬化规则的几何表示

首先假设应变硬化量是关于单位体积总塑功能的函数。这一假设显然与纯弹性应变不会产生应变硬化的事实相一致。如果应变增量张量的塑性部分表示为 $\mathrm{d}\varepsilon_{ij}^{\mathrm{p}} = \dot{\varepsilon}_{ij}^{\mathrm{p}}\mathrm{d}t$，其中 $\dot{\varepsilon}_{ij}^{\mathrm{p}}$ 为塑性变形速率，$\mathrm{d}t$ 为时间单元，则单位体积塑性功增量为

$$\mathrm{d}W_{\mathrm{p}} = \tau_{ij}\,\mathrm{d}\varepsilon_{ij}^{\mathrm{p}} = (\tau_{ij}' + \sigma_{\mathrm{m}}\delta_{ij})\mathrm{d}\varepsilon_{ij}^{\mathrm{p}} = \tau_{ij}'\,\mathrm{d}\varepsilon_{ij}^{\mathrm{p}}$$

其中当 $i = j$，$\tau_{ij} = \sigma_i$，否则 $\tau_{ij}' = \tau_{ij}$，$\mathrm{d}\varepsilon_{ij}^{\mathrm{p}} = 0$ 意味着塑性体积没有变化。金属的塑性不可压缩性与实验观察结果非常一致，也与均匀静水应力不产生塑性应变的事实一致。应变硬化假设在数学上可以表述为

$$\bar{\sigma} = \Phi\left(\int \mathrm{d}\omega_{\mathrm{p}}\right) = \Phi\left(\int \tau_{ij}\,\mathrm{d}\varepsilon_{ij}\right) \tag{7-81}$$

上述方程从某个初始状态开始，对实际应变路径进行积分。函数 Φ 可由单轴拉伸或单轴压缩时的真实应力-应变曲线确定。如果根据应变的塑性部分绘制出真实应力 σ，那么 W_{p} 就等于曲线下部分和纵坐标 σ 之间构成的面积。因为在这种情况下 $\bar{\sigma} = \sigma$，方程(7-81)中的参数可以通过面积得到。

在另一种更常用的假设中，$\bar{\sigma}$ 被认为是总塑性应变的某个表征量的函数。考虑塑性应变增量张量的第二应变增量不变量，等效塑性应变增量定义如下：

$$\mathrm{d}\,\overline{\varepsilon^{\mathrm{p}}} = \sqrt{\frac{2}{3}}\,(\mathrm{d}\varepsilon_{ij}^{\mathrm{p}}\,\mathrm{d}\varepsilon_{ij}^{\mathrm{p}})^{\frac{1}{2}}$$
$$= \sqrt{\frac{2}{3}}\left[(\mathrm{d}\varepsilon_{xx}^{\mathrm{p}})^2 + (\mathrm{d}\varepsilon_{yy}^{\mathrm{p}})^2 + (\mathrm{d}\varepsilon_{zz}^{\mathrm{p}})^2 + 2\,(\mathrm{d}\gamma_{xy}^{\mathrm{p}})^2 + 2\,(\mathrm{d}\gamma_{yz}^{\mathrm{p}})^2 + 2\,(\mathrm{d}\gamma_{zx}^{\mathrm{p}})^2\right]^{\frac{1}{2}} \tag{7-82}$$

该方程只有正根，上述表达式中数学因子的选择是为了单轴拉伸状态下 $\mathrm{d}\,\overline{\varepsilon^{\mathrm{p}}}$ 等于纵向塑性应变增量。事实上，因为各向同性杆状物拉伸试验中，横向压缩塑性应变的大小是纵向拉伸塑性应变的一半，所以应变硬化假设可以表示为

$$\bar{\sigma} = F\left(\int \mathrm{d}\,\overline{\varepsilon^{\mathrm{p}}}\right) = F\left(\sqrt{\frac{2}{3}\mathrm{d}\varepsilon_{ij}^{\mathrm{p}}\,\mathrm{d}\varepsilon_{ij}^{\mathrm{p}}}\right) \tag{7-83}$$

式中，像前文一样沿应变路径进行积分。积分得到的应变，即总等效塑性应变，为塑性变形提供了合适的表征量。函数 F 由单轴拉伸或压缩状态真实应力和塑性应变之间的关系给出。方程(7-83)表明，应变硬化量由导致单元体最终形状的每次微小塑性扭曲变形确定，而不只是由单元体初始形状和最终形状之间的差异决定。

7.19　结　　语

在本章中，我们首先介绍了弹性与塑性性能的差别以及在塑性问题研究中采用的三种不同的解决方法。阐释了多种塑性变形模型。在各向同性、没有包申格效应、塑性变形不发生体积变化、屈服不受流体

静力效应影响的假设下,已经有特雷斯卡和米泽斯两种屈服准则可以用来预测屈服开始。我们解释了米泽斯屈服准则是引起扭曲的弹性能和等八面体剪应力。我们给出了普朗特-罗伊斯和莱维-米泽斯方程,并且说明了莱维-米泽斯方程是普朗特-罗伊斯方程的一种特殊形式。本章也给出了另一种被称为全量理论的关于总应力分量与应变分量之间关系的塑性理论。几种有关正交性问题的证明被提出,并得到了总的应变矢量必须垂直于屈服表面的结论。等效应力和等效应变提供了一种计算塑性功的简便方法。我们也给出了应力偏量矢量的几何解释并且讨论了各向同性应变硬化下的屈服准则。

习题

7-1 证明应力偏量不变量 J_2 等于

$$\frac{1}{6}\left[(\sigma_1-\sigma_2)^2+(\sigma_2-\sigma_3)^2+(\sigma_3-\sigma_1)^2\right]$$

7-2 金属的拉伸屈服强度表示为 Y MPa。如果该材料的试件承受沿 x 和 y 方向上作用的两个正交压应力,分别为 $-\frac{1}{4}Y$ 和 $-\frac{1}{2}Y$,那么在 z 方向上施加的拉伸应力多大时会发生屈服?

(1)根据米泽斯屈服准则求解。

(2)根据特雷斯卡屈服准则求解。

7-3 一个边长为 25 mm 的立方体金属块被施加 250 kN 的压缩荷载时发生屈服。假设加载面处的摩擦力可以忽略不计(即不存在剪应力)。如果另一个相同的立方体在另两个正交面上分别受到 90 kN 和 135 kN 的压缩荷载的约束(同样忽略摩擦力),那么利用米泽斯屈服准则求出第三个坐标方向上的压缩荷载多大时会导致屈服?

7-4 一个由金属制成的端部封闭的薄壁长管,其单轴拉伸屈服强度为 280 MPa。该管长 1500 mm,壁厚为 0.38mm,直径 50mm。该薄壁长管承受轴向拉伸荷载 4.5kN,扭矩为 0.115 kN·m,并内部受压。那么内部压力多大时会发生屈服?

(1)根据特雷斯卡屈服准则求解。

(2)根据米泽斯屈服准则求解。

7-5 一个端部为半球形的圆筒压力容器,其半径为 600 mm,其制造所用金属材料的剪切屈服强度 k 为 560 MPa。使用期间预期的最大内部压力为 35 MPa。如果容器的任何部分都不发生屈服,那么

(1)根据特雷斯卡屈服准则求解容器的最小壁厚是多少?

(2) 根据米泽斯屈服准则求解容器的最小壁厚是多少?

7-6 分别考虑(a)单轴拉伸,(b)纯剪切,(c)三维应力状态下,假设 $\sigma_1>\sigma_2>\sigma_3$,比较每种情况下等效应力 $\bar{\sigma}$ 与最大剪应力 $\tau_{\max}$ 的比值。

7-7 当一个薄壁圆筒(直径 80 mm,壁厚 3.5 mm)仅施加 200 MPa 的均匀轴向应力时发生屈服。如果相同的薄壁圆筒承受了 140 MPa 的最大轴向正应力,则使用米泽斯屈服准则计算导致屈服所需的内部压力是多少。

7-8 由纯剪切屈服强度为 105 MPa 的金属制成的一个端部封闭的薄壁圆管,其中直径为 150 mm,壁厚为 1.9 mm 且长度为 500 mm,弹性模量为 1.4×10^5 MPa 且泊松比为 0.25。如果该薄壁圆管受到大于 3.5 MPa 的内部压力时开始发生屈服,要求根据米泽斯屈服准则确定圆管的最终长度。

7-9 考虑平面应变中的塑性变形($\mathrm{d}\varepsilon_{22}^{\mathrm{p}}=0$),使用米泽斯屈服准则,证明 $\sigma_1-\sigma_3=2\bar{\sigma}/\sqrt{3}$。

7-10 某金属屈服强度为常数,即 $\bar{\sigma}=Y=C$(C 为常数)。如果该金属在单轴拉伸作用下产生应变 ε,且不引起颈缩,证明单位体积的塑性功为 $\bar{\sigma}\bar{\varepsilon}^{\mathrm{p}}$。

7-11 某立方体由屈服强度为 350 MPa 的材料制成,并在一个坐标轴方向上承受拉应力 σ_1,在第二

个坐标轴方向上承受应力 $\sigma_3 = -\sigma_1/2$。

(1)确定主应变增量 $d\varepsilon_{11}^p / d\varepsilon_{22}^p$ 的比值。

(2)使用米泽斯屈服准则,确定屈服开始时 τ_{max} 的大小。

(3)使用米泽斯屈服准则,当 $\sigma_3 = +\dfrac{1}{2}\sigma_1$ 时,确定屈服开始时 τ_{max} 的大小。

(4)使用特雷斯卡屈服准则计算(2)和(3)中的结果。

7-12 应力状态采用 $\sigma_1 = 30$,$\sigma_2 = 15$ 以及 $\sigma_3 = 0$ 来描述(单位:MPa)。

(1)确定 $d\varepsilon_{11}^p / d\varepsilon_{22}^p$ 的比值。

(2)如果在初始应力状态的流体压力上叠加了 20 MPa 的静水压力,那么(1)中的比值如何变化?请做出解释。

7-13 当施加 200 MPa 的均匀轴向应力时,薄壁圆筒(直径为 80 mm,壁厚为 3.5 mm)刚好屈服。如果相同的圆筒承受由弯曲产生的 140 MPa 的最大轴向正应力,则使用特雷斯卡屈服准则计算导致屈服所需的内部压力是多少(参考习题 7-7)。

7-14 当扭转剪应力达到 306 MPa 时,薄壁圆筒刚好屈服。如果相同的圆筒承受 270 MPa 的扭转剪应力,使用特雷斯卡屈服准则,计算屈服时所需的轴向压缩应力是多大?

7-15 一个刚塑性立方体在第一对相对面上承受应力为 σ_1,第二对相对面上承受应力为 $\sigma_2 = 0.2\sigma_1$,在第三对相对面上承受应力为 $\sigma_3 = -0.4\sigma_1$。保持上述比率,应力逐渐增加。使用米泽斯屈服准则($Y = 300$ MPa),计算屈服时的主应力大小以及三个主应变增量的比值。

7-16 米泽斯屈服准则可用于给定金属的屈服预测。在 σ_3 为零的平面应变变形下,在 σ_1-σ_2 应力空间的第一象限中画出米泽斯屈服准则表示的椭圆图,并且

(1)在正交性情况下,说明应变增量 $d\varepsilon_{11}^p$ 和 $d\varepsilon_{22}^p$ 在图上是什么样的。

(2)根据主应力和主应变增量确定塑性功增量。

(3)参考问题(1)中的图,解释问题(2)中的答案。

8 弹塑性分析实例

8.1 引　言

变形体受到不断增大的外部荷载作用时,在临界应力首次达到屈服准则时会出现塑性流动。随着荷载的进一步增大,材料的塑性区不断扩展,其弹性区域与塑性区域由弹塑性边界分隔。关于该边界位置的问题是未知的,并且其形状通常是很复杂的,以至于对于边界值问题的求解常常涉及数值方法。即使变形限制在弹性数量级,也必须对一系列的小应变增量问题求解。确保每个阶段计算的弹性和塑性区域中应力和位移在弹塑性边界处满足连续性条件是必要的。在本章中,我们将主要关注假设在小变形情况下,弹塑性范围内弯曲和扭转问题。

8.2　块体的平面应变压缩问题

作为运用普朗特-罗伊斯理论的一个简单问题,考虑无摩擦受压且在一对刚性叠合板之间的矩形金属块,如图 8-1 所示。金属块体的各面平行于矩形轴,设 x 轴为金属块的压缩方向。

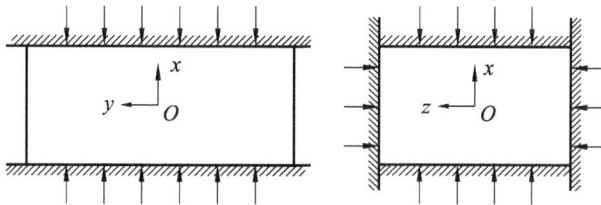

图 8-1　平面应变条件下光滑刚性模板之间块体的塑性压缩

因为刚性模板限制了金属块沿 z 方向的横向膨胀,所以以问题变为平面应变问题。因此,在均匀压缩的情况下,整个变形过程中 $\sigma_y = 0$,当金属块仍处于弹性阶段时,$\sigma_z = \nu\sigma_x$。如果采用特雷斯卡屈服准则,当 $\sigma_x = -Y$ 时,金属块中的每一个单元都开始发生屈服。在塑性阶段相关的应力-应变方程为

$$\begin{cases} \mathrm{d}\varepsilon_{xx} = \dfrac{1}{E}(\mathrm{d}\sigma_x - \nu\mathrm{d}\sigma_z) + \dfrac{1}{3}(2\sigma_x - \sigma_z)\mathrm{d}\lambda \\ \mathrm{d}\varepsilon_{zz} = \dfrac{1}{E}(\mathrm{d}\sigma_z - \nu\mathrm{d}\sigma_x) + \dfrac{1}{3}(2\sigma_z - \sigma_x)\mathrm{d}\lambda = 0 \end{cases} \tag{8-1}$$

如果材料不存在硬化现象,那么在整个塑性压缩过程中都有 $\sigma_x = -Y$。消去方程(8-1)中的 $\mathrm{d}\lambda$,可以得到

$$E\mathrm{d}\varepsilon_{xx} = \left(\frac{1}{2} - \nu\right)\mathrm{d}\sigma_x + \frac{3\mathrm{d}\sigma_z}{2(2\sigma_z + Y)}$$

在发生初始屈服时,有 $\sigma_z = -\nu Y$,$\varepsilon_{xx} = -(1-\nu^2)Y/E$。在这些初始条件下,对上述方程积分有

$$\frac{E}{Y}\varepsilon_{xx} = \left(\frac{1}{2} - \nu\right)\left(\frac{\sigma_z}{Y} + \nu\right) - \frac{3}{4}\ln\frac{1-2\nu}{1+\dfrac{2\sigma_z}{Y}} - (1-\nu^2) \tag{8-2}$$

这就给出了 σ_z 随压缩量的变化规律。随着变形的进行,第一项变得不再重要,而 σ_z 迅速逼近极值 $-(1/2)Y$。例如令 $\nu = 0.3$,可以看出,当 ε_{xx} 为初始屈服时应变的 3.5 倍时,σ_z 的值为 $-0.49Y$。

根据米泽斯屈服准则,当材料屈服时有 $\sigma_x^2 - \sigma_x\sigma_z + \sigma_z^2 = Y^2$,金属块体初始屈服时有 $\sigma_x = \sigma_x^0$,$\sigma_z = \nu\sigma_x^0$,其中

$$\sigma_x^0 = -\frac{Y}{\sqrt{1-\nu+\nu^2}}$$

在随后的压缩过程中,通过用参数 θ 表示如下应力,屈服准则可以恒成立。

$$\begin{cases} \sigma_x = -\dfrac{2Y}{\sqrt{3}}\cos\theta \\[2mm] \sigma_z = -\dfrac{2Y}{\sqrt{3}}\sin\left(\dfrac{\pi}{6} - \theta\right) \end{cases} \tag{8-3}$$

在发生初始屈服时,$\sigma_z = \nu\sigma_x$ 提供初始的 θ 值,即

$$\theta_0 = \arctan\frac{1-2\nu}{\sqrt{3}} \tag{8-4}$$

当 $\nu = 0.3$ 时,我们得到 $\sigma_x^0 \approx -1.127Y$ 以及 $\theta_0 \approx 13°$。将方程(8-3)代入应力-应变关系方程(8-1)中,得

$$\mathrm{d}\varepsilon_{xx} = \frac{2Y}{\sqrt{3}E}\left[\sin\theta - \nu\cos\left(\frac{\pi}{6} - \theta\right)\right]\mathrm{d}\theta - \frac{2Y}{\sqrt{3}}\cos\left(\frac{\pi}{6} - \theta\right)\mathrm{d}\lambda$$

$$0 = \frac{2Y}{\sqrt{3}E}\left[\cos\left(\frac{\pi}{6} - \theta\right) - \nu\sin\theta\right]\mathrm{d}\theta + \frac{2Y}{\sqrt{3}}\sin\theta\mathrm{d}\lambda$$

由于 $\mathrm{d}\lambda$ 必须是正的,第二个方程表明 θ 在压缩过程中会减小。消去以上方程的 $\mathrm{d}\lambda$,我们得到

$$E\mathrm{d}\varepsilon_{xx} = \frac{2Y}{\sqrt{3}}\left[(1-2\nu)\cos\left(\frac{\pi}{6} - \theta\right) + \frac{3}{4}\frac{1}{\sin\theta}\right]\mathrm{d}\theta$$

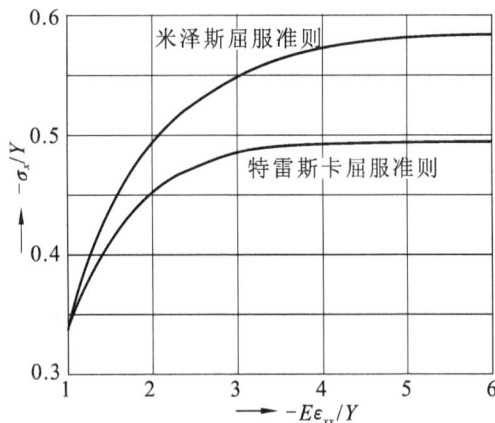

图 8-2 在塑性范围内块体平面应变压缩

利用初始条件,即当 $\theta = \theta_0$ 时,$\varepsilon_{xx} = -(1-\nu^2)\sigma_x^0/E$,对上述方程进行积分可得到

$$-\frac{E}{Y}\varepsilon_{xx} = \frac{2}{\sqrt{3}}(1-2\nu)\sin\left(\frac{\pi}{6} - \theta\right) + \frac{\sqrt{3}}{2}\ln\left(\cot\frac{\theta}{2}\tan\frac{\theta_0}{2}\right) + \sqrt{1-\nu+\nu^2} \tag{8-5}$$

随着变形的继续,方程右边的第一项很快就可以忽略不计了。角 θ 迅速逼近极值零,此时对应的 σ_x 和 σ_z 分别为 $-2Y/\sqrt{3}$ 和 $-Y/\sqrt{3}$。我们可以发现,当 ε_x 仅为初始屈服 $\nu = 0.3$ 时对应应变的 4 倍时,σ_z 值是在极值的 1% 的范围内。由于应力初始变化快速,弹塑性应变增量相当于弹性极限时总应变的 3~4 倍。基于特雷斯卡和米泽斯屈服准则的计算结果图形比较,如图 8-2 所示。

8.3　弯曲梁平面应变问题

两端受到力偶作用的均匀矩形梁在平面应变条件下产生弯曲变形相关问题,如图 8-3 所示。假定弯曲梁的曲率半径大于梁高 $2h$,那么,我们可以忽略其横向应力。中性纤维轴与 Ox 轴重合,弯曲后成半径为 R 的圆弧。在弯曲过程中,中性纤维所在直线以上的所有纤维都被拉伸,以下的所有纤维则被压缩。只要梁仍保持弹性状态,纵向应力 σ_x 则会根据下述关系在梁的高度方向上呈线性分布。

$$\sigma_x = \frac{Ey}{(1-\nu^2)R} = \frac{My}{I_z}$$

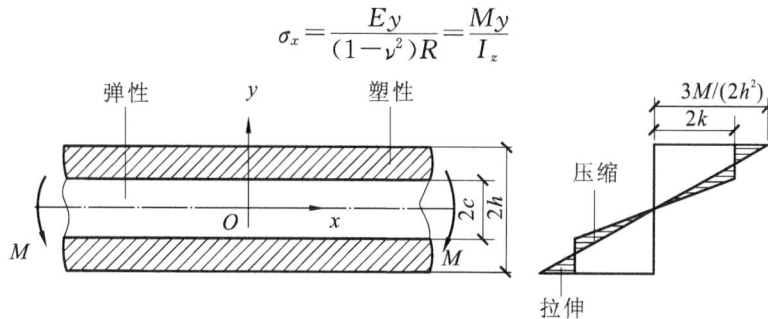

图 8-3　梁在平面应变弯曲中的(近似)几何形状和应力分布(残余应力由阴影三角形绘出)

式中,M 是梁单位宽度上的弯曲力偶,$I_z = (2/3)h^3$ 是梁横截面上每单位宽度绕 z 轴的惯性矩。在弯曲过程中系数 $(1-\nu^2)$ 是根据平面应变条件 $(\varepsilon_{zz}=0)$ 得到的。当纵向应力达到 $\pm Y/\sqrt{(1-\nu+\nu^2)}$ 时,在边界 $y=\pm h$ 处开始出现塑性屈服。那么,发生初始屈服时的弯矩 M_e 以及中性面相应的曲率半径 R_e 为

$$\begin{cases} M_e = \dfrac{2h^2Y}{3}\dfrac{1}{\sqrt{1-\nu+\nu^2}} \\[2ex] R_e = \dfrac{Eh\sqrt{1-\nu+\nu^2}}{Y(1-\nu^2)} \end{cases} \tag{8-6}$$

式中,下标 e 表示弹性极限。当 $\nu=0.3$ 时,M_e 和 R_e 的数值分别为 $0.751Yh^2$ 和 $0.934Eh/Y$。

如果弯矩进一步增大,塑性区会从外表面向内扩展,任何阶段中的弹性部分的高度用 $2c$ 表示。弹性区域的应力为

$$\sigma_x = \frac{Yy}{c\sqrt{1-\nu+\nu^2}}; \quad \sigma_z = \nu\sigma_x \quad (-c \leqslant y \leqslant c) \tag{8-7}$$

整个弯曲过程中,横截面上任意点的纵向应变为 y/R。在弹塑性弯曲过程中,梁弹性部分应用胡克定律,可得

$$R = \frac{Ec\sqrt{1-\nu+\nu^2}}{Y(1-\nu^2)} = \frac{c}{h}R_e$$

在下部塑性区,应力可根据方程(8-3)得到,其中 θ 与 y 有关,且方程(8-5)中 $\varepsilon_{zz}=y/R$。上部塑性区的应力和应变值与下部塑性区的大小相同、符号相反。在单位宽度上作用的力偶为

$$M = 2\int_0^h \sigma_x y\,\mathrm{d}y = \frac{2Yc^2}{3\sqrt{1-\nu+\nu^2}} + \frac{2Y}{\sqrt{3}}\int_c^h y\cos\theta\,\mathrm{d}y \tag{8-8}$$

利用方程(8-5),且用 y/R 代替 $-\varepsilon_{zz}$,我们就可以对上述积分进行数值计算,对任意假定 c/h,可得到 $M/(h^2Y)$ 的值。

为了便于实际应用,用修正的特雷斯卡屈服准则 $\sigma_x = \pm 2Y/\sqrt{3}$ 来代替米泽斯屈服准则可以确保足够精确。那么,纵向应力的大小从中性面处的零增加到弹塑性边界处的 $2Y/\sqrt{3}$。在这种情况下进行积分是

比较简单的,结果是

$$M \approx \frac{2Y}{\sqrt{3}}\left(h^2 - \frac{1}{3}c^2\right) = \frac{1}{2}M_e\left[3 - \left(\frac{R}{R_e}\right)^2\right] \tag{8-9}$$

其中

$$M_e \approx \frac{4Yh^2}{3\sqrt{3}}; \quad R_e \approx \frac{\sqrt{3}Eh}{2Y(1-\nu^2)}$$

这种近似产生的最大误差约为 2%,发生在初始屈服时。弯矩 M 会迅速逼近渐近值$(2/\sqrt{3})h^2Y$ 或$(3/2)M_e$,这是单位宽度梁发生完全塑性或破坏的弯矩。极限塑性状态下通过中性面的应力(应力值大小为 $4Y/\sqrt{3}$)是不连续的。

如果将梁从部分塑性状态下卸载,梁内仍有一定的残余应力分布。假设卸载过程中应力变化是纯弹性的,我们就可以计算出残余应力。因此,叠加弹性应力分布是有必要的,因为反力矩的大小与释放力矩是相等的。将弹塑性梁中存在的应力减去 My/I_z,其中 M 由方程(8-9)得到,我们可计算完全卸载时的残余应力为

$$\begin{cases} \dfrac{\sigma_x}{Y} = \dfrac{2}{\sqrt{3}}\left[\dfrac{y}{c} - \dfrac{y}{2h}\left(3 - \dfrac{c^2}{h^2}\right)\right] & (|y| \leqslant c) \\[3mm] \dfrac{\sigma_x}{Y} = \dfrac{2}{\sqrt{3}}\left[1 - \dfrac{y}{2h}\left(3 - \dfrac{c^2}{h^2}\right)\right] & (|y| \geqslant c) \end{cases} \tag{8-10}$$

该应力分布在图 8-3 中采用阴影三角形表示。在 $c < |y| < h$ 区域内,残余应力符号发生变化,在距离中性面 $2h/(3-c^2/h^2)$ 处消失。当 $c/h \geqslant \sqrt{2}-1$ 时,应力在外表面达到最大值;当 $c/h \leqslant \sqrt{2}-1$ 时,应力在塑性边界处达到最大值。随着梁的塑性增强,在 $y = \pm h$ 处的残余应力接近极限值 $\mp Y/\sqrt{3}$。

卸载后梁的曲率是通过弹塑性曲率 $h/(cR_e)$ 减去弹性回弹量$(1-\nu^2)M/(EI_z)$ 得到的。将 R_e、M、I_z 代入,残余曲率可表示为

$$\frac{1}{R} = \frac{2}{\sqrt{3}}(1-\nu^2)\frac{Y}{Ec}\left(1 - \frac{3c}{2h} + \frac{c^3}{2h^3}\right) \tag{8-11}$$

括号外的系数是梁在卸载时的曲率,括号内的表达式是在 $y = c$ 处的残余应力与平面应变情况下屈服应力 $2Y/\sqrt{3}$ 的比值。对于小的弹塑性弯曲,残余曲率可与弹性回弹量数值相当。

8.4　圆柱体杆件的纯扭转

我们先考虑一个半径为 a 的实心圆柱体杆件,其受到扭矩 T 的作用。只要杆处于弹性状态,那么作用在任何截面上的剪应力都与到中心轴的径向距离 r 成正比。施加的扭矩 T 是围绕该轴的应力分布得到的合力矩。若杆件单位长度的扭转角用 θ 表示,则弹性剪应力可表示为

$$\tau = Gr\theta = \frac{2Tr}{\pi a^4}$$

由于剪应力在 $r = a$ 处最大,当扭矩增加到 T_e 时,杆件在该半径处开始发生屈服,相应的扭转角为 θ_e。设在 $r = a$ 处,对应 $\tau = k$,我们得到

$$T_e = \frac{1}{2}\pi ka^3; \quad \theta_e = \frac{k}{Ga}$$

如果扭矩进一步增大,那么在边界附近会形成一个塑性环,留下一个半径为 c 的中心区域,其中材料仍处于弹性状态,如图 8-4 所示。弹性区域内的应力分布呈线性,在 $r = c$ 处剪应力达到 k 值。对于非应

变硬化材料,整个塑性区内剪应力为常数 k,应力分布为

$$\tau = k\frac{r}{c} \quad (0 \leqslant r \leqslant c)$$

$$\tau = k \quad (c \leqslant r \leqslant a)$$

由于弹性区内的剪应力也等于 $Gr\theta$,因此得到 $\theta = k/(Gc)$。扭矩为

$$T = 2\pi \int_0^a \tau r^2 \mathrm{d}r = \frac{2}{3}\pi k \left(a^3 - \frac{1}{4}c^3\right) = \frac{1}{3}T_e \left[4 - \left(\frac{\theta_e}{\theta}\right)^3\right] \tag{8-12}$$

随着弹塑性扭转继续,扭矩迅速接近完全塑性的值 $(2/3)k\pi a^3$。由于当 c 趋于零时,θ 趋于无穷大,对于所有确定值的扭转角,材料的弹性中心区必须存在。

对于退火材料,其没有明确的屈服点,因此也不存在弹塑性边界。由于任意半径 r 处的工程剪应变为 $\gamma = r\theta$,因此扭矩可表示为

$$T = 2\pi \int_0^a \tau r^2 \mathrm{d}r = \frac{2\pi}{\theta^3}\int_0^{a\theta} \tau \gamma^2 \mathrm{d}\gamma$$

当给定材料的剪应力-剪应变曲线,利用已知的 (τ,γ) 关系,我们可由上述方程计算出扭矩。相反,如果通过实验确定了一个实心杆的扭矩-扭转角关系,我们也能很容易从中得到剪应力-剪应变曲线。将上述方程对 θ 进行微分得到

$$\frac{\mathrm{d}}{\mathrm{d}\theta}(T\theta^3) = 2\pi a^3 \theta^2 \tau_0$$

式中,τ_0 为 τ 在 $r=a$ 处的值,剪应变 $\gamma_0 = a\theta$。因此,τ_0 与 γ_0 的关系式为

$$\tau_0 = \frac{1}{2\pi a^3}\left(\theta \frac{\mathrm{d}T}{\mathrm{d}\theta} + 3T\right) \quad (\gamma_0 = a\theta) \tag{8-13}$$

括号中第一项的几何意义如图 8-5 所示。由于 $\mathrm{d}T/\mathrm{d}\theta$ 必须从测得的 T-θ 曲线中通过数值计算或作图的方式得到,因此在曲线初始部分,用方程(8-13)计算的结果并不是很准确。然而,将剪应力改写为下式则可以提高精度。

$$\tau_0 = \frac{1}{2\pi a^3}\left[\theta^2 \frac{\mathrm{d}}{\mathrm{d}\theta}\left(\frac{T}{\theta}\right) + 4T\right]$$

在弹性范围内 T/θ 为常数,在塑性范围初始阶段会缓慢减小。因此,括号中第一项的贡献很小,方程(8-13)给出的结果会更合适。

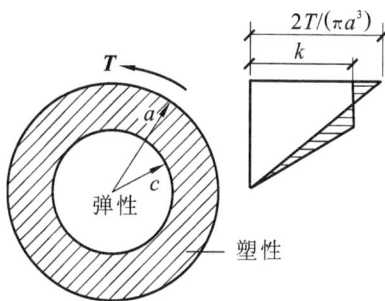

图 8-4 塑性环和 $H=0$ 时的应力分布

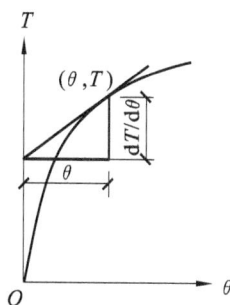

图 8-5 用于退火棒材的纳迪娅本结构

现在,假设半径为 a 的圆杆上有一个半径为 b 的同心圆孔。假设杆件材料为应变硬化材料,单轴应力-塑性应变曲线在相应范围内由斜率为 H 的直线表示。如果圆杆受纯扭转作用,当扭矩和特定扭转角为下述数值时,其将在 $r=a$ 处开始发生屈服

$$T_e = \frac{1}{2}\pi k a^3\left(1 - \frac{b^4}{a^4}\right); \quad \theta_e = \frac{k}{Ga}$$

在弹塑性扭转过程中,弹性区和塑性区中的剪应力都随着半径的增大而增大。由于任意塑性单元的总等

效应变等于$(1/\sqrt{3})(r\theta-\tau/G)$,我们给出假定的应变硬化规律

$$\tau=k+\frac{H}{3}\left(r\theta-\frac{\tau}{G}\right);\quad \theta=\frac{k}{Gc}$$

式中,c 是到弹塑性边界的半径。

将 θ 代入上述方程,我们得到塑性区域中的剪应力为

$$\tau=k\left(\frac{1+\dfrac{Hr}{3Gc}}{1+\dfrac{H}{3G}}\right)\quad(c\leqslant r\leqslant a)$$

在 $b\leqslant r\leqslant c$ 的弹性区域内,剪应力和前文一样为 $\tau=kr/c$。整个截面上的应力分布提供了扭矩 T。直接积分得到

$$\left(1+\frac{H}{3G}\right)T=\frac{2}{3}\pi ka^3\left\{1+\frac{1}{4}\left[-\frac{c^3}{a^3}\left(1+\frac{3b^4}{c^4}\right)+\frac{Ha}{Gc}\left(1-\frac{b^4}{a^4}\right)\right]\right\}\tag{8-14}$$

当 $H=0$, $b=0$ 时,方程(8-14)可简化为方程(8-12)。那么,具有应变硬化的空心圆杆的全塑性扭矩 T_0 为

$$\left(1+\frac{H}{3G}\right)T_0=\frac{2}{3}\pi ka^3\left[1-\frac{b^3}{a^3}+\frac{H}{4G}\left(\frac{a}{b}-\frac{b^3}{a^3}\right)\right]$$

在 $H=0$ 和 $H=0.3G$ 两种情况下,T/T_e 随着 θ/θ_e 的变化情况如图 8-6 所示。杆件的单位长度全塑性扭转角具有确定值 $\theta_0=k/(Gb)$,其与材料的应变硬化速率无关。

从弹塑性状态卸载时留在杆件中的残余应力可以采用像梁弯曲时同样的方法来确定,我们只需将与释放弯矩大小相等的反向扭矩所产生弹性应力分布进行叠加。因此,对于完全卸载的杆件,我们必须将卸载时刻杆件中存在的应力减去 $2Tr/[\pi(a^4-b^4)]$。利用关于扭矩 T 的方程(8-14),则在 $H=0$ 情况下,残余应力分布可表示为

$$\begin{cases}\dfrac{\tau}{k}=\dfrac{r}{c}-\dfrac{r}{3a}\dfrac{4-\dfrac{c^3}{a^3}\left(1+\dfrac{3b^4}{c^4}\right)}{1-\dfrac{b^4}{a^4}}&(b\leqslant r\leqslant c)\\[6mm]\dfrac{\tau}{k}=1-\dfrac{r}{3a}\dfrac{4-\dfrac{c^3}{a^3}\left(1+\dfrac{3b^4}{c^4}\right)}{1-\dfrac{b^4}{a^4}}&(c\leqslant r\leqslant a)\end{cases}\tag{8-15}$$

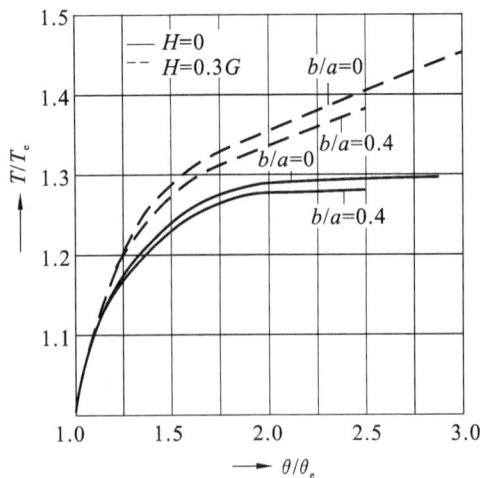

图 8-6　圆柱杆在弹塑性范围内纯扭转的扭矩-扭转的关系

残余应力在塑性环的外侧为负,而在横截面的其余部分为正。在 $b=0$ 情况下,在 $r=c$ 处且 $c/a \leqslant 0.576$ 时以及在 $r=a$ 处且 $c/a \geqslant 0.576$ 时计算得到的残余应力最大。

完全卸载后杆单位长度的弹性扭转角为 $2T/[\pi G(a^4-b^4)]$,其中 T 可由方程(8-14)得到。设 $H=0$,杆单位长度残余扭转角可表示为

$$\theta = \frac{k}{Gc}\left[1 - \frac{c}{3a}\frac{4 - \frac{c^3}{a^3}\left(1+\frac{3b^4}{c^4}\right)}{1 - \frac{b^4}{a^4}}\right] \tag{8-16}$$

方括号外的系数为杆卸载时的 θ 值,其中,方括号内的表达式为杆在弹塑性边界处 τ/k 的残余值。对于给定的弹塑性扭转,残余扭转角会随材料的应变硬化速率的增加而减小。

8.5 棱柱梁的纯弯曲

考虑一个均质棱柱梁,其在两端受到两个大小相等、方向相反的力偶 M 作用,如图 8-7 所示。梁的横截面有一个对称轴 Oy,弯曲力偶的作用轴线平行于 Oz,其中 O 位于中性面上。然后,弯曲平面与 xy 平面平行,中性纤维 Ox 弯曲成半径为 R 的圆弧。在弹性弯曲过程中,点 O 位于截面质心处,且唯一的非零应力 $\sigma_x = \sigma$ 为

$$\sigma = \frac{Ey}{R} = \frac{My}{I_z}$$

式中,E 为材料的杨氏模量;I_z 为截面关于中性轴 Oz 的惯性矩。

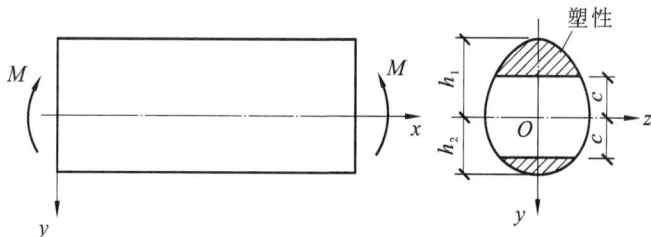

图 8-7 棱柱梁在端部力偶作用下的弯曲

弹性梁的纵向应变 $\varepsilon_{xx} = y/R$,那么伴随的横向应变 $\varepsilon_{yy} = \varepsilon_{zz} = -\nu y/R$,其中 ν 为泊松比。如果位移分量相对于坐标轴用 u、v、w 表示,那么

$$\frac{\partial u}{\partial x} = \frac{y}{R}; \quad \frac{\partial v}{\partial y} = \frac{\partial w}{\partial z} = -\frac{\nu y}{R}$$

$$\frac{\partial u}{\partial y} + \frac{\partial v}{\partial x} = \frac{\partial u}{\partial z} + \frac{\partial w}{\partial x} = \frac{\partial v}{\partial z} + \frac{\partial w}{\partial y} = 0$$

假设 x 轴上的一个单元和 yz 平面上的一个单元被固定在 $x=y=z=0$ 的空间处,那么在坐标原点处根据 $u=v=w=0$ 和 $\partial v/\partial x = \partial w/\partial x = \partial w/\partial y = 0$ 这两个条件可以得到解。结果是

$$\begin{cases} u = \dfrac{xy}{R} \\ v = -\dfrac{x^2 + \nu(y^2 - z^2)}{2R} \\ w = -\dfrac{\nu yz}{R} \end{cases} \tag{8-17}$$

这种变形使横向截面在弯曲过程中仍保持平面。中性面 xz 和每一个平行平面变形后都变为横向曲率为

ν/R 且向上凸的鞍形平面。

当纵向应力数值等于 Y 时,屈服首先发生在离中性面最远的纤维处。如果横截面不关于中性轴 Oz 对称,那么塑性区域就会在截面另一侧发生屈服前从这一侧开始向内扩展。之后,梁弯曲形成两个独立的塑性区,弹塑性边界均位于距中性面等距离 $c=(Y/E)/R$ 处。中性面的位置随弯曲量的变化而改变,并由横截面上纵向合力为零的条件决定,即

$$\int \sigma b(y)\mathrm{d}y = 0$$

式中,b 是距离 Oz 轴为 y 处横截面的宽度。若 Oz 为横截面的对称轴,则在弯曲形成的弹性和塑性范围内,梁的中性轴与质心轴重合。

通常我们假定应力状态是单轴的,即使梁处于部分塑性状态。然而,就像考虑梁的变形时可以看出的那样,严格来说假设并不正确。在小的扭转增量情况下,弹性区域的反向曲率变化量为 $\nu\mathrm{d}(1/R)$,塑性区域的反向曲率变化量为 $\eta\mathrm{d}(1/R)$,其中 η 表示材料超过屈服点后的收缩比。因此,除非在 $\eta=\nu=1/2$ 特殊情况下,否则这些单元将不能在弹塑性界面上结合在一起。因此,只有在材料不可压缩的情况下,前面的弹塑性弯曲理论才严格有效。对于 $\nu<1/2$ 的情况,由于横向应力会影响塑性边界的形状,不引入横向应力不能保证必要的连续性限制。接着,这个问题变得极其复杂。通过简化处理,只要在变形足够小的情况下,ν 取适当的值为 $1/2$,那么弹性和塑性区域的位移可由方程(8-17)得到。

弹性区域的应力 σ 值从中性轴上的零到弹塑性边界上的 Y 呈线性变化。在塑性纤维中,在拉伸或压缩状态下应力具有局部屈服值,并且是应变 $|y/R|$ 给定的函数。任意阶段的弯矩可由下述表达式给出

$$M = \int \sigma y b(y)\mathrm{d}y$$

对于退火材料,则不存在弹塑性界面,但仍然可以对整个截面上所给定的应力-应变定律进行积分计算。对于非应变硬化材料,给定截面的全塑性弯矩与给定截面的初始屈服弯矩之比称为形状系数。

8.6　矩形和圆形截面梁的纯弯曲

作为第一个例子,我们先考虑一个梁的弯曲问题,该梁截面形状是高为 $2h$、宽为 b 的矩形,弯矩作用在梁垂直平面内,如图 8-8(a)所示。根据截面的对称性,中性轴总是经过它的质心,截面关于该轴的转动惯性矩为 $I_z=\frac{2}{3}bh^3$。梁在弯曲过程中对应的弯矩和曲率半径如下时,在 $y\pm h$ 处开始发生塑性屈服:

$$M_e=\frac{2}{3}bh^2Y；\quad R_e=\frac{Eh}{Y}$$

在弹塑性弯曲过程中任意阶段,曲率半径为 $R=Ec/Y$,其中 c 为弹性中心区深度的一半。当 $0\leqslant n<1$ 时,假定材料的应变硬化规律如下:

$$\frac{\sigma}{Y}=\left(\frac{E\varepsilon}{Y}\right)^n \quad \left(\varepsilon\geqslant\frac{Y}{E}\right)$$

显然,σ 和 ε 等于塑性区域内纵向应力和应变的大小。由于 $\varepsilon=|y/R|$,截面受拉一侧的应力分布可表示为

$$\begin{cases}\sigma=Y\left(\dfrac{y}{c}\right) & (0\leqslant y\leqslant c)\\[2mm] \sigma=Y\left(\dfrac{y}{c}\right)^n & (c\leqslant y\leqslant h)\end{cases} \tag{8-18}$$

考虑截面的对称性,任意弹塑性阶段的弯矩为

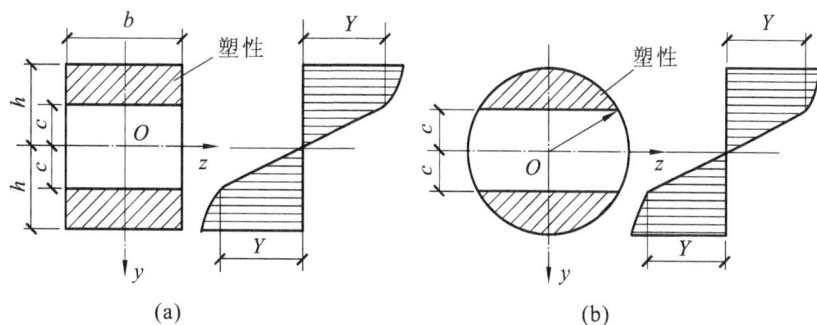

图 8-8 具有应变硬化材料的矩形和圆形截面梁的弹塑性弯曲几何图形和应力分布

$$M = 2b \int_0^h \sigma y \, \mathrm{d}y$$

将方程(8-18)代入上式并积分,弯矩与曲率的关系可表示为

$$\frac{M}{M_\mathrm{e}} = \frac{1}{2+n} \left[3 \left(\frac{R_\mathrm{e}}{R} \right)^n - (1-n) \left(\frac{R}{R_\mathrm{e}} \right)^2 \right] \tag{8-19}$$

对于非应变硬化材料($n=0$),弯矩-曲率关系可简化为方程(8-9)。对于不同数值 n,M/M_e 随 R_e/R 的变化如图 8-9 所示。除了在 $n=0$ 时,存在 $1.5M_\mathrm{e}$ 的极限弯矩外,其他情况下弯矩均随曲率的增加而稳定增大。因此,矩形截面梁的形状系数为 1.5。

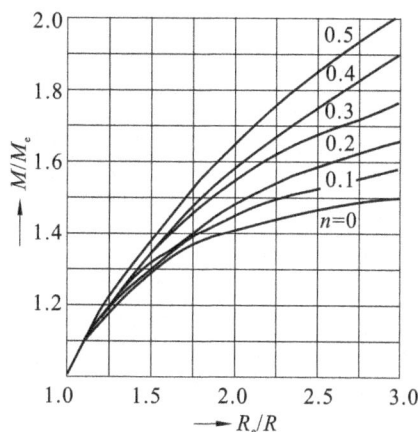

图 8-9 纯弯曲下具有应变硬化矩形截面梁的弯矩-曲率关系

如果从弹塑性梁的弯矩中释放量为 M' 的弯矩,则方程(8-18)叠加了一个 $-M'y/I_z$ 的纯弹性应力。将弹塑性曲率 $1/R$ 减去回弹曲率 $M'/(EI_z)$ 可以得到梁的残余曲率。只要不存在包申格效应,当 M' 继续反向增大时,梁将在压缩区 $y=h$ 处和拉伸区 $y=-h$ 处发生屈服。

$$\frac{M'}{M_\mathrm{e}} = 1 + \left(\frac{h}{c} \right)^n = 1 + \left(\frac{R_\mathrm{e}}{R} \right)^n \tag{8-20}$$

因此,对于非应变硬化材料($n=0$),在弹性阶段弯矩始终为 $2M_\mathrm{e}$,如图 8-10 所示。当弯曲继续反向增大时,非应变硬化梁的总弯矩逐渐趋近于 $-1.5M_\mathrm{e}$。当负曲率的大小大于或等于卸载时的负曲率时,总弯矩与该梁从未变形状态弯曲到该曲率所需的弯矩相同。

单轴拉伸或压缩状态下材料的应力-应变曲线可以由实验确定的弯矩 M 和在长度 l 上测量的弯曲角 α 之间的关系导出。我们可以以下列形式表示弯矩:

$$M = 2bR^2 \int_0^{\varepsilon_0} \sigma \varepsilon \, \mathrm{d}\varepsilon \quad \left(\varepsilon_0 = \frac{h}{R} \right)$$

在关于 M 的表达式两边同乘 α^2,并利用 $R^2 \alpha^2 = l^2$,我们得到导数如下:

$$\frac{\mathrm{d}}{\mathrm{d}\alpha}(M\alpha^2) = 2bh^2\alpha\sigma_0$$

式中，σ_0 为边界 $y=h$ 处应变 ε_0 对应的拉应力。

σ_0 和 ε_0 之间的关系可以写成

$$\begin{cases} \sigma_0 = \dfrac{1}{2bh^2}\Big(\alpha\,\dfrac{\mathrm{d}M}{\mathrm{d}\alpha} + 2M\Big) \\ \varepsilon_0 = \dfrac{h\alpha}{l} \end{cases} \tag{8-21}$$

如果在 M-α 曲线上任意点 P 处画一条切线与 M 轴交于点 Q，那么 PQ 平行于 M 轴的投影表达式即为方程(8-21)括号中的第一项，如图 8-11 所示。因此，基于给定的 M-α 曲线采用作图构造 σ_0-ε_0 曲线是可能的。

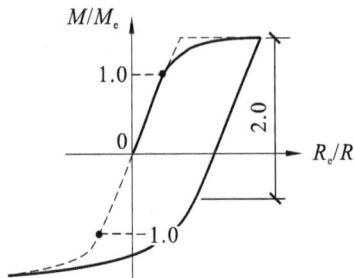

图 8-10 无应变硬化卸载和反向加载的影响，顶部的虚线代表理想状态下的性能

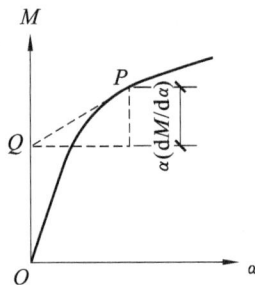

图 8-11 梁的纳迪娅本构图

现在考虑一个圆形截面的梁，其在垂直直径平面上受到纯弯曲。与水平直径重合的中性轴截面的惯性矩等于 $\pi a^4/4$，其中 a 为圆边界半径。当 M 和 R 达到如下值时，垂直直径的上下端，首先发生屈服：

$$M_\mathrm{e} = \frac{\pi}{4}a^3 Y; \quad R_\mathrm{e} = \frac{Ea}{Y}$$

典型的弹塑性阶段用塑性边界到中性轴的距离 c 表示。弹塑性弯矩如下：

$$M = 2\int_0^a \sigma y b(y)\mathrm{d}y = 4\int_0^a \sigma y\sqrt{a^2 - y^2}\mathrm{d}y$$

其中忽略应变硬化，当 $y \leqslant c$ 时，$\sigma = (y/c)\,Y$；当 $y \geqslant c$ 时，$\sigma = Y$。对方程进行积分，结果可以表示为

$$\frac{M}{M_\mathrm{e}} = \frac{2}{\pi}\left[\frac{1}{3}\Big(5 - \frac{2c^2}{a^2}\Big)\sqrt{1 - \frac{c^2}{a^2}} + \frac{a}{c}\arcsin\frac{c}{a}\right] \tag{8-22}$$

弹塑性梁的曲率半径为初始屈服时对应曲率半径的 c/a 倍。随着横截面的塑性增强，比值 M/M_e 迅速逼近渐近值 $16/(3\pi) \approx 1.698$，这是圆形截面的形状系数。

8.7 结　语

随着外荷载逐渐增加，我们关注应力单元第一次满足屈服准则时物体某一阶段的塑性流动问题。我们描述了荷载进一步增加导致弹性材料被弹塑性边界分隔的塑性区扩展情况。本章我们主要介绍了块体的平面应变压缩、弯曲梁平面应变、圆柱体杆件的纯扭转、棱柱梁的纯弯曲及矩形和圆形截面梁的纯弯曲等问题的弹塑性分析实例。

习题

8-1 对一实心圆柱杆进行纯扭转,该杆由初始剪切屈服应力为 k 具有应变硬化的材料制成,施加扭矩直至材料无应变硬化而达到完全塑性状态。假设一个恒定的塑性模量 $H=G/3$,求相关的扭转比 θ/θ_e。当杆从部分塑性状态完全卸载时,根据屈服剪切强度 k,计算外边界处的残余剪应力。

8-2 将外径为 a、内径为 $0.5a$ 的空心圆柱杆扭转至完全塑性,施加的扭矩随后被释放,材料是非应变硬化材料并遵循米泽斯屈服准则。如果被卸载杆受到足够大的轴向拉力,证明屈服将在杆件内半径处重新发生。

8-3 半径为 a 的实心圆柱体在纯扭转情况下,在半径 b 内区域部分出现塑性。当扭转角保持恒定时,向杆施加越来越大的轴向拉力。假设材料是一种不可压缩且非应变硬化普朗特-罗伊斯材料。证明在 $b \leqslant r \leqslant a$ 区域内的应力分布如下:

$$\frac{\sigma}{Y}=\tan\left(\frac{3G}{Y}\varepsilon\right); \quad \frac{\sqrt{3}\tau}{Y}=\sec\left(\frac{3G}{Y}\varepsilon\right)$$

式中,ε 为纵向应变。并证明附加塑性区域内的轴向应力满足的方程为

$$\frac{\sigma}{Y}=\tan\left(\frac{3G}{Y}\varepsilon-\sqrt{1-\frac{r^2}{b^2}}+\arctan\sqrt{1-\frac{r^2}{b^2}}\right)$$

8-4 一根横截面为矩形且非应变硬化的梁,沿着对称轴被弯曲,其弹塑性曲率为 κ_0。然后将梁卸载,并反向重新加载,直到再次发生塑性变形。证明新的弹塑性阶段相关曲率 $\kappa \leqslant \kappa_0 - 2\kappa_e$,其中 κ_e 对应于初始屈服,弯矩-曲率关系为

$$\frac{M}{M_e}=-\frac{1}{2}\left[3+\left(\frac{\kappa_e}{\kappa_0}\right)^2\right]+\left(\frac{2\kappa_e}{\kappa_0-\kappa}\right)^2$$

证明当 $\kappa \leqslant \kappa_0$ 时,梁的性能和初始未进行正向加载时情况一样。

8-5 一根矩形梁的横截面宽度为 b,高度为 $2h$,绕平行于宽度的轴线进行弯曲。该材料是线性应变硬化材料且初始屈服强度为 Y,切线模量为 T。证明弯曲过程中,弹塑性阶段中弯矩-曲率关系可以写成

$$\frac{M}{M_e}=\frac{1}{2}\left(1-\frac{T}{E}\right)\left[3-\left(\frac{R}{R_e}\right)^2\right]+\frac{TR_e}{ER}$$

如果梁从截面一半呈塑性状态时完全卸载,则证明残余曲率是初始屈服曲率的 $5(E-T)/(8E)$ 倍。

8-6 一根由理想的塑性材料制成的方形截面的棱柱梁,在弹塑性范围内沿着对角线进行弯曲。

(1)证明初始屈服弯矩为 $M_e=2\sqrt{2}a^3Y/3$,其中 $2a$ 为截面正方形的边长;

(2)证明该部分塑性梁的弯矩-曲率关系为

$$\frac{M}{M_e}=2-2\left(\frac{R}{R_e}\right)^2+\left(\frac{R}{R_e}\right)^3$$

如果在截面的一半区域变为塑性状态后,将施加的力矩 I 释放,要求找出在不引起进一步塑性流动条件下重新加载的弯矩 M 的极限值。

本书习题答案

参考文献

[1] ROBERT M,CADDELL. Deformation and fracture of solids[M]. New Jersey: Prentice-Hall Inc, 1980.

[2] TIMOSHENKO S P,GOODIER J N. Theory of elasticity[M]. New York: McGraw-Hill Book Company,1970.

[3] SHAMES I H. Mechanics of deformable solids[M]. New Jersey:Prentice-Hall Inc,1964.

[4] LEE E H,SYMONDS P S. Plasticity[M]. New York:Pergamon Press,1960.

[5] LEWIS M H,TAPLIN D M R. Micromechanisms of plasticity and fracture [M]. Waterloo:University of Waterloo Press,1983.

[6] CHAKRABARTY J. Theory of plasticity[M]. New York:McGraw-Hill Book Company,1987.

[7] SLATER R A C. Engineering plasticity[M]. London:The Macmillan Press Ltd,1977.

[8] WESTERGAARD H M. Theory of elasticity and plasticity[M]. New York: Harvard University Press,1964.

[9] SAADA A S. Elasticity:theory and applications[M]. New York:Pergamon Press,1974.

[10] WANG C T. Applied elasticity[M]. New York:McGraw-Hill Inc,1953.

[11] HASHIGUCHI K. Elastoplasticity theory[M]. Berlin:Springer,2002.

[12] MOLOTNIKOV V,MOLOTNIKOVA A. Theory of elasticity and plasticity [M]. Berlin:Springer,2017.

[13] 徐芝纶. 弹性力学[M]. 5 版. 北京:高等教育出版社,2016.

[14] 王仲仁,苑世剑,胡连喜,等. 弹性与塑性力学基础[M]. 2 版. 哈尔滨:哈尔滨工业大学出版社,2007.